Semantic-Based Visual Information Retrieval

Yu-Jin Zhang
Tsinghua University, Beijing, China

IRM Press
Publisher of innovative scholarly and professional information technology titles in the cyberage

Hershey • London • Melbourne • Singapore

Acquisition Editor: Kristin Klinger
Senior Managing Editor: Jennifer Neidig
Managing Editor: Sara Reed
Assistant Managing Editor: Sharon Berger
Development Editor: Kristin Roth
Copy Editor: Sue Vanderhook
Typesetter: Marko Primorac
Cover Design: Lisa Tosheff
Printed at: Yurchak Printing Inc.

Published in the United States of America by
        IRM Press (an imprint of Idea Group Inc.)
        701 E. Chocolate Avenue, Suite 200
        Hershey PA 17033-1240
        Tel: 717-533-8845
        Fax: 717-533-8661
        E-mail: cust@idea-group.com
        Web site: http://www.irm-press.com

and in the United Kingdom by
        IRM Press (an imprint of Idea Group Inc.)
        3 Henrietta Street
        Covent Garden
        London WC2E 8LU
        Tel: 44 20 7240 0856
        Fax: 44 20 7379 0609
        Web site: http://www.eurospanonline.com

Product or company names used in this book are for identification purposes only. Inclusion of the names of the products or companies does not indicate a claim of ownership by IGI of the trademark or registered trademark.

Library of Congress Cataloging-in-Publication Data

Zhang, Yu-Jin, 1954-
  Semantic-based visual information retrieval / by Yu-Jin Zhang.
     p. cm.
  Summary: "This book presents state-of-the-art advancements and developments in the field, and also brings a selection of techniques and algorithms about semantic-based visual information retrieval. It covers many critical issues, such as: multi-level representation and description, scene understanding, semantic modeling, image and video annotation, human-computer interaction, and more"--Provided by publisher.
  Includes bibliographical references and index.
  ISBN 1-59904-370-X (hardcover) -- ISBN 1-59904-371-8 (softcover) -- ISBN 1-59904-372-6 (ebook)
  1. Image processing--Digital techniques. I. Title.
  TA1637.Z528 2006
  621.36'7--dc22

                         2006027731

British Cataloguing in Publication Data
A Cataloguing in Publication record for this book is available from the British Library.

All work contributed to this book is new, previously-unpublished material. The views expressed in this book are those of the authors, but not necessarily of the publisher.

Semantic-Based Visual Information Retrieval

Table of Contents

Preface

Content-based visual information retrieval (CBVIR) is one of the most interesting research topics in the last years for the image and video community. With the progress of electronic equipment and computer techniques for visual information capturing and processing, a huge number of image and video records have been collected. Visual information becomes a well-known information format and a popular element in all aspects of our society. The large visual data make the dynamic research focused on the problem of how efficiently to capture, store, access, process, represent, describe, query, search, and retrieve their contents. In the last years, this field has experienced significant growth and progress, resulting in a virtual explosion of published information.

The research on CBVIR has already a history of more than a dozen years. It was started by using low-level features, such as color, texture, shape, structure, and space relationship, as well as (global and local) motion to represent the information content. Research on feature-based visual information retrieval has made quite a bit, but limited, success. Due to the considerable difference between the users' concerns on the semantic meaning and the appearances described by the aforementioned low-level features, the problem of semantic gap arises. One has to shift the research toward some high levels, and especially the semantic level. So, semantic-based visual information retrieval (SBVIR) began a few years ago and soon became a notable theme of CBVIR.

Research on SBVIR is conducted around various topics, such as (in an alpha-beta list) distributed indexing and retrieval, higher-level interpretation, human-computer interaction, human knowledge and behavior, indexing and retrieval for huge databases, information mining for indexing, machine-learning approaches, news and sport video summarization, object classification and annotation, object recognition and scene understanding, photo album and storybook generation, relevance feedback and association feedback, semantic modeling for retrieval, semantic representation and description, semiotics and symbolic operation, video abstraction and structure analysis, and so forth.

How to bridge the gap between semantic meaning and the perceptual feeling, which also exists between man and computer, has attracted much attention. Many efforts have converged to SBVIR in recent years, though it is still in its commencement. As a consequence, there is a considerable requirement for books like this one, which attempts to make a summary of the past progresses and to bring together a broad selection of the latest results from researchers involved in state-of-the-art work on semantic-based visual information retrieval.

This book is intended for scientists and engineers who are engaged in research and development of visual information (especially image and video content) techniques and who wish to keep their paces with the advances of this field. The objective of this collection is to review and survey new forward-thinking research and development in intelligent, content-based retrieval technologies. A comprehensive coverage of various branches of semantic-based visual information retrieval is provided by more than 30 leading experts around the world.

The book includes 16 chapters that are organized into six sections. They cover several distinctive research fields in semantic-based visual information retrieval and form a solid background for treating the content-based process from low-level to high-level both at static and dynamic aspects. They also provide many state-of-the-art advancements and achievements in filling the semantic gap. Some detailed descriptions for each section and chapter are provided in the following.

Section I, Introduction and Background, consists of one opening and surveying chapter (Chapter I) and provides some surrounding information, various achieved results, and a brief overview of the current research foci.

Chapter I, *Toward High-Level Visual Information Retrieval*, considers content-based visual information retrieval (CBVIR) as a new generation (with new concepts, techniques, mechanisms, etc.) of visual information retrieval, provides a general picture about research and development on this subject. The research on CBVIR starts by using low-level features more than a dozen years ago. The current focus is around capturing high-level semantics (i.e., the so-called semantic-based visual information retrieval [SBVIR]). This chapter conveys the research from feature level to semantic level by treating the problem of semantic gap under the general framework of CBVIR. This chapter first shows some statistics about the research publications on semantic-based retrieval in recent years in order to give an idea about its development status. It then presents some effective approaches for multi-level image retrieval and multi-level video retrieval. This chapter also gives an overview of several current centers of attention by summarizing certain results on those subjects, such as image and video annotation, human-computer interaction, models and tools for semantic retrieval, and miscellaneous techniques in application. In addition, some future research directions such as domain knowledge and learning, relevance feedback and association feedback, as well as research at even higher levels such as cognitive level and affective level are pointed out.

Section II, From Features to Semantics, discusses some techniques on the road. It is recognized that high-level research often relies on low-level investigation, so the development of feature-based techniques would help semantic-based techniques considerably. This section consists of three chapters (Chapters II through IV).

Chapter II, *The Impact of Low-Level Features in Semantic-Based Image Retrieval*, provides a suitable starting point for going into the complex problem of content representation and description. Considering image retrieval (IR) as a collection of techniques for retrieving images on the basis of features (in its general sense), both the low-level (content-based IR) feature and the high-level (semantic-based IR) feature (especially their relations) are discussed. Since semantic-based features rely on low-level ones, this chapter tries to make the reader initially familiarized with the most widely used low-level features. An efficient way to present these features is by means of a statistical tool capable of bearing concrete information, such as the histogram. For use in IR, histograms extracted from the low level features need to be compared by means of a metric. The most popular methods and distance metrics are, thus, opposed. Finally, several IR systems using histograms are presented in a thorough manner, and some experimental results are discussed. The steps in order to develop a custom IR system, along with modern techniques in image feature extraction, also are presented.

Chapter III is titled *Shape-Based Image Retrieval by Alignment.* Among the existing CBIR techniques for still images based on different perceptual features, shape-based methods are particularly challenging due to the intrinsic difficulties in dealing with shape localization and recognition problems. Nevertheless, there is no doubt that shape is one of the most important perceptual features, and successful shape-based techniques would significantly improve the spreading of general-purpose image retrieval systems. In this chapter, a shape-based image retrieval approach that is able to deal efficiently with domain-independent images with possible cluttered backgrounds and partially occluded objects is proposed. It is based on an alignment approach proven to be robust in rigid object recognition, which has been modified in order to deal with inexact matching between the stylized shape input by the user as query and the real shapes represented in the system's database. Results with a database composed of complex real-life images randomly taken from the Web and composed of several objects are provided.

Chapter IV is titled *Statistical Audiovisual Data Fusion for Video Scene Segmentation.* Video has a large data volume and complex structure. Automatic video segmentation into semantic units is important in effectively organizing long videos. In this chapter, the focus is on the problem of video segmentation into narrative units called scenes-aggregates of shots unified by a common dramatic event. A statistical video scene segmentation approach that detects scenes boundaries in one pass by fusing multimodal audiovisual features in a symmetrical and scalable manner is derived. The approach deals properly with the variability of real-valued features and models their conditional dependence on the context. It also integrates prior information concerning the duration of scenes. Two kinds of features, video coherence and audio dissimilarity, extracted both in visual and audio domains are used in the process of scene segmentation. The results of experimental evaluations carried out with ground truth video are reported. They show that this approach effectively fuses multiple modalities with higher performance, compared with an alternative rule-based fusion technique.

Section III focuses on Image and Video Annotation, which gets a lot of attention from the SBVIR research community. The text could be considered as a compact medium that expresses more abstract sense than do by image and video. So, by annotating image and video with characteristic textural entities, their semantic contents would be represented efficiently and would be used in retrieval. This section consists of three chapters (Chapters V through VII).

Chapter V is titled *A Novel Framework for Image Categorization and Automatic Annotation.* Image classification and automatic annotation could be treated as effective solutions to enable keyword-based semantic image retrieval. Traditionally, image classification and automatic annotation are investigated in different models separately. In this chapter, a novel framework combining image classification and automatic annotation by learning semantic concepts of image categories is proposed. In order to choose representative features for obtaining information from image, a feature selection strategy is proposed, and visual keywords are constructed by using both discrete and continuous methods. Based on the selected features, the integrated patch (IP) model is proposed to describe the properties of different image categories. As a generative model, the IP model describes the appearance of the mixture of visual keywords in considering the diversity of the object composition. The parameters of the IP model then are estimated by EM algorithm. Some experimental results on Corel image dataset and Getty Image Archive demonstrate that the proposed feature selection and image description model are effective in image categorization and automatic image annotation, respectively.

Chapter VI is titled *Automatic and Semi-Automatic Techniques for Image Annotation*. When retrieving images, users may find that it is easier to express the desired semantic content with keywords than with visual features. Accurate keyword retrieval can occur only when images are described completely and accurately. This can be achieved either through laborious manual effort or complex automated approaches. Current methods for automatically extracting semantic information from images can be classified into two classes. One is text-based methods, which use metadata such as ontological descriptions and/or texts associated with images to assign and/or refine annotations. Although highly specialized in domain- (context-) specific image annotations, the text-based methods are usually semi-automatic. Another is image-based methods, which focus on extracting semantic information directly and automatically from image content, though they are domain-independent and could deliver arbitrarily poor annotation performance for certain applications. The focus of this chapter is to create an awareness and understanding of research and advances in this field by introducing basic concepts and theories and then by classifying, summarizing, and describing works with a variety of solutions from the published literature. By identifying some currently unsolved problems, several suggestions for future research directions are pointed out.

Chapter VII is titled *Adaptive Metadata Generation for Integration of Visual and Semantic Information*. The principal concern of this chapter is to provide those in the visual information retrieval community with a methodology that allows them to integrate the results of content analysis of visual information (i.e., the content descriptors and their text-based representation) in order to attain the semantically precise results of keyword-based image retrieval operations. The main visual objects under discussion are images that do not have any semantic representations therein. Those images demand textual annotation of precise semantics, which is to be based on the results of automatic content analysis but not on the results of time-consuming manual annotation processes. The technical background and literature review on a variety of annotation techniques for visual information retrieval is outlined first. The proposed method and its implemented system for generating metadata or textual indexes to visual objects by using content analysis techniques that can bridge the gaps between content descriptors and textual information then are described.

Section IV is devoted to the topic of Human-Computer Interaction. Human beings play an important role in semantic level procedures. By putting human into the loop of computer routine, it is quite convenient to introduce the domain knowledge into description module and to greatly improve the performance of the retrieval system. This chapter consists of three chapters (Chapters VIII through X).

Chapter VIII is titled *Interaction Models and Relevance Feedback in Image Retrieval*. Human-computer interaction increasingly is recognized as an indispensable component of image retrieval systems. A typical form of interaction is that of relevance feedback, whereby users supply relevant information on the retrieved images. This information subsequently can be used to optimize retrieval parameters and to enhance retrieval performance. The first part of the chapter provides a comprehensive review of existing relevance feedback techniques and also discusses a number of limitations that can be addressed more successfully in a browsing framework. Browsing models form the focus of the second part of this chapter, in which the merit of hierarchical structures and networks for interactive image search are evaluated. This exposition aims to provide enough detail to enable the practitioner to implement many of the techniques and to find numerous pointers to the relevant literature otherwise.

Chapter IX, *Semi-Automatic Ground Truth Annotation for Benchmarking of Face Detection in Video*, presents a method of semi-automatic ground truth annotation for benchmarking of face detection in video. It aims to illustrate the solution to the issue in which an image processing and pattern recognition expert is able to label and annotate facial patterns in video sequences at the rate of 7,500 frames per hour. These ideas are extended to the semi-automatic face annotation methodology in which all object patterns are categorized into four classes in order to increase flexibility of evaluation results analysis. A strict guide on how to speed up manual annotation process by 30 times is presented and is illustrated with sample test video sequences that consist of more than 100,000 frames, including 950 individuals and 75,000 facial images. Experimental evaluation of face detection using ground truth data that were semi-automatically labeled demonstrates the effectiveness of the current approach for both learning and testing stages.

Chapter X is titled *An Ontology-Based Framework for Semantic Image Analysis and Retrieval*. In order to overcome the limitations of keyword- and content-based visual information access, an ontology-driven framework is developed in this chapter. Under the proposed framework, an appropriately defined ontology infrastructure is used to drive the generation of manual and automatic image annotations and to enable semantic retrieval by exploiting the formal semantics of ontology. In this way, the descriptions considered in the tedious task of manual annotation are constrained to named entities (e.g., location names, person names, etc.), since the ontology-driven analysis module can automatically generate annotations concerning common domain objects of interest (e.g., sunset, trees, sea, etc.). Experiments in the domain of outdoor images show that such an ontology-based scheme realizes efficient visual information access with respect to its semantics.

Section V is intended to present some Models and Tools for Semantic Retrieval, which continuously have been incorporated in recent years. As for other disciplines or subjects, the progress of research on SBVIR also should get support from a variety of mathematic models and technique tools. Several models and tools utilized in SBVIR thus are introduced, and some pleasing results also are presented. This section consists of three chapters (Chapters XI through XIII).

Chapter XI is titled *A Machine Learning-Based Model for Content-Based Image Retrieval*. Index is an important component of a retrieval system. A suitable index makes it possible to group data according to similarity criteria. Traditional index structures frequently are based on trees and use the k-nearest neighbors (k-NN) approach to retrieve databases. Due to some disadvantages of such an approach, the use of neighborhood graphs was proposed. This approach is interesting, but it also has some disadvantages consisting mainly in its complexity. This chapter first proposes an effective method for locally updating neighborhood graphs that constitute the required index. This structure then is exploited in order to make the retrieval process using queries in an image form more easy and effective. In addition, the indexing structure is used to annotate images in order to describe their semantics. The proposed approach is based on an intelligent manner for locating points in a multidimensional space. Promising results are obtained after experimentations on various databases.

Chapter XII, *Neural Networks for Content-Based Image Retrieval*, introduces the use of neural networks for content-based image retrieval (CBIR) systems. It presents a critical literature review of both the traditional and neural network-based techniques that are used in retrieving images based on their content. It shows how neural networks and fuzzy logic can be used in various retrieval tasks, such as interpretation of queries, feature extraction, and classification of features by describing a detailed research methodology. It investigates a neural network-based technique in conjunction with fuzzy logic in order to improve the overall performance of the CBIR systems. The results of the investigation on a benchmark database with a comparative analysis are presented in this chapter. The methodologies and results presented in this chapter will allow researchers to improve and compare their methods and also will allow system developers to understand and implement the neural network and fuzzy logic-based techniques for content-based image retrieval.

Chapter XIII is titled *Semantic-Based Video Scene Retrieval Using Evolutionary Computing*. A new emotion-based video scene retrieval method is proposed in this chapter. Five features extracted from a video are represented in a genetic chromosome, and target videos that users have in mind are retrieved by the interactive genetic algorithm through the feedback iteration. After selecting the videos that contain the corresponding emotion from the initial population of videos by the proposed algorithm, the feature vectors extracted from them are regarded as chromosomes, and a genetic crossover is applied to those feature vectors. Next, new chromosomes after crossover and feature vectors in the database videos are compared based on a similarity function in order to obtain the most similar videos as solutions of the next generation. By iterating this process, a new population of videos that users have in mind is retrieved. In order to show the validity of the proposed method, six example categories—action, excitement, suspense, quietness, relaxation, and happiness—are used as emotions for experiments. This method of retrieval shows 70% of effectiveness on the average of more than 300 commercial videos.

Section VI brings together several Miscellaneous Techniques in Applications of semantic retrieval. Research on SBVIR is still in its infancy, a large number of special ideas and exceptional techniques have been applied and also are going to apply to SBVIR. These works provide new sights and fresh views from various sides and enrich the technique pool for treating the process on semantics. This chapter consists of three chapters (Chapters XIV through XVI).

Chapter XIV, *Managing Uncertainties in Image Databases*, focuses on those functionalities of multimedia databases that are not present in traditional databases but are needed when dealing with multimedia information. Multimedia data are inherently subjective; for example, the association of a meaning and the corresponding content description to an image as well as the evaluation of the difference between two images usually depend on the user, who is involved in the evaluation process. For retrieval purposes, such subjective information needs to be combined with objective information such as image color histograms obtained through (generally imprecise) data analysis processes. Therefore, the inherently fuzzy nature of multimedia data, at both the subjective and objective sides, may lead to multiple, possibly inconsistent interpretations of data. In this chapter, a fuzzy, nonfirst, normal form

(FNF^2) data model that is an extension of the relational models is presented. It takes into account subjectivity and fuzziness. It is intuitive and enables user-friendly information access and manipulation mechanisms. A prototype system based on the FNF2 model has been implemented.

Chapter XV is titled *A Hierarchical Classification Technique for Semantics-Based Image Retrieval*. A new approach with multiple steps for improving image retrieval accuracy by integrating semantic concepts is presented in this chapter. First, images are represented according to different abstraction levels. At the lowest level, they are represented with visual features. At the upper level, they are represented with a set of very specific keywords. At the subsequent higher levels, they are represented with more general keywords. Second, visual content together with keywords are used to create a hierarchical index. A probabilistic classification approach is proposed, which allows grouping similar images into the same class. Finally, this index is exploited to define three retrieval mechanisms: text-based, content-based, and a combination of both. Experiments show that such a combination allows nicely narrowing the semantic gap encountered by most current image retrieval systems. Furthermore, it is shown that the proposed method helps to reduce retrieval time and improve retrieval accuracy.

Chapter XVI is titled *Semantic Multimedia Information Analysis for Retrieval Applications*. Most of the research works in multimedia retrieval applications have focused on retrieval by content or retrieval by example. Since the beginning of the century, a new interest has grown immensely in the multimedia information retrieval community: retrieval by semantics. This exciting new research area arises as a combination of multimedia understanding, information extraction, information retrieval, and digital libraries. This chapter presents a comprehensive review of analysis algorithms in order to extract semantic information from multimedia content. Some statistical approaches to analyze image and video contents are described and discussed.

Overall, the 16 chapters in six sections with hundreds of pictures and several dozen tables offer a comprehensive image about the current advancements of semantic-based visual information retrieval.

Yu-Jin Zhang

Editor

Tsinghua University, Beijing, China

Acknowledgments

It is always with great pleasure to write this page.

First, credit goes to the senior academic editor of Idea Group Inc., Mehdi Khosrow-Pour, for the invitation made to me to organize and edit this book.

Sincere thanks goes to the 32 authors, coming from 11 countries and regions (4 from Asia, 20 from Europe, 6 from North America, and 2 from Oceania), who made great contributions to this project by submitting chapters and reviewing chapters, among other things.

Special gratitude goes to all the staff at Idea Group Inc. for their valuable communication, guidance and suggestion along with the development process.

Last, but not least, I am indebted to my wife, my daughter and my parents for their encouragement, patience, support, tolerance and understanding during the last two years.

Yu-Jin Zhang
Editor
Tsinghua University, Beijing, China

Section I

Introduction
and Background

Chapter I

Toward High-Level Visual Information Retrieval

Yu-Jin Zhang, Tsinghua University, Beijing, China

Abstract

Content-based visual information retrieval (CBVIR) as a new generation (with new concepts, techniques, mechanisms, etc.) of visual information retrieval has attracted many interests from the database community. The research starts by using a low-level feature from more than a dozen years ago. The current focus has shifted to capture high-level semantics of visual information. This chapter will convey the research from the feature level to the semantic level by treating the problem of semantic gap under the general framework of CBVIR. This high-level research is the so-called semantic-based visual information retrieval (SBVIR). This chapter first shows some statistics about the research publications on semantic-based retrieval in recent years; it then presents some existing approaches based on multi-level image retrieval and multi-level video retrieval. It also gives an overview of several current centers of attention by summarizing certain results on subjects such as image and video annotation, human-computer interaction, models and tools for semantic retrieval, and miscellaneous techniques in application. Before finishing, some future research directions, such as domain knowledge and learning, relevance feedback and association feedback, as well as research at even a high level such as cognitive level, are pointed out.

Introduction

It is said that "a picture is worth a thousand words." Human beings obtain the majority of information from the real world by visual sense. This could include all entities that can be visualized, such as image and video (a chain/sequence of images) in a narrow sense, as well as animation, charts, drawings, graphs, multi-dimensional signals, text (in fact, many documents are used in image form, as indicated by Doermann, 1998), and so forth in a more general sense.

With the fast technique progress of computer science, electronics, medium capturing, and so forth, and the rapidly rising use of the Internet and the growing capability of data storage, the quantity of visual information expands dramatically and results in many huge visual information databases. In addition, many data are created and collected by amateurs, which is quite different than by professional people (Luo, Boutell, & Brown, 2006). In addition, visual media become a widespread information format in the World Wide Web (WWW) in which data are dispersed in various locations. All these make the search of required visual information more complex and time-consuming (Zhang, 2006). Along with the quickly increasing demands to create and store visual information comes the need for a richer set of search facilities. Providing tools for effective access, retrieval, and management of huge visual information data, especially images and videos, has attracted significant research efforts. Several generations of techniques and systems have been developed.

Traditionally, textual features such as captions, file names, and especially keywords have been used in searching required visual information. However, the use of keywords in the search is not only cumbersome but also inadequate to represent the riche content of visual information. Images are snapshots of the real world. Due to the complexity of scene content, there are many images for which no words can exactly express their implications. Image is beyond words, so it has to be seen and must be searched as image by content (i.e., object, purpose, scene, style, subject, etc.).

Content-based visual information retrieval has attracted many interests, from image engineering, computer vision, and database community. A large number of researches, especially on feature-based techniques, have been developed and have achieved plentiful and substantial results (Bimbo, 1999; Rui, Huang, & Chang, 1999; Smeulders et al., 2000; Zhang, 2003). However, in light of the complexity of the real world, low-level perceptive cues/indexes are not enough to provide suitable interpretation. To probe further, some higher-level researches and techniques for content understanding are mandatory. Among three broad categories of high-level techniques—synthetic, semantic, and semiotic—the semantic approach is quite natural from the understanding point of view. Nevertheless, from feature to semantic, there is a semantic gap. Solving this problem has been a focal point in content-based visual information retrieval. This chapter will summarize some recent research results and promote several new directions in higher-level researches.

Background

In the following, visual information will refer mainly to image and video, although other media also will be covered with some specific considerations.

Content-Based Visual Information Retrieval

More than 10 years ago, the term *content-based image retrieval* made its first appearance (Kato, 1992). Content-based image retrieval (CBIR) could be described as a process framework for efficiently retrieving images from a collection by similarity. The retrieval relies on extracting the appropriate characteristic quantities describing the desired contents of images. Shortly thereafter, content-based video retrieval (CBVR) also made its appearance in treating video in similar means as CBIR-treating images. Content-based visual information retrieval (CBVIR) soon combines CBIR and CBVR together. Also, the MPEG-7 standard was initiated with the intention to allow for efficient searching, indexing, filtering, and accessing of multimedia (especially image and video) contents (Zhang, 2003).

A general scheme for content-based visual information retrieval is shown in Figure 1. The system for retrieval is located between the information user and the information database. It consists of five modules: Querying, for supporting the user to make an inquiry; Description, for capturing the essential content/meaning of inquest and transferring it to internal representation; Matching, for searching required information in a database; Retrieval, for extracting required information from a database; and Verification, for ensuring/confirming the retrieval results.

Research on CBVIR today is a lively discipline that is expanding in breadth. Numerous papers have appeared in the literature (see the next section), and several monographs have been published, such as Bimbo (1999) and Zhang (2003). As happens during the maturation process of many disciplines, after early successes in a few applications, research then concentrates on deeper problems, challenging the hard problems at the crossroads of the

Figure 1. General scheme for content-based visual information retrieval

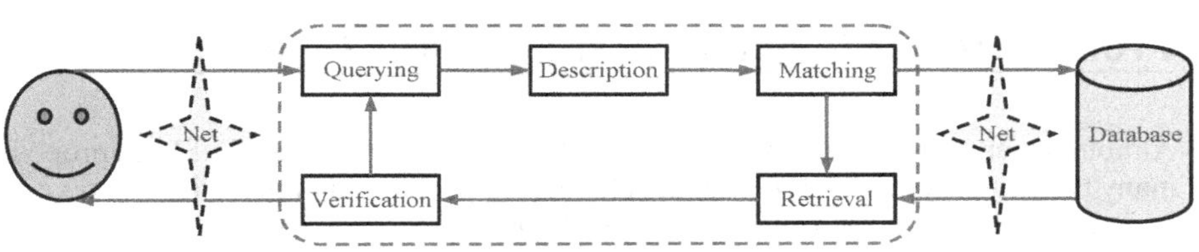

discipline from which it was born: image processing, image analysis, image understanding, databases, and information retrieval.

Feature-Based Visual Information Retrieval

Early CBVIR approaches often rely on the low-level visual features of image and video, such as color, texture, shape, spatial relation, and motion. These features are used widely in the description module of Figure 1 for representing and describing the contents. It is generally assumed that the perceptual similarity indicates the content similarity. Such techniques are called feature-based techniques in visual information retrieval.

The color feature is one of the most widely used visual features in image retrieval. It is relatively robust to background complication and independent of image size and orientation. Texture refers to the visual patterns that have properties of homogeneity that do not result from the presence of only a single color or intensity. It is an innate property of virtually all surfaces, including clouds, trees, bricks, hair, and fabric. The shape feature is at some higher level than that of color and texture, as the focus is now on interesting objects. Some required properties for shape features are translation, rotation, and scaling invariants. Once interesting objects are determined, their spatial relationship will provide more information about the significance of the whole scene, such as the structural arrangement of object surfaces and the association with the surrounding environment.

The aforementioned features often are combined in order to provide more complete coverage of various aspects of image and video. For example, the global color feature is simple to calculate and can provide reasonable discriminating power, but using the color layout feature (both color feature and spatial relation feature) is a better solution to reduce false positives when treating large databases. A number of feature-based techniques and examples (using single features and/or composite features) can be found in Zhang (2003).

There are several early surveys on this subject. In Rui et al. (1999), more than 100 papers covering three fundamental bases of CBIR—image feature representation and extraction, multidimensional indexing, and system design—are reviewed. In Smeulders et al. (2000), a review of 200 references in content-based image retrieval published in the last century has been made. It starts with discussing the working conditions of content-based retrieval: patterns of use, types of pictures, the role of semantics, and the sensory gap. The discussion on feature-based retrieval is divided into two parts: (1) the features could be extracted from the pixel-level, such as color, texture, and local geometry; and (2) the features need to be extracted from the pixel-group level (more related to objects), such as accumulative and global features (of pixels), salient points, object and shape features, signs, and structural combinations.

From Features to Semantics: Semantic Gap

Although many efforts have been put on CBVIR, many techniques have been proposed, and many prototype systems have been developed, the problems in retrieving images according to image content are far from being solved. One problem of the feature-based technique is that there is a considerable difference between users' interest in reality and the image contents described by using only the previously mentioned low-level perceptive features (Zhang, Gao, & Luo, 2004), although all current techniques assume certain mutual information between the similarity measure and the semantics of images and videos. In other words, there is a large gap between content description based on low-level features and that of human beings' understanding. As a result, these feature-based approaches often lead to unsatisfying querying results in many cases. In Smeulders et al. (2000), some discussions on the semantic gap also are presented.

One solution to fill the semantic gap is to make the retrieval system work with low-level features while the user puts in high-level knowledge so as to map low-level visual features to high-level semantics (Zhou & Huang, 2002). Two typical early methods are to optimize a query request by using relevance feedback and a semantic visual template (Chang, 1998) and to interpret progressively the content of images by using interactive interface (Castelli, Bergman, Kontoyiannis, et al., 1998).

Nowadays, the mainstream of the research converges to retrieval based on semantic meaning, which tries to extract the cognitive concept of a human by combining the low-level features in some way. However, semantic meaning extraction based on feature vectors is difficult, because feature vectors indeed cannot capture the perception of human beings. For example, when looking at a colorful image, an ordinary user hardly can figure out the color histogram from that image but rather is concerned about what particular color is contained.

The methods toward semantic image retrieval have been categorized roughly into the following classes (Cheng, Chen, Meng, Sundaram, & Zhong, 2005): (1) automatic scene classification in whole images by statistically based techniques; (2) methods for learning and propagating labels assigned by human users; (3) automatic object classification using knowledge-based or statistical techniques; (4) retrieval methods with relevance feedback during a retrieval session. An indexing algorithm may be a combination of two or more of the aforementioned classes.

Main Thrust

First, some statistics about research publication are provided and analyzed. Then, demonstrative approaches for multi-level image and video retrievals are presented. Finally, an overview of recent research works is made by a brief survey of representative papers.

Statistics about Research Publications

The study for CBVIR has been conducted for more than 10 years. As a new branch of CBVIR, SBVIR recently has attracted many research efforts. To get a rough idea about the scale and progress of research on (general) image retrieval and semantic-based image retrieval for the past years, several searches in EI Compendex database (http://www.ei.org) for papers published from 1995 through 2004 have been made. In Table 1, the results of two searches

Table 1. List of records found in the title field of EI Compendex

Searching Terms	1995	1996	1997	1998	1999	2000	2001	2002	2003	2004	Total
(1) Image Retrieval	70	89	81	135	157	166	212	237	260	388	1795
(2) Semantic Image Retrieval	0	1	1	2	4	5	5	9	12	20	59
Ratio of (2) over (1)	0	1.12	1.23	1.48	2.55	3.01	2.36	3.80	4.62	5.15	3.29

in the title field of EI Compendex for the numbers of published papers (records) are listed; one term used is *image retrieval* and other is *semantic image retrieval*. The papers found out by the second term should be a subset of the papers found out by the first term.

It is seen from Table 1 that both numbers (of papers) are increasing in that period, and the number of published papers with the term *semantic image retrieval* is just a small set of papers with the term *image retrieval*.

Other searches take the same terms as used for Table 1, but are performed in the field of title/abstract/subject. The results are shown in Table 2.

The numbers of records in Table 2 for both terms are augmented during that period in comparison with that of Table 1. This is expected, as the fields under search now also include abstract field and subject field in addition to only title field.

Comparing the ratios of SIR over IR for 10 years in two tables, these ratios in Table 2 are much higher than those ratios in Table 1. This difference indicates that the research for semantic retrieval is still in an early stage (many papers have not put the word *semantic* in the title of papers), but this concept starts to get numerous considerations or attract much attention (*semantic* appeared already in abstract and/or subject parts of these papers).

In order to have a closer comparison, these ratios in Table 1 and Table 2 are plotted together in Figure 2, in which light bars represent ratios from Table 1, and dark bars represent ratios

Table 2. List of records found in the subject/title/abstract field of EI Compendex

Searching Terms	1995	1996	1997	1998	1999	2000	2001	2002	2003	2004	Total
(1) Image retrieval	423	581	540	649	729	889	1122	1182	1312	2258	9685
(2) Semantic image retrieval	11	25	27	46	62	80	114	124	155	335	979
Ratio of (2) over (1)	2.60	4.30	5.00	7.09	8.50	9.00	10.16	10.49	11.81	14.84	10.11

Figure 2. Comparison of two groups of ratios

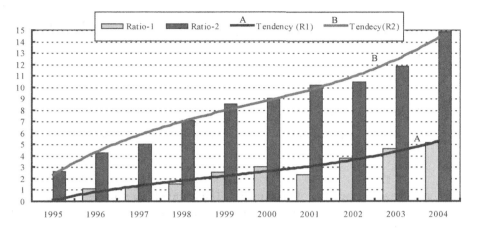

from Table 2. In addition, the tendencies of ratio developments are approximated by third-order polynomial. It is clear that many papers have the semantic concept in mind, although they do not always use the word *semantic* in the title. It is also clear that both ratios have the tendency to increase, with the second ratio going up even faster.

Multi-Level Image Retrieval

Content-based image retrieval requires capturing the content of images, which, in turn, is a challenging task. To improve the performance of image retrieval, there is a strong trend to analyze images in a hierarchical way so as to represent and describe image contents progressively toward the semantic level.

Multi-Level Representation

Many approaches have been proposed to represent and describe the content of images in a higher level than the feature level, which should be more corresponding to human beings' mechanisms for understanding and which also reflects the fuzzy characteristics of image contents. Several representation schemes have been proposed to represent the contents of images in different levels (Amir & Lindenbaum, 1998), such as the three-level content representation, including feature-level content, object-level content, and scene-level content (Hong, Wu, & Singh, 1999); and the five-level representation, including region level, perceptual region level, object part level, object level, and scene level (Jaimes & Chang, 1999).

On the other side, people distinguish three layers of abstraction when talking about image database: (1) raw data layer, (2) feature layer, and (3) semantic layer. The raw data are

Figure 3. A general paradigm for object-based image retrieval

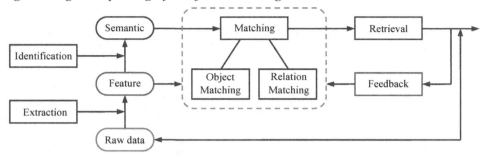

original images in the form of pixel matrix. The feature layer shows some significant characteristics of the pixel patterns of the image. The semantic layer describes the meaning of identified object in images. Note that the semantic level also should describe the meaning of an image as a whole. Such a meaning could be obtained by the analysis of objects and the understanding of images.

According to the previous discussions, a multi-layer approach should be used in which the object characterization plays an important role. The basic idea behind such an approach is that images in each object class have similar semantic meanings and visual perceptions (Dai & Zhang, 2005). A general paradigm is shown in Figure 3 (Zhang, 2005a). Two important tasks are:

1. **Extracting meaningful regions:** In order to be able to base image retrieval on objects, the interesting regions related to objects should be extracted first. This process relates the raw data layer to the feature layer.
2. **Identification of interesting objects:** Based on the extracted regions, (perceptual) features should be taken out, and those required objects could be identified. This corresponds to the step from the feature layer to the object layer.

Multi-Level Description

A typical realization of the aforementioned general paradigm consists of four layers: original image layer, meaningful region layer, visual perception layer, and object layer, as depicted in Figure 4 (Gao, Zhang, & Merzlyakov, 2000). One note here is that instead of deriving accurate measures of object properties for further identification or classification, the primary concerns in CBIR are to separate required regions and to obtain more information related to the semantic of these regions (Zhang, 2005b).

In Figure 4, the left part shows the multi-level model, while the right part gives some presentation examples. The description for a higher layer could be generated on the basis of the description obtained from the adjacent lower layer by using the following various techniques: presegmentation from the original image layer to the meaningful region layer (Luo et al.,

Figure 4. Multi-level image description

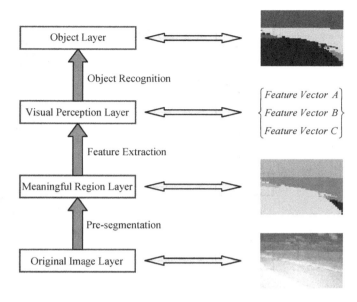

2001), feature extraction from the meaningful region layer to the visual perception layer, and object recognition from the visual perception layer to the object layer (Zhang, Gao, & Merzlyakov, 2002a). At the object layer, retrieval also can be performed with the help of self-adaptive relevance feedback (Gao, Zhang, & Yu, 2001). Such a progressive procedure could be considered as a synchronization of the procedure for progressive understanding of image contents. These various layers could provide distinct information of the image content, so this model is suitable to access from different levels. More details can be found in Zhang et al. (2004).

From the point of view of image understanding, the next stage would go beyond objects. The actions and interactions of objects and, thus, generated events (or scenes) are important in order to understand fully the contents of images. The images in this case would be described by some metadata.

Multi-Level Video Retrieval

Due to the great length and rich content of video data, quickly grasping a global picture or effectively retrieving pertinent information from such data becomes an challenging task. Organization and summarization of video data are the main approaches taken by content-based video retrieval.

Figure 5. Multi-level video organization

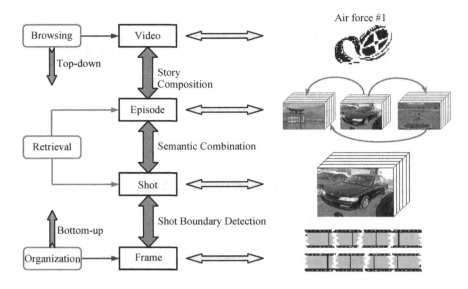

Video Organization

Video organization is a process of connecting and assembling the components of video in a predefined structure in order to provide a fast and flexible ability to browse and/or retrieve. In order to effectively organize the video for efficient browsing and querying, multi-level representation of video data is adopted. A typical hierarchical structure (scheme) for organization consists of four layers: video, episode, shot, and stream, as shown in Figure 5 (Zhang & Lu, 2002b). Three video operations—organization, browsing, and retrieval—are enabled by this scheme.

In contrast to normal browsing, which is a top-down process, organization is a bottom-up process. It starts from the lowest layer—the frame layer. This layer corresponds to the original video data that consist of a time sequence of frames. By shot boundary detection, the frames are grouped into shots. A shot is a basic unit of a video program. It consists of a number of frames that temporally are connected and spatially neighboring; it contains a continuous action in space. By using some high-level knowledge, several shots are combined to form an episode. Episode is a semantic unit that describes an act or a story. In other words, shots in an episode are content-related but can be separated temporally and/or disconnected spatially. A video program (e.g., movie) is built by a number of episodes that form a meaningful story. Retrieval can be conducted either in shot layer or episode layer by using various cues, such as global camera motion (frequently conveys the semantic implication of the video creator) and object motion vector (often represents intended action), as indicated by Yu and Zhang (2001a, 2001b).

Video Summarization

Numerous techniques used in video organization (shot detection, shot clustering) also find their applications in video summarization, which is aimed at providing an abstraction and capturing essential subject matter with a compact representation of complex information. In addition, the object-based approach also could be employed. It first imposes spatial-temporal segmentation to get individual regions in video frames. Low-level features then are extracted for analysis from these objects instead of from frames directly. Therefore, semantic information can be expressed explicitly via the object features (Jiang & Zhang, 2005a).

With the help of these techniques, video summarization can be achieved. One framework for hierarchical abstraction of news programs with emphases on key persons (i.e., Main Speaker Close-Up, MSC, denoting camera-focused talking heads in center of the screen) has been proposed (Jiang & Zhang, 2005b). It can answer user queries such as to find all the key persons in a program or to highlight specific scenes with particular persons. The framework consists of three levels (from low physical level to high semantic level): shot level, item level, and item-group level. One actual example is shown in Figure 6 to provide an explanation.

Suppose a user is interested in two persons appearing in the news program (e.g., Bill Clinton and Jorge Sampaio). The user can find both of them from the previously constituted human table by name or by representative images and can check the corresponding boxes. Then, multi-level abstractions can be acquired: abstraction at shot level contains all the MSC shots of these two persons; abstraction at item level is made up of news items, including those MSC shots; abstraction at item-group level extends to every news item group with news items

Figure 6. Multi-level video abstraction

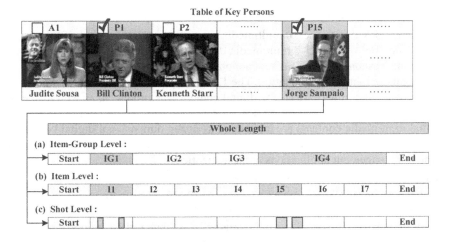

contained at item level. To obtain a summarization of preview sequence, frames belonging to all the video segments (at shot/item/item-group level) are concatenated together. Finally, abstractions at three levels with different durations are obtained in order to adaptively fulfill users' needs, from independent close-ups to complete video clips.

Overview of Current Focus

Current research works on SBVIR could be classified in several directions. A brief overview of some tracks is made in the following.

Image and Video Annotation

Image classification has been used to provide access to large image collections in a more efficient manner because the classification can reduce search space by filtering out the images in an unrelated category (Hirata et al., 2000). Image and video annotation significantly can enhance the performance of content-based visual retrieval systems by filtering out images from irrelevant classes during a search so as to reduce the processing time and the probability of error matching. One typical example is SIMPLIcity system, which uses integrated region matching based on image segmentation (Wang, Li, & Wiederhold, 2001). This system classifies images into various semantic categories such as textured/non-textured graph photograph. A series of statistical image classification methods also is designed during the development of this system for the purpose of searching images.

Inspired from SIMPLIcity system, a novel strategy by using feature selection in learning semantic concepts of image categories has been proposed (Xu & Zhang, 2006). For every image category, the salient patches on each image are detected by scale invariant feature transform (SIFT) introduced by Lowe (2004), and the 10-dimensional feature vectors are formed by PCA. Then, the DENCLUE (DENsity-based CLUstEring) clustering algorithm (Hinneburg & Keim, 1998) is applied to construct the continuous visual keyword dictionary.

A semantic description tool of multimedia content constructed with the Structured-Annotation Basic Tool of MPEG-7 multimedia description schemes (MDS) is presented in Kuo, Aoki, and Yasuda (2004). This tool annotates multimedia data with 12 main attributes regarding its semantic representation. The 12 attributes include answers to who, what, when, where, why, and how (5W1H) the digital content was produced, as well as the respective direction, distance, and duration (3D) information. With the proposed semantic attributes, digital multimedia contents are embedded as dozen dimensional digital content (DDDC). The establishment of DDDC would provide an interoperable methodology for multimedia content management applications at the semantic level.

A new algorithm for the automatic recognition of object classes from images (categorization) has been proposed (Winn, Criminisi, & Minka, 2005). It classifies a region according to the proportions of various visual words (clusters in feature space) by taking into account all pixels in images. The specific visual words and the characteristic proportions for each object are learned automatically by a supervised algorithm from a segmented training set, which gives the compact and yet discriminative appearance-based models for the object class.

According to a hypothesis that images dropped into the same text-cluster can be described with common visual features of those images, a strategy is described that combines textual and visual clustering results to retrieve images using semantic keywords and auto-annotate images based on similarity with existing keywords (Celebi & Alpkocak, 2005). Images first are clustered according to their text annotations using the C^3M clustering technique. The images also are segmented into regions and then clustered based on low-level visual features using the k-means clustering algorithm. The feature vector of the images then is mapped to a dimension equal to the number of visual clusters in which each entry of the new feature vector signifies the contribution of the image to that visual cluster. A feature vector also is created for the query image. Images in the textual cluster that give the highest matching score with the query image are determined by the matching of feature vectors.

A mechanism called Weblog-Style Video Annotation and Syndication to distribute and advertise video contents and to form communities centered on those contents effectively on the Internet is described in Yamamoto, Ohira, and Nagao (2005). This mechanism can tightly connect video contents, Weblogs, and users. It can be used for applications based on annotation by extracting the video annotations from the contents. In contrast to the machine processing of both voice- and image-recognition technology, which is difficult to apply to general contents, this mechanism has a low cost for annotation, as a variety of annotations is acquired from many users.

Human-Computer Interaction

In order to alleviate the problems that arise because of user subjectivity and the semantic gap, interactive retrieval systems have been proposed that place the user in the loop during retrievals. Such relevance feedback (RF) approaches aim to learn intended high-level query concepts and adjust for subjectivity in judgment by exploiting user input on successive iterations. In fact, since humans are much better than computers at extracting semantic information from images, relevance feedback has proved to be an effective tool for taking the user's judgment into account (Zhou & Huang, 2003). Generally, the user provides some quality assessment of the retrieval results to the system by indicating the degree of satisfaction with each of the retrieved results. The system then uses this feedback to adjust its query and/or the similarity measure in order to improve the next set of results.

Image retrieval is a complex processing task. In order to simplify the procedure, most current approaches for interactive retrieval make several restrictive assumptions (Kushki et al., 2004). One is that images considered similar according to some high-level concepts also fall close to each other in the low-level feature space. This generally is not true, as high-level semantic concepts may not be mapped directly to elements in a low-level feature space. Another is that the ideal conceptual query in the mind of a user can be represented in the low-level space and used to determine the region of this space that corresponds to images relevant to the query. However, as the mapping between low-level features and the conceptual space often is unclear, it is not possible to represent a high-level query as a single point in the low-level feature space. A novel approach has been proposed for interactive content-based image retrieval that provides user-centered image retrieval by lifting the previously mentioned restrictive assumptions imposed on existing CBIR systems while maintaining accurate retrieval performance (Kushki et al., 2004). This approach exploits user feedback to generate multiple query images for similarity calculations. The final similarity between a given image and a high-level user concept then is obtained as a fusion of similarity of these images to a set of low-level query representations.

A new RF framework based on a feature selection algorithm that nicely combines the advantages of a probabilistic formulation with those of discriminative learning methods, using both the positive example (PE) and the negative example (NE), has been proposed (Kherfi & Ziou, 2006). It tries through interaction with the user to learn the weights the user assigns to image features according to their importance and then to apply the results obtained to define similarity measures that correspond better to the user's judgment for retrieval.

Models and Tools for Semantic Retrieval

Many mathematic models have been developed, and many useful tools from other fields have been applied for semantic retrieval, such as machine learning (Dai & Zhang, 2004; Lim & Jin, 2005). One powerful tool for such a work is data mining, especially for Web mining. The huge amounts of multivariate information offered by the Web have opened up new possibilities for many areas of research. Web mining refers to the use of data mining techniques to automatically retrieve, extract, and evaluate (generalize/analyze) information for knowledge discovery from Web documents and services (Arotaritei & Mitra, 2004). Web data typically are unlabeled, distributed, heterogeneous, semi-structured, time varying, and high-dimensional. Hence, any human interface needs to handle context-sensitive and imprecise queries and provide for summarization, deduction, personalization, and learning. A survey on Web mining involving fuzzy sets and their hybridization with other soft computing tools is presented in Arotaritei and Mitra (2004). The Web-mining taxonomy has been described. The individual functions like Web clustering, association rule mining, Web navigation, Web personalization, and Semantic Web have been discussed in the fuzzy framework.

Pattern recognition plays a significant role in content-based recognition. A list of pattern recognition methods developed for providing content-based access to visual information (both image and video) has been reviewed (Antani, Kasturi, & Jain, 2002). Here, the term *pattern recognition methods* refer to their applicability in feature extraction, feature clustering, generation of database indices, and determining similarity between the contents of the query and database elements.

A semantic learning method for content-based image retrieval using the analytic hierarchical process (AHP) has been proposed (Cheng et al., 2005). The AHP provides a good way to evaluate the fitness of a semantic description that is used to represent an image object. The idea behind this work is that the problem of assigning semantic descriptions to the objects of an image can be formulated as a multi-criteria preference problem, while AHP is a powerful tool for solving multi-criteria preference problems. In this approach, a semantic vector consisting of the values of fitness of semantics of a given image is used to represent the semantic content of the image according to a predefined concept hierarchy, and a method for ranking retrieved images according to their similarity measurements is made by integrating the high-level semantic distance and the low-level feature distance.

A new technique, cluster-based retrieval of images by unsupervised learning (CLUE), which exploits similarities among database images, for improving user interaction with image retrieval systems has been proposed (Chen, Wang, & Krovetz, 2005). The major difference between a cluster-based image retrieval system and traditional CBIR systems lies in the two processing stages: selecting neighboring target images and image clustering, which are the major components of CLUE. CLUE retrieves image clusters by applying a graph-theoretic clustering algorithm to a collection of images in the vicinity of the query. In CLUE, the clustering process is dynamic, and the clusters formed depend on which images are retrieved in response to the query. CLUE can be combined with any real-valued symmetric similarity measure (metric or nonmetric). Thus, it may be embedded in many current CBIR systems, including relevance feedback systems.

A retrieval approach that uses concept languages to deal with nonverbally expressed information in multimedia is described in Lay and Gua (2006). The notion of concept language here is a rather loose one. It covers all other conventional methods of communication, including the systems of signs and rules, such as body language, painterly language, and the artificial languages of chess and the solitaire game. In operation, a finite number of elemental concepts of a concept language is identified and used to index multimedia documents. The elemental concepts then allow a large number of compound semantic queries to be expressed and operated as sentences of elemental concepts. Managing semantics by concept languages not only extends an intuitive query regime in which semantic queries can be specified more expressively and extensively but also allows concept detection to be restricted to a more manageable sum of semantic classes.

Miscellaneous Techniques in Application

Still other techniques could have potential in semantic-based visual information retrieval. A cascading framework for combining intra-image and interclass similarities in image retrieval motivated from probabilistic Bayesian principles has been proposed (Lim & Jin, 2005). Support vector machines are employed to learn local view-based semantics based on just-in-time fusion of color and texture features. A new detection-driven, block-based segmentation algorithm is designed to extract semantic features from images. The detection-based indexes also serve as input for support vector learning of image classifiers in order to generate class-relative indexes. During image retrieval, both intra-image and interclass similarities are combined to rank images. Such an approach would be suitable for unconstrained consumer photo images in which the objects are often ill-posed, occluded, and cluttered with poor lighting, focus, and exposure.

A novel, contents-based video retrieval system called DEV (Discussion Embedded Video), which combines video and an electronic bulletin board system (BBS), has been proposed (Haga & Kaneda, 2005). A BBS is used as an index of video data in the DEV system. The comments written in the BBS are arranged not according to the time that comments were submitted but by the playing time of video footage whose contents correspond to the contents of submitted comments. The main ideas behind this work are twofold: (1) since a participant's comments would be a summarization of the contents of a part of video, so it can be used as an index of it; and (2) as a user of this system can search part of the video by retrieving it via keywords in the BBS comments, so by detecting the part on which the comments from participants are concentrating it is possible to ascertain the topic of the video.

A two-stage retrieval process is described in Vogel and Schiele (2006). Users perform querying through image description by using a set of local semantic concepts and the size of the image area to be covered by the particular concept. In first stage, only small patches of the image are analyzed, whereas in the second stage, the patch information is processed, and the relevant images are retrieved. In this two-stage retrieval system, the precision and recall of retrieval can be modeled statistically. Based on the model, closed-form expressions that allow for the prediction as well as the optimization of the retrieval performance are designed.

Future Trends

There are several promising directions for future research.

Domain Knowledge and Learning

A semantic multimedia retrieval system consists of two components (Naphade & Huang, 2002). The first component links low-level physical attributes of multimedia data to high-level semantic class labels. The second component is domain knowledge or any such information apart from the semantic labels themselves that makes the system more competent to handle the semantics of the query. Many research works related to the first component have been conducted. For the second component, which can be in the form of rules, heuristics, or constraints, much more effort is required to automatically capture long-term human experience of human experts.

Various learning techniques such as active learning and multiple-instance learning are efficient to alleviate the cost of annotation (Naphade & Huang, 2002). One important direction is related to the development of an intelligent dialogue mechanism to increase the effectiveness of the user's feedback. The challenge is to attain performance that is considered useful by the end users of the systems. To this end, an active role of statistical learning in medium analysis will bridge the gap between the user's desire and the system's reply and will prevail in future media applications that involve semantic understanding.

What can be learned from human beings is not only domain knowledge but also the comportment of human beings. A technique employing the concept of small-world theory, which mimics the way in which humans keep track of descriptions of their friends and acquaintances, has been proposed (Androutsos, Androutsos, & Venetsanopoulos, 2006). Such a social networking behavior is extremely general to be applied to both low-level and semantic descriptors. This mirroring of the characteristics of social acquaintance provides an intelligent way to incorporate human knowledge into the retrieval process. Extending this procedure to other kinds of knowledge would be promoting.

Relevance Feedback and Association Feedback

Domain knowledge could be used in combination with relevance feedback to approach the optimal retrieval results. However, feedback is just a method for refining the results so it cannot totally determine the performance of retrieval systems. How to make this refining process fast following the user's aspiration in the course of retrieval is interesting (Zhang et al., 2004). When the system is based on lower-level features, relevance feedback only could improve the retrieval results to some degree. The use of relevance feedback based on high-level content description in the object level could further improve the performance.

On the other side, a potential direction would be to use association feedback based on feature elements (Xu & Zhang, 2001, 2002). Association feedback tries to find out the associated parts between the existing interest (user intent) and the new target (related to the current retrieval results), which usually is a subset of the demand feature element set; that is, the bridge to the new retrieval.

Even Higher-Level Exploration

Semantic retrieval requires the use of a cognitive model—a feature element construction model that tries to enhance the view-based model—while importing some useful inferences from image-based theory has been proposed (Xu & Zhang, 2003). A feature element is the discrete unit extracted from low-level data that represent the distinct or discriminating visual characteristics that may be related to the essence of the objects.

The semantic level is higher than the feature level, while the affective level is higher than the semantic level (Hanjalic, 2001). Affection is associated with some abstract attributes that are quite subjective. For example, atmosphere is an important abstract attribute for film frames. Atmosphere serves an important role in generating the scene's topic or in conveying the message behind the scene's story. Five typical categories of atmosphere semantics have been studied: vigor and strength, mystery or ghastfulness, victory and brightness, peace or desolation, and lack unity and appears disjoint (Xu & Zhang, 2005).

Conclusion

There is no clear image yet about the development of semantic-based visual information retrieval. Research made from a low feature level to a high semantic level in visual information retrieval already has obtained the result of numerous noteworthy progresses and quite a number of publications. However, many approaches have obtained only limited success, and various investigations are still pursued on the way. For example, still very little work has been done to address the issue of human perception of visual data content (Vogel & Schiele, 2006). Several practical approaches, such as multi-level model, classification and annotation, machine-learning techniques, human-computer interaction, as well as various models and tools, have been discussed in this chapter. Few potential research directions, such as the domain knowledge and learning, relevance feedback and association feedback, as well as research at an even high level, are discussed. It is apparent that the future for content-based visual information retrieval will rely on high-level research.

Acknowledgments

This work has been supported by Grants NNSF-60573148 and SRFDP-20050003013.

References

Amir, A., & Lindenbaum, M. (1998). A generic grouping algorithm and its quantitative analysis. *IEEE PAMI, 20*(2), 168–185.

Androutsos, P., Androutsos, D., & Venetsanopoulos, A. N. (2006). Small world distributed access of multimedia data. *IEEE Signal Processing Magazine, 23*(2), 142–153.

Antani, S., Kasturi, R., & Jain, R. (2002). A survey on the use of pattern recognition methods for abstraction, indexing and retrieval of images and video. *Pattern Recognition, 35*(4), 945–965.

Arotaritei, D., & Mitra, S. (2004). Web mining: A survey in the fuzzy framework. *Fuzzy Sets and Systems, 148*, 5–19.

Bimbo, A. (1999). *Visual information retrieval*. Morgan Kaufmann.

Castelli, V., Bergman, L. D., Kontoyiannis, I., et al. (1998). Progressive search and retrieval in large image archives. *IBM J. Res. Develop., 42*(2), 253–268.

Celebi, E., & Alpkocak, A. (2005). Combining textual and visual clusters for semantic image retrieval and auto-annotation. In *Proceedings of the 2nd European Workshop on the Integration of Knowledge, Semantics and Digital Media Technology* (pp. 219–225).

Chang, S. F., Chen, W., Meng, H. J., Sundaram, H., & Zhong, D. (1998). A fully automated content-based video search engine supporting spatiotemporal queries. *IEEE CSVT, 8*(5), 602–615.

Chen, Y. X., Wang, J. Z., & Krovetz, R. (2005). CLUE: Cluster-based retrieval of images by unsupervised learning. *IEEE IP, 14*(8), 1187–1201.

Cheng, S. C., Chou, T. C., Yang, C. L., et al. (2005). A semantic learning for content-based image retrieval using analytical hierarchy process. *Expert Systems with Applications, 28*(3), 495–505.

Dai, S. Y., & Zhang, Y. J. (2004). Adaboost in region-based image retrieval. In *Proceedings of the International Conference on Acoustic, Speech, and Signal Processing* (pp. 429–432).

Dai, S. Y., & Zhang, Y. J. (2005). Unbalanced region matching based on two-level description for image retrieval. *Pattern Recognition Letters, 26*(5), 565–580.

Doermann, D. (1998). The indexing and retrieval of document images: A survey. *Computer Vision and Image Understanding, 70*(3), 287–298.

Gao, Y. Y., Zhang, Y. J., & Merzlyakov, N. S. (2000). Semantic-based image description model and its implementation for image retrieval. In *Proceedings of the First International Conference on Image and Graphics* (pp. 657–660).

Gao, Y. Y., Zhang, Y. J., & Yu, F. (2001). Self-adaptive relevance feedback based on multi-level image content analysis. *SPIE, 4315*, 449–459.

Haga, H., & Kaneda, H. (2005). A usability survey of a contents-based video retrieval system by combining digital video and an electronic bulletin board. *Internet and Higher Education, 8*(3), 251–262.

Hanjalic, A. (2001). Video and image retrieval beyond the cognitive level: The needs and possibilities. *SPIE, 4315*, 130–140.

Hinneburg, A., & Keim, D. A. (1998). An efficient approach to clustering in large multimedia databases with noise. *KDD '98*, 58–65.

Hirata, K., et al. (2000). Integration of image matching and classification for multimedia navigation. *Multimedia Tools and Applications, 11*(3), 295–309.

Hong, D. Z., Wu, J. K., & Singh, S. S. (1999). Refining image retrieval based on context-driven method. *SPIE, 3656*, 581–593.

Jaimes, A., & Chang, S. F. (1999). Model-based classification of visual information for content-based retrieval. *SPIE, 3656*, 402–414.

Jiang, F., & Zhang, Y. J. (2005a). Camera attention weighted strategy for video shot grouping. *SPIE, 5960*, 428–436.

Jiang, F., & Zhang, Y. J. (2005b). News video indexing and abstraction by specific visual cues: MSC and news caption. In S. Deb (Ed.), *Video data management and information retrieval* (pp. 254–281). IRM Press.

Kato, T. (1992). Database architecture for content-based image retrieval. *SPIE, 1662*, 112–123.

Kherfi, M. L., & Ziou, D. (2006). Relevance feedback for CBIR: A new approach based on probabilistic feature weighting with positive and negative examples. *IEEE IP, 15*(4), 1014–1033.

Kuo, P. J., Aoki, T., & Yasuda, H. (2004). MPEG-7 based dozen dimensional digital content architecture for semantic image retrieval services. In *Proceedings of the 2004 IEEE International Conference on E-Technology, E-Commerce and E-Service (EEE '04)* (pp. 517–524).

Kushki, A., et al. (2004). Query feedback for interactive image retrieval. *IEEE CSVT, 14*(5), 644–655.

Lay, J. A., & Gua, L. (2006). Semantic retrieval of multimedia by concept languages. *IEEE Signal Processing Magazine, 23*(2), 115–123.

Lim, J. H., & Jin, J. S. (2005). Combining intra-image and inter-class semantics for consumer image retrieval. *Pattern Recognition, 38*(6), 847–864.

Lowe, D. G. (2004). Distinctive image features from scale-invariant key-points. *International Journal of Computer Vision, 60*(2), 91–110.

Luo, J. B., Boutell, M., & Brown, C. (2006). Pictures are not taken in a vacuum: An overview of exploiting context for semantic scene content understanding. *IEEE Signal Processing Magazine, 23*(2), 101–114.

Luo, Y., et al. (2001). Extracting meaningful region for content-based retrieval of image and video. *SPIE, 4310,* 455–464

Naphade, M. R., & Huang, T. S. (2002). Extracting semantics from audiovisual content: The final frontier in multimedia retrieval. *IEEE NN, 13*(4), 793–810.

Rui, Y., Huang, T. S., & Chang, S.F. (1999). Image retrieval: Current techniques, promising directions, and open issues. *Journal of Visual Communication and Image Representation, 10*(1), 39–62.

Smeulders, A. W. M., et al. (2000). Content-based image retrieval at the end of the early years. *IEEE PAMI, 22*(12), 1349–1380.

Vogel, J., & Schiele, B. (2006). Performance evaluation and optimization for content-based image retrieval. *Pattern Recognition, 39,* 897–909.

Wang, J. Z., Li, J., & Wiederhold, G. (2001). SIMPLIcity: Semantics-sensitive integrated matching for picture libraries. *IEEE PAMI, 23*(9), 947–963.

Winn, J., Criminisi, A., & Minka, T. (2005). Object categorization by learned universal visual dictionary. In *Proceedings of the 10th International Conference on Computer Vision (ICCV'05)* (Vol. 2, pp. 1800–1807).

Xu, Y., & Zhang, Y. J. (2001). Association feedback: A novel tool for feature elements based image retrieval. In *Proceedings of the Second IEEE Pacific Rim Conference on Multimedia* (pp. 506–513).

Xu, Y., & Zhang, Y. J. (2002). Feature element theory for image recognition and retrieval. *SPIE, 4676,* 126–137

Xu, Y., & Zhang, Y. J. (2003). Semantic retrieval based on feature element constructional model and bias competition mechanism. *SPIE, 5021,* 77–88.

Xu, F., & Zhang, Y. J. (2005). Atmosphere-based image classification through illumination and hue. *SPIE, 5960,* 596–603.

Xu, F., & Zhang, Y. J. (2006). Feature selection for image categorization. In *Proceedings of the Seventh Asian Conference on Computer Vision* (pp. 653–662).

Yamamoto, D., Ohira, S., & Nagao, K. (2005). Weblog-style video annotation and syndication. In *Proceedings of the First International Conference on Automated Production of Cross Media Content for Multi-Channel Distribution (AXMEDIS'05)* (pp. 1–4).

Yu, T. L., & Zhang, Y. J. (2001a). Motion feature extraction for content-based video sequence retrieval. *SPIE, 4311,* 378–388.

Yu, T. L., & Zhang, Y. J. (2001b). Retrieval of video clips using global motion information. *IEE Electronics Letters, 37*(14), 893–895.

Zhang, Y. J. (2003). *Content-based visual information retrieval.* Beijing, China: Science Publisher.

Zhang, Y. J. (2005a). Advanced techniques for object-based image retrieval. In M. Khosrow-Pour (Ed.), *Encyclopedia of information science and technology* (Vol. 1, pp. 68–73). Hershey, PA: Idea Group Reference.

Zhang, Y. J. (2005b). New advancements in image segmentation for CBIR. In M. Khos-row-Pour (Ed.), *Encyclopedia of information science and technology* (Vol. 4, pp. 2105–2109). Hershey, PA: Idea Group Reference.

Zhang, Y. J. (2006). Mining for image classification based on feature elements. In J. Wang (Ed.), *Encyclopedia of Data Warehousing and Mining* (Vol. 1, pp. 773–778).

Zhang, Y. J., Gao, Y. Y., & Luo, Y. (2004). Object-based techniques for image retrieval. In S. Deb (Ed.), *Multimedia systems and content-based image retrieval* (pp. 156–181). Hershey, PA: Idea Group Publishing.

Zhang, Y. J., Gao, Y. Y., & Merzlyakov, N. S. (2002a). Object recognition and matching for image retrieval. *SPIE, 4875*, 1083–1089.

Zhang, Y. J., & Lu, H. B. (2002b). A hierarchical organization scheme for video data. *Pattern Recognition, 35*(11), 2381–2387.

Zhou, X. S., & Huang, T. S. (2003). Relevance feedback for image retrieval: A comprehensive review. *Multimedia Systems, 8*(6), 536–544.

Section II

From Features
to Semantics

Chapter II

The Impact of Low-Level Features in Semantic-Based Image Retrieval

Konstantinos Konstantinidis, Democritus University of Thrace, Greece

Antonios Gasteratos, Democritus University of Thrace, Greece

Ioannis Andreadis, Democritus University of Thrace, Greece

Abstract

Image retrieval (IR) is generally known as a collection of techniques for retrieving images on the basis of features, either low-level (content-based IR) or high-level (semantic-based IR). Since semantic-based features rely on low-level ones, in this chapter the reader initially is familiarized with the most widely used low-level features. An efficient way to present these features is by means of a statistical tool that is capable of bearing concrete information, such as the histogram. For use in IR, the histograms extracted from the previously mentioned features need to be compared by means of a metric. The most popular methods and distances are thus apposed. Finally, a number of IR systems using histograms are presented in a thorough manner, and their experimental results are discussed. The steps in order to develop a custom IR system along with modern techniques in image feature extraction also are presented.

Introduction

Research in color imaging has recently emerged in a number of different applications, including military, industrial, and civilian, that generates gigabytes of color images per day. Moreover, recent improvements in information and communication technology have led to higher data transmission rates and, consequently, to a boom in networking. Therefore, more and more people have access to an increasing number of images. It is obvious that this will lead to a chaotic predicament, unless the enormous amount of available visual information is organized (Gagliardi & Schettini, 1997). Organization here means that appropriate indexing is available in order to allow efficient browsing, searching, and retrieving as in keyword searches of text databases. Associating a text to each image is one of the most popular and straightforward ways to index (Rowe, 2005). However, this means that prior to submitting each image into a database, a human agent must accompany it with a caption, thus leading to a lack of system automization. In many applications, such as in digital photography, area surveillance, remote sensing, and so forth, the images are labeled with automatically produced computerized names that are totally irrelevant to their semantic content. The best solution to such cases is the extraction and storage of meaningful features from each image for indexing purposes. In order to retrieve these images, a procedure known as query by example is performed; that is, the user has to present an image to the system, and the latter retrieves others alike by extracting features from the query image and comparing them to the ones stored in the database. The extraction of meaningful features, both content (Del Bimbo, 1999) and semantic (Zhang & Chen, 2003), is critical in IR and, therefore, an active field of research (Eakins, 2002; Smeulders, Worring, Santini, Gupta, & Jain, 2000). Nevertheless, while considering a semantic query (e.g., A Red Round Dirty Car), the descriptive components are based on low-level features; red on color, round on shape, and dirty on texture. Hence, in order for a semantic-based IR system to perform effectively, its lower features

Figure 1. Block diagram of the basic structure of a generic IR system

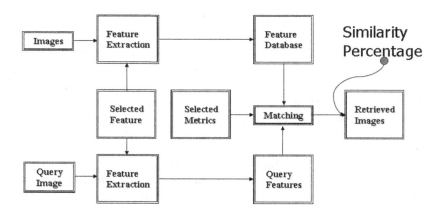

must be extracted and indexed accordingly. The low-level features most commonly used by researchers are color (Castelli & Bergman, 2002, Del Bimbo, 1999, Gagliardi & Schettini, 1997; Konstantinidis & Andreadis, 2005; Konstantinidis, Gasteratos, & Andreadis, 2005; Liang, Zhai, & Chavel, 2002; Pass, Zabih, & Miller, 1996; Swain & Ballard, 1991), texture (Castelli & Bergman, 2002; Del Bimbo, 1999; Gasteratos, Zafeiridis, & Andreadis, 2004; Howarth & Rüger, 2004; Wang & Wiederhold, 2001), and shape (Brandt, Laaksonen, & Oja, 2002; El Badawy & Kamel, 2002; Jain & Vailaya, 1996). Color possesses a dominant function in vision, thus allowing the performance of complex tasks such as the discrimination between objects with similar shape characteristics but different color features, the tracking of moving objects as well as scene property analysis. Therefore, the use of color as a meaningful feature in IR is straightforward. Also, due to variations in viewing angles, texture patterns may vary in scale and orientation from image to image or even in different parts of the same image. Therefore, texture provides a useful cue for IR. Finally the role of shape is essential in IR, since there are examples in which two images with similar color and texture characteristics may exhibit highly dissimilar semantic content.

A useful tool in color image analysis is the histogram (Gonzalez & Woods, 1992)—a global statistical low-level descriptor that represents the color distribution for a given image. In color image processing methods, histograms usually enclose information of three components that correspond to the three components of the color space used. A histogram-based retrieval system (Figure 1) requires the following components: a suitable color space such as HSV, CIELAB or CIELUV; a feature representation such as classic, joint, or fuzzy histograms; and a similarity metric such as the Euclidean Distance, the Matusita distance, or the Histogram Intersection method. Similar to color, an image also can be indexed using a textural energy histogram. To this end, the image is processed by a method that may include convolution with Gabor filters (Zhang & Lu, 2000), wavelets (Bartolini, Ciaccia, & Patella, 1999), and Laws' masks (Laws, 1980a, 1980b).

IR using color histograms has both advantages and limitations (Konstantinidis & Andreadis, 2005):

- It is robust, since color histograms are rotation- and scale-invariant.
- Histograms are straightforward to implement.
- It is fast. The histogram computation has $O(M^2)$ complexity for an MxM image, while a histogram comparison has $O(n)$, where n is the number of histogram bins, or quantization levels, of the colors used.
- It has low storage requirements. The color histogram size is much smaller than the size of the image itself.
- However, no spatial information of the color distribution is included, so the problem of two completely different images having similar histograms may arise.
- It is not immune to lighting variations.

As stated before, the efficiency of the semantic representation depends heavily on how compact and robust the indexing of the low-level features is. Therefore, in this chapter, we approach the role of color and texture in IR by constructing compact, easy-to-handle, meaningful histograms. We also present examples in which color, texture, and spatial information are combined in generalized histograms for indexing. Methods (e.g., joint, fuzzy, etc.) and tradeoffs in histogram making are also discussed as well as the metrics most frequently encountered in a histogram-based IR system. Recent advances are presented through exemplar IR systems. These are based on the aforementioned methods and are comparatively apposed. At the end of the chapter, the user will have a sufficient background with which to compose a custom IR system.

Background and Related Work

Color Histograms

In order to define a color histogram, let I be an $n \times n$ image. Each pixel, $p=(x, y)$, of the image may have one of m colors of the set $\{c_1, c_2, ..., c_m\}$; that is, $I(p) \in \{c_1, c_2, ..., c_m\}$. Let $I^c \overset{\Delta}{=} \{p \in nxn | I(p) = c\}$ be the set of pixels of image I, which are of color c.

Using this notation a histogram, $HI(.)$, for an image I is given by:

$$H_I(i) \overset{\Delta}{=} n^2 \Pr[p \in I^{c_i}] \tag{1}$$

In most applications of the color histogram, the term $\Pr[p \in I^{c_i}]$ is estimated as the fractional number of pixels with color c_i.

The meaning of this is that, given a discrete color space, a color histogram is a statistic that counts the frequency with which each color appears in the image.

Color Histogram Extraction Techniques

In the past, various scientists have attempted to tackle the problem of real-time IR. In view of the fact that content-based IR has been an active area of research since the 1990s, quite a few IR systems, both commercial and research, have been developed. The most popular systems are query by image content (QBIC) and MIT's Photobook and its new version of FourEyes. Other popular systems are VisualSEEK, MetaSEEK and WebSEEK, Netra, Multimedia Analysis and Retrieval Systems (MARS), and Chabot.

In IR systems, image content is stored in visual features that can be divided into three classes according to the properties they describe: color, texture, and shape. Color and texture contain vital information, but nevertheless, shape-describing features are also essential in an efficient content-based IR system.

In addition to the complex real-time systems mentioned previously, several researchers also tried to present simpler and, therefore, computationally wise and much lighter solutions to the problem.

For example, using the approach in Swain and Ballard (1991), a histogram is created from the opponent color space by subdividing the three derived components: rg, by, and wb into 16, 16, and 8 sections, respectively, resulting in a 2048-bin histogram.

Furthermore, Liang et al. (2002) presented a system that uses crisp values produced by a Gaussian membership function in order to characterize a similarity fuzzy set centered at a given color vector in the RGB color space. Simple histograms are created using the previously mentioned Gaussian membership function.

Lu and Phillips (1998) proposed the use of perceptually weighted histograms (PWH). Instead of dividing each color channel by a constant (quantization step) when obtaining a histogram, they find representative colors in the CIELUV color space. The number of the representative colors equals the required number of bins.

Pass et al. (1996) created a joint histogram by selecting a set of local pixel features, subsequently constructing a multidimensional histogram. Each entry in this histogram contains the number of pixels that are described by a particular combination of feature values. Just as a color histogram approximates the density of pixel color, a joint histogram approximates the joint density of several pixel features. The features used were selected empirically and can be implemented time efficiently: color, edge density, textureness, gradient magnitude, and rank.

The idea of normalizing color images separately in each band as a reasonable approach to color constancy preprocessing in the context of indexing an image database was adopted by Drew, Wei, and Li (1998). The information extracted from the images is transformed into a 2D representation by using histograms of chromaticity. Then, regarding the 2D feature space histograms as images, they apply a wavelet-based image reduction transformation for low-pass filtering, a square root operation, and discrete cosine transform (DCT) and truncation. The resulting indexing scheme uses only 36 or 72 values to index the image database.

The weak spot of the indexing methods described earlier is the lack of spatial information in the histograms. For example, the two pictures shown in Figure 2 have identical histograms but different spatial distributions. Their semantic content is noticeably different, so evidently, one cannot assume that color distribution alone is always sufficient to represent the pictorial content of an image, since it comprises an abstract representation of it.

So, although the histograms become computationally heavier, in order to avoid false negatives, several methods have been introduced that enrich the color feature with geometric or spatial information. Stricker and Swain (1994) used boundary histograms to encode the lengths of the boundaries between different discrete colors in order to take into account geometric information in color image indexing. But this method may produce a large feature space (for a discrete color space of 256 elements, a boundary histogram of 32,768 bins) and is not robust enough to deal with textured color images. Gagliardi and Schettini (1997) investigated the use and integration of various color information descriptions and similarity measurements in order to enhance the system's effectiveness. In their method, both query and database images are described in the CIELAB color space with two limited palettes of perceptual importance of 256 and 13 colors, respectively. A histogram of the finer color quantization and another of the boundary lengths between two discrete colors

Figure 2. Two images with different semantic content, but with identical histograms

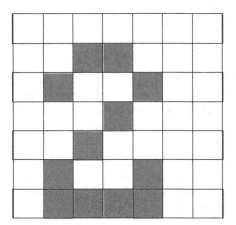 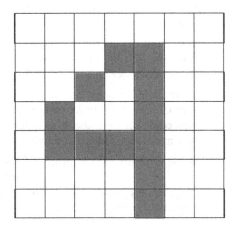

of the coarser quantization are used as indices of the image. While the former contains no spatial information (it describes only the color content in the image), the latter provides a brief description of the color spatial arrangement.

Pass and Zabih (1999) proposed another color histogram enhancement that also augments color histograms with spatial information. Color coherence vectors (CCVs) classify each pixel as either coherent or incoherent; based on whether it is part of a large color homo-geneous region in the image or not. After the classification, the histogram is constructed (as in the original color histogram formulation), and the value of each bin is the number of coherent pixels. One shortcoming of this method is that it neglects the relationship of a connected component to its background. It also fails to capture the shape of the component. Thus, Zachary and Iyengar (1999) developed the so-called threshold edge vector (TEV) (an extension to the CCV method), which addresses these two issues, by storing an additional vector containing edge information.

Heidemann (2004) presented an approach to represent spatial color distributions using local principal component analysis (PCA). The representation is based on image windows, which are selected by two complementary data-driven attentive mechanisms: a symmetry-based saliency map and an edge/corner detector. The eigenvectors obtained from the local PCA of the selected windows form color patterns that capture both low and high spatial frequencies, so they are well-suited for shape as well as texture representation.

Overall, color is the most widely used feature in IR; sometimes enriched with spatial in-formation or else with multiple feature vectors. In a number of applications however, even these enrichments are insufficient to provide the exact semantic content (e.g., the Red Round Dirty Car). Thus, a synergy of features becomes a necessity.

Texture Histogram Extraction Techniques

Liapis and Tziritas (2004) approached the IR problem based on a combination of texture and color features. Texture features are extracted using the discrete wavelet frame analysis, whereas histograms of the CIELAB chromaticity coordinates are used as color features.

A method for color texture classification using self-relative histogram ratio features was presented by Paschos and Petrou (2003). The method utilizes the 3-D xyY color histogram of a given image (xyY is derived from the CIEXYZ color space, where xy is chrominance and Y is luminance). The chrominance component (xy) turns out to contain sufficient information for the proposed method to adequately classify the set of 164 VisTex color textures. When any of the previously described histogram extraction phases comes to an end and when an adequate image descriptor has been produced for every image in the database, a way to compare the latter to the one from the query image is needed.

Histogram Comparison Methods (Metrics)

The histogram comparison methods presented in this chapter are the ones most frequently met in the literature for the purpose of IR. In the mathematical expressions presented next, H_Q is the query histogram, H_C is the histogram to be compared, and *(i)* is the number of bins.

The Bhattacharyya distance (Fukunaga, 1990) measures the statistical separability of spectral classes, giving an estimate of the probability of correct classification. This distance overpasses zero histogram entries. For highly structured histograms (i.e., those that are not populated uniformly), this can lead to the selection of matches in which there is a strong similarity between the structure of query and database histogram.

$$B(H_Q, H_C) = -\ln \sum_i \sqrt{H_Q(i) \times H_C(i)} \tag{2}$$

The divergence factor measures the compactness of the color distribution in the query histogram with respect to the histograms of the images in the database.

$$D(H_Q, H_C) = \sum_i \left[\left(H_Q(i) - H_C(i) \right) \ln \frac{H_Q(i)}{H_C(i)} \right] \tag{3}$$

The Euclidean distance is one of the oldest distances used for IR and can be expressed through equation (4).

$$L_2(H_Q, H_C) = \sqrt{\sum_i \left(H_Q(i) - H_C(i) \right)^2} \tag{4}$$

The Matusita distance (Fukunaga, 1990) is a separability measure that provides a reliable criterion apparently because, as a function of class separability, it behaves much more like the probability of correct classification. It is expressed as:

$$M(H_Q, H_C) = \sqrt{\sum_i \left(\sqrt{H_Q(i)} - \sqrt{H_C(i)}\right)^2} \qquad (5)$$

Swain and Ballard (1991) introduced the histogram intersection method, which is robust in respect to changes in image resolution, histogram size, occlusion, depth, and viewing point. The similarity ratio that belongs to the interval [0, 1] is compared to a given threshold. It can be described by:

$$H(H_Q, H_C) = \frac{\sum_i \min\left(H_Q(i), H_C(i)\right)}{\min\left(\sum_i H_Q(i), \sum_i H_C(i)\right)} \qquad (6)$$

A New Perspective to Image Retrieval

In this section, a novel perspective to the problem of IR is presented, and a number of solutions are proposed. In order to tackle the problem of IR, four exemplar systems are presented that deal with all the aspects of the matter except for shape, which is used in much more application-dependent systems. The features addressed are those of color, texture, and spatial information. The first example shows that the color spaces used can be addressed as fuzzy sets to construct a more efficient and compact image descriptor. Due to its robustness, it could easily be included as a basis for a semantic feature. In the second system, it is explained that although color is sufficient enough to perform efficient retrieval, there are times when dissimilar images have very similar color distributions, thus producing false positives. These can be avoided by embedding spatial information into the feature at hand. Textureness is introduced in the third system, which is an example of an application-specific IR system for aerial images. Nonetheless, it is a fair example of the synergy between color and texture. In order to avoid the dependency, the features of color and color-texture are interlaced, as presented in the fourth system.

Fuzzy Histogram Image Retrieval

One might say that the largest image database in the world is that of the Worldwide Web as a whole, in which, since new sites are introduced daily, the images become truly countless. Thus, IR systems are needed that are fast and light but also robust and efficient. The most robust feature of all is that of color, and the easiest way to index it by also preserving memory is with histograms. However, a straightforward histogram IR system may be accompanied with a series of drawbacks; the number of bins required can be neither large, which presents the issue of perceptually similar colors belonging to adjacent bins and, thus, seemingly different, nor too small, since a few bins will result in a loss of resolution and, thus, of efficiency. In order to overcome this obstacle, a fuzzy histogram creation system is introduced that uses the CIELAB color space and produces a 10-bin histogram (Konstantinidis et al., 2005). The CIELAB color space is selected since it is a perceptually uniform color space, which approximates the way that humans perceive color. However, the main reason is that CIELAB was found to perform better than other color spaces in strenuous retrieval tests performed in the laboratory for this exact purpose. In CIELAB, the luminance is represented as L*, the relative greenness-redness as a*, and the relative blueness-yellowness as b*. The L* component does not contribute in providing any unique colors but for

Figure 3. Membership functions of L, a*, and b**

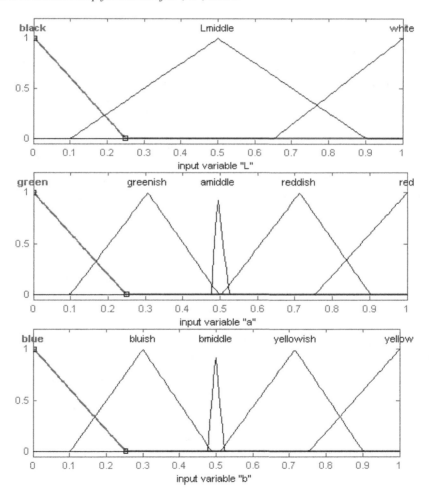

Figure 4. Membership functions of the output of the fuzzy system

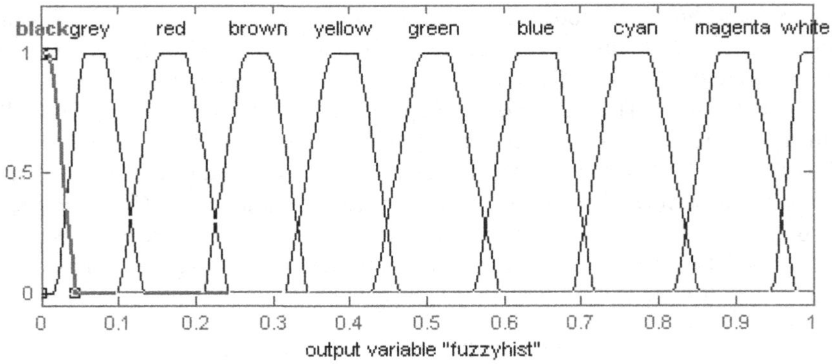

the shades of colors white, black, and grey. Thus, the L* component receives a lower weight with respect to the other two components of the triplet that represent the coloration. The tests performed on the regions of the CIELAB color space prove that in order for the IR system to work effectively the L*, a*, and b* components should be subdivided roughly into three, five, and five regions, respectively. The fuzzification of the input is accomplished by using triangular-shaped, built-in membership functions for the three input components, which represent the regions as shown in Figure 3. The reason for which the middle membership function even exists both in a* and b* is that in order to represent black, grey, and white, as seen in L*, then a* and b* must be very close to the middle of their regions.

The Mamdani type of fuzzy inference was used. The implication factor that determines the process of shaping the fuzzy set in the output membership functions based on the results of the input membership functions is set to min, and the OR and AND operators are set to max and min, respectively. The output of the system has 10 equally divided membership functions, as shown in Figure 4.

The defuzzification phase is best performed using the lom (largest of maximum) method along with 10 trapezoidal membership functions, thus producing 2,500 clustered bin values (all images were resized to a 50x50 pixel size to increase performance), which lead to the 10-bin final fuzzy histogram. The fuzzy linking of the three components is made according to 27 fuzzy rules, which leads to the output of the system. The rules were established through empirical conclusions that arose through thorough examination of the properties of a series of colors and images in the CIELAB color space.

In Figure 5, a query image (the well-known mandrill) and its respective fuzzy histogram are presented. The bins of the histogram shown in this Figure are in respect to (1) black, (2) dark grey, (3) red, (4) brown, (5) yellow, (6) green, (7) blue, (8) cyan, (9) magenta, and (10) white. The dominant colors in the image are easily noticed; bins 3, 4, 6, and 7 are activated mostly because of its fur and nose. The histogram in the proposed scheme, though apparently rough, has proven to be an efficient tool for accurate IR, as presented in the experimental results section.

Figure 5. (a) Query image 1 and (b) the respective 10-bin fuzzy linked histogram

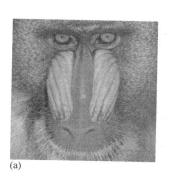

(a)

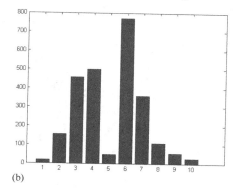

(b)

Once the histograms of the images in the database have been extracted, they are compared to the histogram derived from the query image by means of intersection (Swain & Ballard, 1991). This method was selected through a large number of tests performed in order to find the most suitable metric for this kind of histogram.

Spatially Biased Histogram Image Retrieval

Most color histogram creation methods like the one previously described contain global attributes; that is, the histogram in hand describes the overall statistics of the color in the images. Although such systems have proved to be efficient in most cases and are insensitive to rotation and scaling of the images, they also can present deficiencies in cases where the images have similar colors but are spatially distributed differently. This problem can be overcome with the use of local histograms (i.e., splitting each image into smaller regions) (Konstantinidis & Andreadis, 2005). These, on the other hand, suffer from a severe lack in speed due to the rapid increase in computational burden, which results from the repetitiveness needed to produce them. This led to the need of adopting global histograms with embedded local characteristics, such as a spatially biased histogram. The main gist of this method is to create a histogram by taking into consideration the color distribution in the image along with the concentrations of the dominant colors. The suggested histogram creation method has a straightforward algorithm, and only the hue component is enriched with spatial information so as to maintain the original histogram speed. The main reason the HSV color space was selected is because it reflects human vision quite accurately and because it mainly uses only one of its components (hue) to describe color in an image. The other two components (i.e., saturation and value) are significant only when describing black, white, gray, and the various shades of the colors. Thereby, in this method, hue is used mostly, being divided into eight regions, whereas saturation and value are divided into four each. The three color components then are linked, thus creating a (8x4x4) histogram of 128 bins. During the reading of every pixel's value in the hue frame from left to right and top to bottom, the algorithm also searches in the manner of a cross having 15 pixels width and 15 pixels height around

Figure 6. Graphical presentation of the presented method: A 50x50 pixel image I with the cross scanning the neighboring pixels of I (8,8) (The gray dots are the pixels enlarged for presentation purposes, and the black dot is the pixel at hand.)

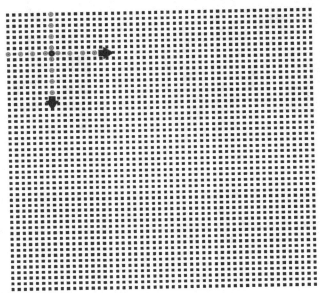

the current pixel for pixels with similar colors, as shown in Figure 6. If any pixel that is included in the vicinity of the full length of the cross possesses a color similar to the one of the central pixel, then instead of increasing the value of the specified bin of the particular color by one, it increases it by two, thus considerably enlarging the number of pixels in the bins that contain colors with significant concentrations in the image. The similarity of any pixel on the cross and the central pixel is measured by whether or not they belong to the same color region (bin).

Nonetheless, the pattern of the neighborhood for color vicinity may be replaced by any regular geometrical shape, such as a rectangle, a circle, or a rhombus, or even by an arbitrary geometrical shape. However, the cross was chosen due to its simplicity and also due to the gain in computational burden compared to using other shapes. The rest of the histogram, including saturation and value, is created in a straightforward manner so that even the slightest of information is included. The key idea of the method is that an 8x4x4 histogram is created, and then extra values are added to bins whose colors have significant concentrations anywhere in the image, thus solving the problem of noticeably different images having approximately the same histograms. The respective histogram of the mandrill for this method is depicted in Figure 7. One might notice the four distinct areas of peaks, which represent the four colors of the image that dominate both in volume and concentration. In this case, the peaks have slightly spread out due to the number of adjacent bins representing similar colors.

As a final instruction, when using this method, experimental results show that the size of the image need not be larger than 50x50 pixels in order to produce satisfying results, thus

Figure 7. The respective spatially biased histogram of the mandrill image

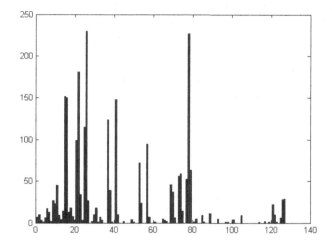

preserving disk space and speed, but increasing the size might produce a slight accuracy boost.

Having concluded the histogram extraction process and following numerous tests, the supposition was made that the similarity metric to be used in this method was that of the Bhattacharyya distance (equation (2)). As presented in the experimental results section in Table 2, one might notice the increase in accuracy compared to that of straightforward histogram creation methods.

Nevertheless, in some cases, as mentioned in the previous section, color alone is not sufficient to provide an efficient IR system, and it might seem more sensible to combine all the existing features into a single, possibly semantic, strong one. Yet this would lead to a nonrobust, computationally wise, heavy image descriptor, which most probably would be application-dependent. In order to overcome this predicament, the features need to be combined, but in a selective manner.

In the following sections, two systems are presented that are based on the synergy of color and texture.

Fuzzy Color-Texture Classification and Retrieval

The method presented is an example of semantic-based application dependency, as stated in the previous section. It produces a very efficient indexing and retrieval system, but it is confined in the range of classes with which it has been trained (i.e., deserts, clouds, sea, plantation). Overall, it is a fuzzy system in which the features of color and texture are combined via a least mean square (LMS) technique. The texture features of the images are extracted using Laws' (1980a, 1980b) convolution method. However, instead of extracting a new image in which each of its pixels describes the local texture energy, a single descriptor is proposed for the whole image. Each class of scenes corresponds to a certain band in the descriptor space. The color similarity is examined by means of its characteristic colors (Scharcanski, Hovis, & Shen, 1994) in the RGB color space. The same feature set also can be used for image classification by its semantic content. The classification is performed by a fuzzy system. The membership functions of the proposed method are constructed by statistical analysis of the training features. As an example, a system that classifies aerial images is described, through which it can be noticed that the redundancy of texture information decreases the classification uncertainty of the system.

Specifically, the texture feature extraction of the presented system relies on Laws' (1980a) texture measures in which where the notion of "local texture energy" is introduced. The idea is to convolve the image with 5x5 kernels and then to apply a nonlinear windowing operation to the convolved image. In this way, a new image results in which each pixel represents the local texture energy of the corresponding pixel of the original image. Laws (1980a) proposed 25 individual zero-summing kernels, each describing a different aspect of the local texture energy. These kernels are generated by the one-dimensional kernels, as shown in Table 1. As an example of how the two-dimensional kernels are generated, L5S5 results by multiplying the one-dimensional kernel L5 with S5. Experiments with all the 25 kernels showed that, as far as our application is concerned, the most potent ones are R5R5, E5S5, L5S5, and E5L5. More specifically, by applying each of these four masks to images of a certain class (sea, forest, etc.), the global texture descriptors were more concentrated than was the rest of the masks. These kernels were used to extract the four texture descriptors of the proposed system.

The first texture descriptor of the image is extracted by convolving it with the first kernel (R5R5). The descriptor is the absolute average of the convolved image pixel values. Thus, instead of measuring local texture descriptors by averaging over local windows (typically 15x15), as proposed in Laws' (1980a) original work, we keep one global texture descriptor by averaging the whole image. This descriptor is normalized by the maximum average found among a database of 150 training images. If, for a query image, the absolute average of the convolved image is greater than the maximum value, then the descriptor is 1.

Table 2. 1-D kernels: The mnemonics stand for level, edge, spot, wave and ripple, respectively

L5 =	[	1	4	6	4	1	]
E5 =	[	-1	-2	0	2	1	]
S5 =	[	-1	0	2	0	-1	]
W5=	[	-1	2	0	-2	1	]
R5 =	[	1	-4	6	-4	1	]

$$d_1 = \begin{cases} \dfrac{\left(\dfrac{1}{m \times n}\sum\limits_{i=1}^{m}\sum\limits_{j=1}^{n}|I(i,j)*R5R5|\right)}{d_{1\max}} & \text{if } \left(\dfrac{1}{m \times n}\sum\limits_{i=1}^{m}\sum\limits_{j=1}^{n}|I(i,j)*R5R5|\right) \leq d_{1\max} \\ 1 & \text{otherwise} \end{cases} \tag{7}$$

The same procedure is followed in order to extract the other three texture descriptors, d_2, d_3, and d_4, by replacing in equation (7) kernel R5R5 with the kernels E5S5, L5S5, and E5L5, respectively.

According to Scharcanski et al. (1994), in order to extract the characteristic colors of an image, the frequency of appearance of each color is assigned. Next, the colors are sorted in descending order according to their frequency of appearance. Given a color and a certain radius, a spherical volume is constructed in the RGB color space. The first color in the descending order comprises the first characteristic color of the image. Starting with the second color, it is examined whether it lies within the volume of any color above it. If so, then the examined color is merged with the color in the volume in which it lies. Otherwise, it comprises a new characteristic color of the image.

Considering the set of the characteristic colors as a vector, the color similarity of two images is computed by means of the angle between these two vectors. More specifically, the ratio of the inner product to the product of the measures of the two vectors corresponds to the cosine of the angle of these two vectors. The greater the value of the cosine, the smaller the angle and the more similar the two images (in terms of their color prospect). Therefore, the cosine could be used as the color descriptor of similarity. However, because the angle is the absolute descriptor and the cosine is a nonlinear function, the descriptor used in the proposed system is:

$$d_5 = \frac{2}{\pi}\arccos\frac{\bar{C_1}\cdot\bar{C_2}}{\|\bar{C_1}\|\cdot\|\bar{C_2}\|} \tag{8}$$

where $\bar{C_1}$ and $\bar{C_2}$ are the sets of the characteristic colors of images 1 and 2, respectively.

After extracting the descriptors both for the query and the database images, retrieval is performed by minimizing the following distance:

$$m = \frac{1}{\sum\limits_{i=1}^{5}w_i}\sqrt{\sum\limits_{i=1}^{4}w_i\left(din_i - ds_i\right)^2 + w_5(d_5)^2} \tag{9}$$

where din_i (i=1,…4) are the four texture descriptors of the input image, resulting according to equation (7); ds_i is the corresponding texture descriptor of the sought image; d_5 is the color descriptor according to equation (8), and w_i is a weight tuning the retrieval process according to the importance of each descriptor. By comparing equations (7) and (8), it can be observed that although d_5 is a differential descriptor (i.e., it presents the difference of

Figure 8. The histogram of the global texture energy distribution for the training images belonging to the class of sea

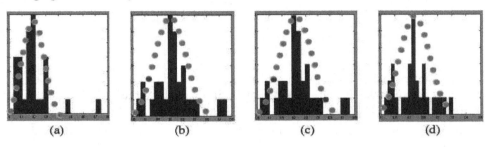

(a)   (b)   (c)   (d)

two images by means of their color aspect) $d_1,...d_4$ are absolute ones. This is the reason the difference of the latter appears in equation (9).

The same feature set described previously and used for IR may be used to classify images according to their texture and color properties by fusion of the descriptors previously presented. The system is tailored to meet the needs of the target application (i.e., the categorization of aerial images into five classes. However, with slight variations, it might be applied to other applications of image classification as well. The inputs of the fuzzy system are the five descriptors presented beforehand. In order to construct the membership functions for the inputs, a statistical analysis was carried out. Distinctively, five classes of photographs were used; namely, sea, clouds, desert, forests, and plantations. As training data, 100 images of each class were used, and for each image, the four texture descriptors were extracted. In

Figure 9. The first input of the fuzzy system is the descriptor d_1. The membership functions from left to right are: clouds, desert, sea, plantations, and forests.

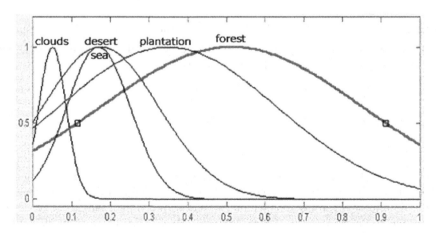

Figure 8, the histograms of the distribution of the four descriptors for the class of the sea are presented.

As might be noticed, the distribution can be approximated by a trapezoidal or even a triangular membership function. However, a Gaussian function is also a good approximation, far better than the two latter ones. The reason is that its curve is not as steep as those of a triangular or a trapezoidal one, and therefore, it also includes the sided values. Experiments with several membership functions proved this intuition. For each descriptor and for each image class, the mean value and the standard deviation were calculated. The membership functions were computed as the normal distribution for the previous values (see Figure 8). The membership functions for the d_1 descriptor are depicted in Figure 9 as an example of the four first inputs of the fuzzy system.

For the color descriptor, five inputs were used. The characteristic colors of the 100 training images of each class were merged in the same way as already described for a single image. The result is a color codebook containing the characteristic colors of the whole image class. Equation (8) is used to compute the similarity between the characteristic colors of the input image and the codebook of each of the classes. The result of each of the color similarity values is used as an input to the fuzzy system (inputs from five to 10). Similarly, five sigmoid membership function outputs (one for each class) were used. In order to evaluate the performance of both the retrieval and the classification systems, several tests were carried out. The first experiments were run in order to acquire the weights of equation (9) that give optimum results. Each time, six more relevant images were asked to be retrieved. The precision (i.e., the ratio of the correctly retrieved images) over the total retrieved images was used to measure the efficiency of the retrieval system. It was observed that the best results occurred when $w_1=w_2=w_3=w_4=1$ and $w_5=11$. In particular, the retrieval precision was measured in the range of 35% to 100%, while no other combination of weights ever had resulted to 100% precision. This is to say that the color information plays a dominant role in the IR process, as the ratio of color to texture coefficients in the optimization of equation (9) is 11/4. Therefore, when the classification system is used in cascade to the retrieval system, it elevates the precision of the latter, as it rejects the majority of the scenes that do not belong to the same class as the retrieved one.

Color-Texture Histograms

The second color-texture-based system, unlike the previous one, is robust enough for usage in a nonclassified image database. It is a histogram creation method that uses both the CIELAB and HSV color spaces. The color-texture feature, which is extracted from the a* and b* components in synergy with the Hue component from the HSV color space, produce a single 256-bin histogram as the image descriptor.

In the first part of the method, all the images are transformed into the CIELAB color space, and only the a* and b* components are reserved. The same is performed for the HSV color space from which only the Hue component is stored. The color-texture feature is extracted from the image by means of convolution with Laws' (1980a, 1980b) energy masks similar to the previously presented system. The a* and b* components are convolved with the 5x5 masks, hence producing the respective ENa* and ENb* components, which represent the chromaticity texture energy of the image. The reason of having 25x2 color-texture components for each image is that only the components with the greatest energy are reserved. The energy volume for each convolved image is computed by equation (10). Experimental results with all the 25 kernels show that the mask that most commonly results in the greatest energy is the L5L5.

$$E_{Vol} = \sum \left(\sum_{x=1,y=1}^{m,n} ENa^*(x,y), \sum_{x=1,y=1}^{m,n} ENb^*(x,y) \right) \qquad (10)$$

When the components with the greatest energy are found, the discretization process is activated during which all three components (ENa*, ENb*, and Hue) are divided into sections. The color-texture components are split up into eight parts each and the hue component into four. This way, they all are interlinked to each other, thus creating a histogram of (8*8*4) 256 bins. The respective histogram for the mandrill image is presented in Figure 10, where, in some cases, the dominant color peaks have been strengthened by the color-texture insertion, and in others, they have been weakened, thus producing a very strong peak and four other weak ones. The strong peak at around 110 represents the color textureness of the image, which is very strong for all colors.

Experiments with different metrics showed that the one performing best in conjunction with the histogram at hand was the Bhattacharyya distance. Comparative results are presented in the experimental results section.

Figure 10. The respective color-texture histogram for the mandrill image

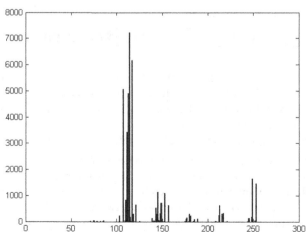

Table 3. Experimental results

Image Set	Classic CIELAB Histogram	Swain & Ballard	Fuzzy Histogram	Spatially Biased Histogram	Color-Texture Histogram
1	65%	75%	90%	100%	100%
2	75%	70%	80%	80%	85%
3	90%	90%	95%	100%	100%
Overall Time(sec)	226.95	511.80	525.54	364.36	703.25

Experimental Results

The systems introduced have been implemented and tested on a database of 1,100 images for comparison purposes. The images in the collection are representative of the general requirements of an IR system over the Internet. Therefore, the range of topics present in the image database is quite wide and varies from several different landscapes, flowers, sunsets, and everyday life images to sports, concerts, and computer graphics, which usually confuse IR systems. (The images are online, available at the following URL: http://utopia.duth.gr/~konkonst). The presented systems' effectiveness and efficiency, their precision (Muller, Muller, Squire, McG., & Marchand-Maillet, 2001), and time cost performances for three representative image sets are displayed next to Swain and Ballard's (1991) and the global histogram extraction method in the CIELAB color space in Table 2 with the intension of giving the reader an idea of the increase in accuracy together with the respective increase in time cost (i.e., computational burden).

Future Trends and Conclusion

Past and recent trends in content-based IR with the use of histograms were reported here as well as new heuristic methods to create histograms. One of the reasons just the low-level image descriptor of histograms was taken into consideration is that it constitutes the cornerstone of any semantic-based IR system. Moreover, the field of IR has become so broad that in order to cover all methods and techniques for all the features and their combinations using all kinds of descriptors would produce a very lengthy survey. However, the main reason for choosing color and texture histograms is that they are very easy to produce, compute, and store, thus making them an attractive feature descriptor to whomever is interested in constructing a custom IR system. To this end, it could be said that having read this chapter, the reader will have acquired sufficient knowledge to select an adequate color space, the tools with which to create a robust low-level image descriptor, and the know-how of which comparison method to use in order to create a custom IR system.

In conclusion, the reader has been presented with four systems that cover all the aspects of a low-level feature-based IR system. At first, by presenting fuzzy histograms, the reader is walked through a color space discretization phase, which is one of the most important steps in IR. Moreover, it is shown that a histogram need not have a large number of bins in order to be effective. In addition, spatially biased histograms are presented, introducing the idea of embedding supplementary information into the straightforward histogram and, therefore, making the system even more efficient. Then, the importance of combining the features of color and texture is stressed out and discussed by means of an application-dependent system that makes use of color and texture as a whole, as well as a robust system that combines only selected attributes from the latter features.

With reference to the methods apposed, it is fair to say that they perform satisfactorily in terms of speed and accuracy.

Finally, general instructions to produce an IR system are as follows:

- The histogram should consist of a number of bins as small as possible so that the computational burden will be as light as possible, while at the same time preserving all the significant information.

- The color-space selection depends very much on the application, but uniform and human color perception spaces result in enhanced retrievals.

- The metric selection also depends on the application and only should be selected after running exhaustive tests by use of the extracted feature.

- The textural information somewhat improves the performance of the IR system but also significantly increases its computational burden; therefore, the method applied should be made as simple as possible. This can be solved by a weighted combination of the features.

Future work should be based either on implementing some new way of representing the features (as histograms once were) or on producing an optimal combination of the low-level features at hand (e.g., color, texture) in order to keep the systems as robust as possible; it is noted that specific applications (e.g., face recognition) do not belong in this area.

Additionally, as computer speed and memory, including disk space, become greater and cheaper, evaluations and experiments on databases of a much larger scale become feasible.

Acknowledgments

Mr. Konstantinidis is funded partially by: (1) the Greek-Slovenian Bilateral Project "Development of New Techniques for Recognition and Categorization" and (2) the Archimedes-II Project of the EU and the Greek Ministry of Education EPEAEK (code 2.2.3.z, subprogram 10 in the ATEI of Thessaloniki).

References

Bartolini, I., Ciaccia, P., & Patella, M. (1999). WINDSURF: A region-based image retrieval system. In *Proceedings of the 10ᵗʰ International Workshop on Database & Expert Systems Applications* (pp. 167–173).

Brandt, S., Laaksonen, J., & Oja, E. (2002). Statistical shape features for content-based image retrieval. *Journal of Mathematical Imaging and Vision, 17*(2), 187–198.

Castelli, V., & Bergman, L.D. (2002). *Image databases: Search and retrieval of digital imagery*. John Wiley & Sons.

Del Bimbo, A. (1999). *Visual information retrieval*. San Francisco: Morgan Kaufman Publishing.

Drew, M. S., Wei, J., & Li, Z. N. (1998). Illumination-invariant color object recognition via compressed chromaticity histograms of color-channel-normalized images. In *Proceedings of the IEEE International Conference on Computer Vision (ICCV'98)* (pp. 533–540).

Eakins, J. P. (2002). Towards intelligent image retrieval. *Pattern Recognition, 35*, 3–14.

El Badawy, O., & Kamel, M. (2002). Shape-based image retrieval applied to trademark images. *International Journal of Image and Graphics, 2*(3), 375–393.

Fukunaga, K. (1990). *Introduction to statistical pattern recognition* (2ⁿᵈ ed.). Academic Press.

Gagliardi, I., & Schettini, R. (1997) A method for the automatic indexing of color images for effective image retrieval. *The New Review of Hypermedia and Multimedia, 3*, 201–224.

Gasteratos, A., Zafeiridis, P., & Andreadis, I. (2004). An intelligent system for aerial image retrieval and classification. In G. A. Vouros & T. Panayiotopoulos (Eds.), *An Intelligent System for Aerial Image Retrieval and Classification* (LNCS 3025, pp. 63–71). Berlin: Springer-Verlag.

Gonzalez, R. C., & Woods, R. E. (1992). *Digital image processing*. Reading, MA: Addison-Wesley.

Heidemann, G. (2004). Combining spatial and color information for content based image retrieval. *Computer Vision and Image Understanding, 94*, 234–270.

Howarth, P., & Rüger, S. (2004). Evaluation of texture features for content-based image retrieval. In P. Enser et al. (Eds.), *CIVR 2004* (LNCS 3115, pp. 326–334). Berlin: Springer-Verlag.

Jain, A. K., & Vailaya, A. (1996). Image retrieval using color and shape. *Pattern Recognition, 29*, 1233–1244.

Konstantinidis, K., & Andreadis, I. (2005). Performance and computational burden of histogram based color image retrieval techniques. *Journal of Computational Methods in Sciences and Engineering (JCMSE), 5*, 141–147.

Konstantinidis, K., Gasteratos, A., & Andreadis, I. (2005). Image retrieval based on fuzzy color histogram processing. *Optics Communications, 248*, 375–386.

Laws, K. (1980a). *Textured image segmentation* [doctoral dissertation]. Los Angeles, CA: University of Southern California.

Laws, K. (1980b). Rapid texture identification. *SPIE, Image Processing for Missile Guidance, 238*, 376–380.

Liang, Y., Zhai, H., & Chavel, P. (2002). Fuzzy color image retrieval. *Optics Communications, 212*, 247–250.

Liapis, S., & Tziritas, G. (2004). Color and texture image retrieval using chromaticity histograms and wavelet frames. *IEEE Transactions on Multimedia, 6*(5), 676–686.

Lu, G., & Phillips, J. (1998). Using perceptually weighted histograms for color-based image retrieval. In *Proceedings of the 4th International Conference on Signal Processing, China* (pp. 1150–1153).

Muller, H., Muller, W., Squire, D., McG., & Marchand-Maillet, S. (2001). Performance evaluation in content-based image retrieval: Overview and proposals. *Pattern Recognition Letters, 22*(5), 593–601.

Paschos, G., & Petrou, M. (2003). Histogram ratio features for color texture classification. *Pattern Recognition Letters, 24*, 309–314.

Pass, G., & Zabih, R. (1999). Comparing images using joint histograms. *Multimedia Systems, 7*(3), 234–240.

Pass, G., Zabih, R., & Miller, J. (1996). Comparing images using color coherence vectors. In *Proceedings of the 4th ACM Multimedia Conference* (pp. 65–73).

Rowe, N. C. (2005). Exploiting captions for Web data mining. In A. Scime (Ed.), *Web mining: applications and techniques* (pp. 119-144). Hershey, PA: Idea Group Publishing.

Scharcanski, J., Hovis, J. K., & Shen, H.C. (1994). Representing the color aspect of texture images. *Pattern Recognition Letters, 15*, 191–197.

Smeulders, A.W.M., Worring, M., Santini, S., Gupta, A., & Jain, R. (2000). Content-based image retrieval: The end of the early years. *IEEE Transactions on Pattern Analysis and Machine Intelligence, 22*(12), 1349–1380.

Stricker, M., & Swain, M. (1994). The capacity of color histogram indexing. In *Proceedings of the IEEE Conference on Computer Vision and Pattern Recognition* (pp. 704–708).

Swain, M. J., & Ballard, D. H. (1991). Color indexing. *International Journal of Computer Vision, 7*, 11–32.

Wang, J. Z., Li, J., & Wiederhold, G. (2001). SIMPLIcity: Semantics-sensitive integrated matching for picture libraries. *IEEE Transactions on PAMI, 23*, 947–963.

Zachary Jr., J. M., & Iyengar, S. S. (1999). Content-based image retrieval systems. In *Proceedings of the IEEE Symposium on Application-Specific Systems and Software Engineering and Technology* (pp. 136–143).

Zhang, C., & Chen, T. (2003). From low level features to high level semantics. In O. Marques & B. Furht (Eds.), *Handbook of video database design and applications* (pp. 613–624). CRC Press.

Zhang, D. S., & Lu, G. (2000). Content-based image retrieval using gabor texture features. In *Proceedings of the First IEEE Pacific-Rim Conference on Multimedia*, Sydney, Australia (pp. 392–395).

Chapter III

Shape-Based Image Retrieval by Alignment

Enver Sangineto, University of Rome "La Sapienza," Italy

Abstract

Among the existing content-based image retrieval (CBIR) techniques for still images based on different perceptual features (e.g., color, texture, etc.), shape-based methods are particularly challenging due to the intrinsic difficulties in dealing with shape localization and recognition problems. Nevertheless, there is no doubt that shape is one of the most important perceptual feature and that successful shape-based techniques would significantly improve the spreading of general-purpose image retrieval systems. In this chapter, we present a shape-based image retrieval approach that is able to deal efficiently with domain-independent images with possible cluttered backgrounds and partially occluded objects. It is based on an alignment approach that is proved to be robust in rigid object recognition that we have modified in order to deal with inexact matching between the stylized shape input by the user as query and the real shapes represented in the system's database. Results with a database composed of difficult real-life images are shown.

Introduction

A typical shape-based image retrieval system accepts as input an image provided by the user and outputs a set of (possibly ranked) images of the system's database, each of which should contain shapes similar to the query. There are two main types of possible queries: query by example and query by sketch (Colombo & Del Bimbo, 2002; Forsyth & Ponce, 2003). In the former, the user submits an image that is an example of what he or she is looking for. For instance, we could be interested in searching a painting having only the a digital image of a part of the whole painting (Schmid, & Mohr, 1997). A typical application of such a search is copyright protection, in which the user is interested in looking for pictures used in Web documents without permission (Forsyth & Ponce, 2003).

The second type of query is usually more interesting and is thought for those cases in which the user has only a vague idea of what he or she is looking for. For this reason, the system accepts a drawing that represents a stylized sketch of the shape the user is seeking in the system's repository. Then, it searches for those images that contain shapes that are similar to the sketch but not necessarily identical. Figure 1(b) shows an example of a user-drawn sketch, which is compared in Figure 1(c) with the edge map of a real image containing a horse (Figure 1(a)). The real and stylized shapes do not match exactly, because the second is intended to be a (fast) representation of a generic quadruped. It is worth noticing that in the example of Figure 1, the user's sketch also implicitly defines the pose of the searched horse. For instance, the well-known David's painting, *Napoleon Crossing the Alps*, representing a reared-up horse cannot be retrieved by the system unless the user provides a sketch with a horse in a similar pose.

In this chapter, we deal only with query by sketch systems because of their greater practical interest. Moreover, query by example based on shape sometimes can be viewed as a special case of query by sketch in which the system accepts only an exact matching between the template of the query and the images.

There are two main problems that make query by sketch systems particularly challenging. The first is the need to localize the shape of interest in the whole image. For instance, in Figure 1(c), only a portion of the image (the horse) is similar to the shape input by the user, which should not be compared with the trees and the other objects in the background. While, for instance, in color-based image retrieval the whole color content of the image often can be compared with the query color content without selecting a specific part of the image, in shape-based retrieval, nonisolated objects are difficult to deal with because they need to be localized in the image before in order to be compared with the query. Shape localization is a nontrivial problem, since it involves high-level scene segmentation capabilities. How to separate interesting objects from the background is still an open and difficult research problem in computer vision.

The second problem is the necessity to deal with inexact matching between a stylized sketch and a real, possibly detailed, shape contained in the image. We need to take into account possible differences between the two shapes when we compare them.

The differences in the two shapes make it difficult to apply exact-matching techniques such as the alignment approaches (Ayache, 1986; Huttenlocher & Ullman, 1990; Ullman, 1996), which are based on an exact description of the target shape. Alignment techniques perform a search in the pose space in order to find a correct rigid geometric transformation that is capable of aligning the model's description with the image data (see the next subsection). These techniques are used widely in object recognition tasks and usually can deal with cluttered scenes and unsegmented images but are not suitable to deal with inexact matching issues.

According to Smeulders, Worring, Santini, Gupta, and Jain (2000), most of the approaches to shape-based image retrieval can be broadly classified into three main types: statistical methods, scale-space approaches, and deformable template-matching techniques. The first types of methods are based on the extraction of statistical features from either the internal or the boundary pixels of the object's silhouette (Kimia, 2002). Examples of features are the Fourier descriptors of the silhouette's contour (Folkers & Samet, 2002), the digital moments, the compactness (i.e., the ratio between the squared perimeter and the area) of the silhouette, the edge histograms (Brand, Laaksonen, & Oja, 2000), and so forth. The values of a fixed number (N) of features describing a given shape are represented by means of a point in an N-dimensional feature space (Duda, Hart, & Strorck, 2000). We can retrieve a shape similar to a sketch S searching for those points in the feature space that are the closest to the point representing S. This allows the system to do a quick indexing of large-sized visual databases in which the image representations are organized exploiting the spatial proximity of the corresponding points in the feature space (multidimensional point access methods, PAMs) (Del Bimbo, 1999). In fact, feature space representation is the only approach that actually allows for an efficient database indexing. Almost all the other representation methodologies need to sequentially process all the repository's images and match them with the user's sketch. The approach presented in this chapter has the same drawback.

Although these methods often perform good recognition skills, especially if they are applied to well-defined domains (Flickner et al., 1995), they only can be used with isolated shapes. Scenes with complex backgrounds and/or possible occluding objects are not suitable for statistical methods, since the silhouettes of a group of contiguous objects are merged together, and it is not clear what pixels have to be used in the feature extraction process. For these reasons, statistical methods do not deal with localization issues, and they usually are applied to simple databases in which every image contains a single, isolated silhouette.

We find an analogous situation in scale-space approaches (Del Bimbo & Pala, 1999; Mokhtarian, 1995; Mokhtarian & Abbasi, 2002; Tek & Kimia, 2001). These methods are based on

Figure 1. (a) An example of image of the system's database with a difficult background; (b) a sketch representing a quadruped; (c) the same sketch superimposed upon the horse contained in the image. The difference between (b) and (c) is evident.

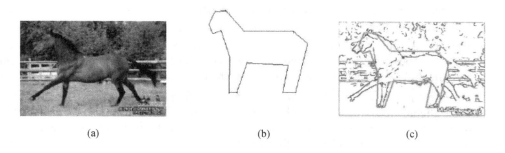

(a)   (b)   (c)

a shape representation whose details vary depending on the chosen scale value. The aim of the scale-space representation is the simulation of those important pre-attentive human vision features that allow us to focus our attentions on different details of the same object in different situations. Nevertheless, scale-space approaches usually are based on different levels of smoothing of the object's silhouette. Consequently, like statistical methods, they also need a previous (possibly manual) segmentation of the scene in order to extract the silhouettes' boundaries of the interesting shapes.

A more accurate approach is deformable template matching based on active contours (also called *snakes*) (Del Bimbo & Pala, 1997; Jain, Zhong, & Lakshmanan, 1996; Xue, Li, & Teoh, 2002). The user sketch S is superimposed upon each image I of the archive and progressively deformed until it matches with the image edge pixels. The deformation usually is controlled by means of an iterative process of local modifications of S, which aims to maximize the matching function $M_{I,S}(O,D)$, where O expresses the overlapping rate between S and I, and D is the deformation energy. $M_{I,S}(O,D)$ represents the similarity measure between S and I: the larger O is, the larger is $M_{I,S}(O,D)$, but the larger D is, the smaller is $M_{I,S}(O,D)$. Jain, Zhong, and Lakshmanan (1996) and Del Bimbo and Pala (1997) present some working examples exploiting this idea with different formalizations of both the function M and the maximization process.

Deformable template matching is robust to occlusions and can be applied to images with a nonuniform background. Indeed, it is not based on the analysis of an isolated object silhouette but rather on the whole image edge map, and noise and extraneous objects usually do not prevent S from being progressively attracted by the true shape possibly present in I. Nevertheless, the deformation process is strictly dependent on the initial position in which S is superimposed upon I, because the method is usually not invariant with respect to rigid geometric transformations (i.e., translation, rotation, and scaling). This problem has been addressed either by iterating the deformation process for different initial positions (Jain, Zhong, & Lakshmanan, 1996) or off-line manually selecting candidate portions of each image of the system's repository in which the online search subsequently is performed (Del Bimbo & Pala, 1997). It is worth noticing that the former solution is usually slow, while the latter involves a human intervention in the segmentation process as in statistical and scale-space approaches.

As this brief overview of the existing approaches has shown, most of the proposed techniques for inexact matching are not able to accurately and efficiently perform in domains in which localization capabilities are requested. For this reason, the (few) existing benchmarks used to evaluate shape-based image retrieval systems, such as those proposed by the MPEG-7 work group (Jeannin, 1999), are collections of images that show isolated objects, usually represented by means of their binarized silhouettes. However, images with isolated objects and uniform backgrounds are not very common in real-life applications. On the other hand, the need for an off-line manual selection of the interesting objects in each image of the repository leads to the reduction or to the loosing of the advantages of an automatic image retrieval tool.

In this chapter, we aim to present a method that has both inexact matching and localization capabilities for a shape-based retrieval problem with a low computational complexity. Since our approach utilizes alignment techniques for localization issues, in the next subsection, we give a brief overview of the alignment paradigm, and we sketch the main ideas of our method.

Extending Alignment-Based Techniques for Inexact Matching

If M is a geometric model of the object's shape for which we are looking and I is the binarized edge map of the input image, a typical alignment approach is composed of two main phases. In the former (hypothesis phase), the system hypothesizes a set of geometric transformations, either 2D or 3D: T_1, T_2, ... each T_i usually being a similarity transformation composed of a translation, a rotation, and a scaling. $T_i(M)$ (the transformed model) is then compared in the test phase with I, taking into account the edge pixels of I close to the points of $T_i(M)$.

The object is recognized in the input image if there is a transformation T_i such that a sufficient number of points of $T_i(M)$ can be matched with points of I. The advantage of these techniques is that they do not need a previous segmentation of the scene (I), since searching for a possible transformation between M and I corresponds to localize M in I.

Despite this fact, alignment techniques have never been applied to image retrieval until now, because they assume to deal with an exact description (M) and an exact matching of the target shape. In this work, we propose an alignment technique that is able to deal efficiently with both localization and inexact matching issues for a query by sketch CBIR system.

First, our system automatically groups the image edge pixels in line segments, maximizing the probability that pixels belonging to a same object are grouped together. Line segments then are used both in the hypothesis and in the test phase to control matching. In the hypothesis phase, line segments are used in order to select those similarity transformations of the sketch (S), which are locally consistent with some segment of the image (I). Candidate hypotheses are represented in a pose space conceptually similar to a Hough accumulator. The pose space then is scanned, and the hypotheses are clustered. The clustering process selects those transformations that correspond to the largest coherent sets of image segments. Moreover, it allows the system to formulate new transformations represented by the centers of the clusters, which compensates for minor differences among the elements of a given cluster. Finally, the transformation hypotheses that correspond to the biggest clusters are verified by using them to project S into I and by using the line segments in order to check the global consistency of each alignment hypothesis.

Feature Extraction

We show in this section the preprocessing operations performed on the user's sketch and on the database images. For each image of the system's repository, we perform a standard edge detection by means of the Canny edge detector (Canny, 1986). Moreover, we suppose that the user's sketch is represented by a set of points (see Figure 2(b)). In the rest of this chapter, we will use S to indicate the user's sketch and I for the edge map of the currently analyzed image of the system's repository.

The points of S and I are merged in order to obtain line segments. The aim is to group the pixels into sets such that the probability is high that elements of a same set represent points of the same object in the image. The merging criteria we use are inspired to some of the well-known Gestalt principles: pixels proximity and curvature uniformity.

The line segments creation process is composed of three steps. In the first step we merge edge pixels that are adjacent, stopping in line junctions.

In the second step, we merge the line segments previously obtained, taking into account the distances and the tangent orientations between their endpoints (Huttenlocher & Ullman, 1990). For each segment s_i, we look for the segment s_j, which minimizes the function $D(s_i, s_j)$ defined as follows:

$$D(s_i, s_j) = |e_1 - e_2|(\alpha + \beta\theta) \tag{1}$$

where e_1 is one of the two endpoints of s_i, and e_2 is one of the endpoints of s_j, $|e_1 - e_2|$ is the Euclidean distance between e_1 and e_2, and θ is defined as below:

$$\theta = 180 - (\mathbf{v_1} - \mathbf{v_2}), \tag{2}$$

being $\mathbf{v_1}$, $\mathbf{v_2}$, respectively, the orientations of the unit tangent vectors in e_1 and e_2. If s_j is the segment minimizing $D(s_i, s_j)$ with respect to s_i, and, vice versa, s_i is the segment minimizing $D(s_j, s_i)$ with respect to s_j, then s_i and s_j are merged in a new segment s. The segment-merging process is performed iteratively, choosing at each step the mutually favorite pair of segments, as suggested in Huttenlocher and Ullman (1990), stopping when no more merging is possible.

Finally, each segment so obtained is split in one or more separate segments using a polygonal approximation process as follows. We use a standard split algorithm (Shapiro & Stockman, 2001) for line polygonal approximation to break each segment into a set of line segments with a low curvature, each one delimited by the breakpoints of the approximation process. Huttenlocher and Ullman (1990) use the extreme curvature values of the segments for the splitting step. However, we have found experimentally that polygonal approximation produces more stable results and a greater number of final segments, which are needed in the hypothesis phase of our matching process (see the next section).

The output of this processing phase is the sets: $SEG_S = \{s_1, ..., s_m\}$ and $SEG_I = \{r_1, ..., r_n\}$ of line segments respectively representing S and I. We exclude from I and S all the segments composed of few points. SEG_I is extracted off-line when the image I is stored in the system's database. SEG_S is extracted online when S is input by the user. Figures 3(a) and 3(b) respectively show the segments SEG_I and SEG_S for the image of Figure 2(a) and the sketch of Figure 2(b).

The last data structure built in this phase is the *NeighbouringSegments* array, used to efficiently check the distance of a point with respect to the segments of SEG_I (see the next two sections). *NeighbouringSegments* and SEG_I are stored in the system's database together with I and used online when the user inputs a query. The construction of *NeighbouringSegments* will be explained in the next section.

Figure 2. (a) Example of real image of the system's database; (b) a user-drawn sketch of a guitar

(a)   (b)

Hypothesis Phase

In this section, we show how we iteratively match elements of SEG_S and SEG_I in order to hypothesize candidate transformations, which then are clustered and verified.

Searching for Candidate Transformations

We are looking for an object in I whose shape S' is similar, even if it is not necessarily identical to S. Both S and S' (if present) have been segmented in a set of line segments, depending on the curvature of their contours, as shown in the second section of this chapter. If S and S' are similar with similar corners and composed of lines with a similar curvature, and if a sufficiently large part of S' is non-occluded in I, then the segmentation process usually produces a sufficient number of segments in S and S' whose relative positions with respect to S and to S' are similar. For instance, the segment r_1 of Figure 3(a) approximately corresponds to the segment s_1 of Figure 3(b). Even if this situation were not true for all the elements of SEG_S, in the hypothesis formulation process, we only need this to be true for a few of them.

Suppose $r \in SEG_I$ is a segment of S' corresponding to $S \in SEG_S$ and that $p_1 = (x_1, y_1)$, $p_2 = (x_2, y_2)$ are the endpoints of r and $p'_1 = (x'_1, y'_1)$, $p'_2 = (x'_2, y'_2)$ are the endpoints of s. We can use p_1, p_2, p'_1, and p'_2 in order to find the parameter values of the two-dimensional similarity transformation T mapping p'_1 in p_1 and p'_2 in p_2. T is composed of a translation (represented by the parameters t_x and t_y), a rotation (θ) and a scale factor (k): $T = (\theta, k, t_x, t_y)$. The parameters of T can be found easily (Ayache, 1986) by solving the following system of equations:

$$x_1 = t_x + x'_1 \times k \times \cos\theta - y'_1 \times k \times \sin\theta \tag{3}$$

$$y_1 = t_y + x'_1 \times k \times \sin\theta + y'_1 \times k \times \cos\theta \tag{4}$$

$$x_2 = t_x + x'_2 \times k \times \cos\theta - y'_2 \times k \times \sin\theta \tag{5}$$

$$y_2 = t_y + x'_2 \times k \times \sin\theta + y'_2 \times k \times \cos\theta. \tag{6}$$

Given a segment $r \in SEG_I$, we a priori do not know if it actually corresponds to any segment s of SEG_S. For this reason we exhaustively formulate transformation hypotheses for every couple $(r, s)SEG_I \times SEG_S$, and we subsequently verify what hypotheses are correct (see the next section). Note that if we had not grouped the points of I and S in line segments (respectively, SEG_I and SEG_S), then every pair of points $(p_1, p_2) \in I^2$ should be matched with every pair of points $(p'_1, p'_2) \in S^2$ in the hypothesis generation process and, of course, $|I^2 \times S^2| >> |SEG_I \times SEG_S|$.

For efficiency reasons, the hypothesis process is performed using only the largest segments of SEG_I. False negatives will be rejected in the verification phase (see the next section) using all the elements of SEG_I. Let SEG'_I be the set of the largest elements of SEG_I (i.e., those segments whose number of points is greater than a given threshold). The transformation hypotheses are generated as follows:

Hypothesis Generation (S, I, SEG_S, SEG'_I)

0 $H := \emptyset$

1 For each $r_i \in SEG'_I$ and each $s_j \in SEG_S$, do:

2   Let p_1, p_2 be the endpoints of r_i and p'_1, p'_2 the endpoints of s_j

3    $T := ComputeTransform(p_1, p_2, p'_1, p'_2)$

4   If *GoodCandidate* (S, I, T) then:

5   If *LocallySimilar*(r_i, s_j, T) then:

6          $H := H \cup \{<T, |s_j|>\}$

7    $T := ComputeTransform(p_1, p_2, p'_2, p'_1)$

8   If *GoodCandidate*(S, I, T) then:

9   If *LocallySimilar*(r_i, s_j, T) then:

10         $H := H \cup \{<T, |s_j|>\}$

11   Return H

H is the set of all the transformation hypotheses. In Step 3, a candidate transformation is computed using equations (3) through (6) and matching the couple of points (p_1, p_2) with the couple (p'_1, p'_2). The process is repeated in Step 7 by matching (p_1, p_2) with (p'_2, p'_1).

In Steps 4 and 8, a candidate transformation T is checked in order to reject those scale, rotation, and translation parameters that produce either very small- or very large-scale differences between S and I or project the most of S out of I. In *GoodCandidate*, we check that $th_{k_1} \leq k \leq th_{k_2}$, being th_{k_1} and th_{k_2} two prefixed thresholds defining the minimum and the maximum accepted scale difference between S and S'. Currently, we have fixed: $th_{k_1} = 1/3$, $th_{k_2} = 4$. Moreover, if c is the centroid of S, we check that $T(c) \in I$ (i.e., we reject those candidate transformations which project c out of I).

A new hypothesis T is added to H (Steps 6 and 10) if in Step 5 (respectively, Step 9) a sufficient number of points of s_j is close to r_i when transformed using T. This is checked by means of the Boolean function *LocallySimilar*, which uses the bidimensional array *NeighbouringSegments* computed off-line when the image I is stored in the system's database. *NeighbouringSegments* aims at representing a tolerance area surrounding each line segment of I; this tolerance area will be exploited by the function *LocallySimilar* to check if the points of the set $t(s_j)$ are close to r_i. *NeighbouringSegments* is an array of the same dimensions of I. Each element *NeighbouringSegments*(x,y) corresponds to the point $p = (x,y)$ of I and contains the indexes of the segments of SEG_I which are close to the point p. *NeighbouringSegments* is obtained performing a sort of dilatation operation on the edge pixels of I as shown next.

NeighbouringSegments Construction (I, SEG_I)

1   For all the elements (x, y) of I, *NeighbouringSegments*$(x,y) := \varnothing$

2   For each $r_i \in SEG_I$ and each point $p \in r_i$, do:

3   For all the elements $p' \in W(p)$ do:

4   *NeighbouringSegments*$(p') := $ *NeighbouringSegments*$(p') \cup \{i\}$

Figure 3. (a) The set of segments SEGI corresponding to the image of Figure 2a; (b) the set of segments SEGS corresponding to the sketch represented in Figure 2b

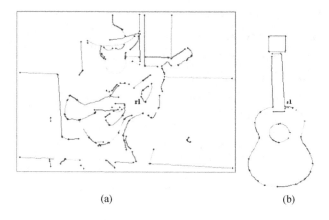

(a)   (b)

Figure 4. (a) A graphical representation of NeighbouringSegments: in the figure, each point is associated with only one segment (instead of a set of segments), and different segments are represented by different gray levels; (b) an example of system output; the sketch of Figure 2(b) is superimposed upon the image of Figure 2(a) using the best transformation found by the system (see the Test algorithm)

(a)   (b)

In Step 3, $W(p)$ is a square box centered in p, and in Step 4, each point p' of $W(p)$ adds the segment index i to the set of its neighboring segments represented in *NeighbouringSegments(p')*. This construction is performed off-line in the preprocessing phase of each image I after the extraction of the segments of the set SEG_I (see the second section). Figure 4(a) shows a graphical representation of *NeighbouringSegments* for the example of Figure 3(a). Online, *NeighbouringSegments* is used in the function *LocallySimilar* as follows:

LocallySimilar (r_i, s_j, T)

0    $n:=0$

1   For each point $p \in s_j$, do:

2   If $i \in NeighbouringSegments(T(p))$ then:

3                $n:=n+1$

4   Return $n \geq th_l \times |s_j|$

In Step 4, th_l is a threshold, and $|s|$ is the number of points of s. The function *LocallySimilar* checks if the segments s_j and r_i matched in Step 1 of the *Hypothesis Generation* procedure are sufficiently similar to each other and help to largely reduce the number of hypotheses to be dealt with subsequently (see "Clustering" and "Verification Phase" sections) by rejecting those that are clearly inconsistent.

Finally, in Steps 6 and 10 of the *Hypothesis Generation* procedure, T is stored in H together with the cardinality of the segment s_j ($|s|$) which has generated T. $|s|$ is a weight of the importance of T and will be used in the cluster selection process (see the next subsection).

Clustering

The hypotheses contained in the set H, obtained as shown previously, are reduced further by means of a clustering process. Moreover, the clustering process is necessary also because we are looking for a shape S' in I, which is similar to S but possibly not identical. For instance, suppose S' is composed of a set of segments $r_1, ..., r_p$ and S is composed of the segments $s_1, ..., s_q$. Furthermore, suppose that in the *Hypothesis Generation* algorithm the elements r_i and r_k of S' have been matched, respectively, with s_j and s_h of S, producing the transformation hypotheses $T^{(i,j)}$ and $T^{(k,h)}$. If $S = S'$, then we expect that $T^{(i,j)} - T^{(k,h)}$ (possibly modulo some minor errors due to noise effects and/or common quantization problems of Hough-like accumulators). But if S and S' are not exactly equal, then the relative position of r_i with respect to r_k can be different from the relative position of s_j with respect to s_h. For instance, the relative positions of the legs and the head of the real horse in Figure 1(a) is different from the relative positions of the same body elements of the sketch of Figure 1(b). For this reason, we need to average on the parameter values of $T^{(i,j)}$ and $T^{(k,h)}$. The clustering process makes it possible, and it is realized as follows.

Each hypothesis $T = (\theta, k, t_x, t_y)$ such that $<T, n> \in H$ is mapped into a point $p = (x, y, z, w)$ of the four-dimensional transformation space TS using:

$$x = \alpha_1\theta, \tag{7}$$

$$y = \alpha_2 k, \tag{8}$$

$$z = \alpha_3 t_x, \tag{9}$$

$$w = \alpha_4 t_y, \tag{10}$$

where $\alpha_1 - \alpha_4$ are normalization constants chosen in order to make TS a uniform space in which the Euclidean distance between two points is a good indicator of the significance of the difference between the corresponding transformation hypotheses.

The points in TS are clustered using a standard agglomerative clustering algorithm (Duda, Hart, & Strorck, 2000; Forsyth & Ponce, 2003), consisting of the following steps. In the initialization step, we make each point a separate cluster. Then we iteratively merge the two clusters c_1, c_2. with the smallest intercluster distance $dist(c_1, c_2)$ until such a distance is smaller then a given threshold. $dist(c_1, c_2)$ is defined as the Euclidean distance between the centroid of c_1 and the centroid of c_2.

If c is one of the clusters produced by the clustering process and $p = (x, y, z, w)$ is its centroid, then we can compute the transformation corresponding to p: $\overline{T} = (x/\alpha_1, y/\alpha_2, z/\alpha_3, w/\alpha_4)$. Moreover, if $<T_1, n_1>, ..., <T_q, n_q>$ are the hypotheses of H clustered into c, then:

- The candidate hypotheses $T_1, ..., T_q$ have been generated by matching pairs of segments $(r_{i_1}, s_{j_1}), ..., (r_{i_q}, s_{j_q})$ in the *Hypothesis Generation* procedure. Since they are physically close to each other in the transformation space *TS*, the average value $\overline{T}$ of T_1, ..., T_q is likely to be a transformation of *S* into *I* such that $\overline{T}(S)$ matches with different segments of *I*. This is conceptually similar to Hough-like approaches in which the parameter space (or pose space) representing all the possible similarity transformations model-to-image is quantized and a voting process leads to select the most likely parameter values. Nevertheless, using a clustering process, we do not need to explore either all the parameter space, which can be huge for a problem with four dimensions, or to deal with the well-known quantization problems of common Hough accumulators.

• $n = n_1 + ... n_q$ is a weight representing the importance of the cluster *c*.

We order all the clusters in decreasing order with respect to their associated weight, and we select the set of the *N* most important clusters c_1, ..., c_N. Let $H' = \{\overline{T}_1, ... \overline{T}_N\}$ be the set of transformations corresponding to the centroids of c_1, ..., c_N. *H'* is the final set of transformation hypotheses, which will be verified (see the next section). Currently, we set *N* = 100.

Verification Phase

In this section, we show how the hypotheses represented by the elements of *H'* are verified. Let $T \in H'$. We project every point of *S* into *I* using *T*, and we look for those segments in SEG_I that are close enough to *T(S)*. A segment $r \in SEG_I$ is valid if the ratio between the portion of *T(S)* lying close to *r* and the length of *r* are close to 1. We use the valid segments to check the portion of *T(S)* that can be matched with segments of *I*, and we use the size of this portion for estimating the goodness of the whole match. In matching *T(S)* with *I*, we take into account only valid segments excluding all the other segments of *I*. For instance, we do not take into account those segments that lie close to *T(S)* for only a minor part of their length. In fact, such segments likely belong to the background or to other objects rather than to the searched object *S*.

Test(H', SEG$_I$, S, NeighbouringSegments)

Let *MatchedPoints* be an array of integers indexed by segment indexes.

0   *MaxMatched*:=0; *BestT*:=nil

    *** Main loop ***

1   For each $T \in H'$ do:

2   For each *i* such that $r_i \in SEG_I$, do: *MatchedPoints(i)*:=0

3      $SEG_V = \varnothing$; *Matched*:=0

      *** Valid segment construction ***

4   For each $p \in S$, do:

5   For each $i \in NeighbouringSegments(T(p))$, do:

6                  $MatchedPoints(i):=MatchedPoints(i)+1$

7   For each $r_i \in SEG_I$, do:

8   If $MatchedPoints\ s(i)\ /\ |r_i| \geq th_v$ then:

9              $SEG_V := SEG_V \cup \{r_i\}$

         *** Matching I and $T(S)$ using valid segments ***

10   For each $p \in S$, do:

11   If $\exists i \in NeighbouringSegments(T(p))$ such that $r_i \in SEG_V$, then:

12              $Matched:=Matched+1$

13   If $MaxMatched<Matched$ then:

14           $MaxMatched:=Matched;\ BestT:=T$

     *** End main loop ***

15   Return $<MaxMatched,BestT>$

In Steps 4 through 6, S is projected into I, and *NeighbouringSegments* is used to retrieve the segments close to $T(S)$. *MatchedPoints*(i) is used to take trace of the number of points of $T(S)$ found to be close to the segment r_i of SEG_I. This information is used in Step 8 in order to decide whether to include r_i in the set of valid segments SEG_V. $|r_i|$ is the number of points of r_i, and th_v is a prefixed threshold (currently, $th_v = 0.7$).

In Steps 10 through 14, S is projected into I once again. This time, we use the elements of SEG_V (and only these elements) to check the correspondence between $T(S)$ and I. In Steps 11 and 12, the current number of matched points of $T(S)$ is incremented if $T(p)$ belongs to the tolerance area of any valid segment (see the previous section). It is worth noticing that the number of average segment indexes stored in each entry of *NeighbouringSegments* is close to 1; thus, both Step 11 and the loop in Steps 5 and 6 are executed in almost constant time.

Finally, the value of the best match, together with the generating best hypothesis, is returned in Step 15. In Figure 4(b), we show the sketch of Figure 2(b) superimposed over the image of Figure 2(a) using the best transformation hypothesis returned by our system. A different system output with the same sketch but another database image is shown in Figure 5, while Figure 1(c) shows the sketch of a quadruped superimposed on the edge map of a database image representing a horse. In all the mentioned cases, the sketches have been positioned on the images using the corresponding transformations returned by the system.

Finally, Figure 6(b) shows a case in which the system fails in detecting the correct position of the guitar in the image. Although the low part of the guitar has been matched correctly with the image, the large amount of noise present in the neighboring segment map (i.e., in the set *NeighbouringSegments*, see Figure 6(a)) causes an incorrect selection of the best transformation in the *Test* algorithm.

This problem can be limited, including orientation information in *NeighbouringSegments(p)* and comparing it with the orientation of the point $T(p)$ (Step 11 of the *Test* algorithm).

The images of the database are processed sequentially and ranked accordingly with the value of *MaxMatched* obtained. The best 10 images are shown as output to the user. In Figure 7(a), we show the first five images chosen by the system for the sketch of Figure 2(b).

Figure 5. (a) The neighboring segment map (NeighbouringSegments) of an image of the system's database representing a guitar; (b) the map is used by the system to compute the best position for the sketch of Figure 2(b)

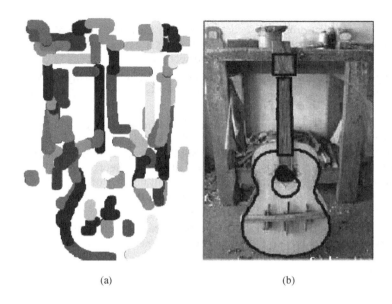

(a)   (b)

Figure 6. (a) The neighboring segment map of an image of the system's database representing two people and two guitars; (b) the sketch of Figure 2(b) is superimposed over the real image corresponding to the map

(a)   (b)

Computational Complexity

In this section, we analyze the computational complexity of the online phases of the proposed method.

The feature extraction process is executed off-line for every image of the database when this is stored in the repository. The user sketch does not need edge detection, but its points have to be grouped in line segments (second section).

Merging points using adjacency information can be performed easily in $O(M)$, being M the number of points of S. Suppose that the output of this step produces m_1 segments. Usually, the magnitude order of m_1 is a few dozens.

The second step in the line segment creation is the iterative process of "mutually favorite pairs" merging, whose worst case computational cost is $O(m_1^2 \lg m_1)$ (we refer to Huttenlocher & Ullman, 1990) for more details). Nevertheless, as observed in Huttenlocher and Ullman (1990), this worst case occurs only if all the segments have an endpoint near the same place in the image. In general, there is a small number of mutually favorite pairs of segments, especially in S; thus, the expected running time is effectively linear in m_1 rather then quadratic (Huttenlocher & Ullman, 1990).

Finally, suppose that the split-based polygonal approximation of a line segment s_j with M_j points produces b_j new breakpoints. It can be shown easily that an upper bound of the worst case computational cost is $(b_j + 1)M_j$, even if the average case is actually lower. Hence, for m_1 segments, we have that:

$$\sum_{j=1}^{m1} M_j (b_j + 1) \le (b_{max} + 1)M \le (m - m_1 + 1)M,$$
(11)

where b_{max} is the maximum number of new segments produced by the polygonal approximation process for a given segment and $m = |SEG_S|$. The whole polygonal approximation phase is then $O((m - m_1)M) = O(mM)$.

In conclusion, the whole feature extraction process for S is:

$$O(M + m_1^2 \lg m_1 + mM) = O(m_1^2 \lg m_1 + mM).$$

Let us now analyze the computational costs of the *Hypothesis Generation* algorithm. The loop of Steps 1 through 10 is repeated for every pair of segments belonging to $SEG'_I \times SEG_S$. If n' and m are, respectively, the cardinality of the sets SEG'_I and SEG_S, then the total number of iterations is $n'm$.

In the *GoodCandidate* function, we verify a constant number of (trivial) geometric constraints. Also the solutions of the few equations of the system defined in equations (3) through (6) can be found in constant time in the procedure *ComputeTransform*.

Conversely, the function *LocallySimilar* is linear with the number of points of the input segment s_j. Hence, if N_l is the cardinality of the largest segment $s_i \in SEG_S$, the worst case computational cost of the whole *Hypothesis Generation* algorithm is $O(n'mN_l)$.

Figure 7. (a) The first five images chosen by the system for the sketch of Figure 2(b); (b) a snapshot of the images contained in the database used for testing the system

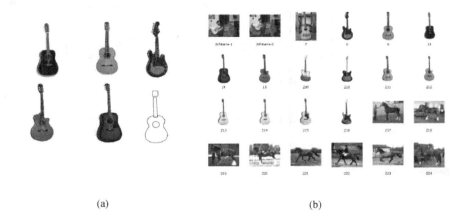

(a)   (b)

The clustering process is $O(N''^2 k)$ (Duda, Hart, & Strorck, 2000), being N'' the cardinality of H and k the number of merge operations performed by the algorithm. The parameter k depends on the maximum possible width of a cluster, which is a prefixed threshold. In practical cases, the magnitude order of k is a few dozen, because TS is a scattered set of geometrically meaningless hypotheses, only a few of which are sufficiently close one to the other to be clustered. Moreover, if C is the number of final clusters, then they can be ordered in $O(C \lg C)$ for the selection of the elements of H'. Since $C < N''$, then $O(C \lg C) = O(N''^2 k)$.

Finally, the computational complexity of the *Test* algorithm is the following. Suppose that N, n, and M are, respectively, the cardinality of H', SEG_l and S. Moreover, suppose that c is the maximum number of elements of *NeighbouringSegments(p)* ($\forall p \in I$). As mentioned in the "Verification Phase" section, c is usually close to 1.

The initialization process performed in Step 2 is trivially $O(n)$. Steps 4 through 6 are performed in $O(cM)$. If SEG_V is realized using a Boolean vector, then the loops in Steps 7 through 9 and 10 through 12 are, respectively, $O(n)$ and $O(cM)$. Hence, the whole computational complexity of the *Test* algorithm is $O(N(n+cM))=O(cNM)$, the last equation deriving from the fact that $M >> n$.

In conclusion, the worst case computational cost of all the online phases of our method is:

$$O(m_1^2 \lg m_1 + mM + n'mN_1 + N''^2 k + cNM). \tag{12}$$

As we will see in the next section, all the involved computations are, in fact, very fast.

Figure 8. Other examples of test images

(a)   (b)

Results

We have implemented our method in Java and tested it on a Pentium M, 1.6 Ghz, 1GB RAM. The average execution time is about one second for processing each single image of the database, with images from 200×200 up to 400×300 pixels.

The test database is composed of 284 images taken randomly from the Web. They contain a great variety of subjects, as shown in Table 1. No simplifying assumption has been made about the images; they often have nonuniform backgrounds, and the represented objects sometimes are occluded by other objects. The background often is composed of trees, rocks, wooden walls, or other objects with thick textures. No manual segmentation has been performed on the images in order to separate the interesting objects from their background or from other adjacent or occluding objects. Also, the lighting conditions and the noise degree are not fixed; we simply have taken our images by randomly browsing the Web.

Figures 7(a) and 8 show three snapshots of the database directory. Table 1 shows the composition of the whole database by grouping the images with respect to the main object(s) they represent (human figures represented together with other objects as well as landscapes or minor objects are not taken into account).

We asked a nonprofessional drawer to draw a sketch for each object category of the set shown in Table 1. The voluntary drawer had not seen the database images when he drew the sketches used as queries. In such a way, he was not influenced by the shapes present in the database. We asked him for a stylized representation for each object category (see Figures 1(b) and 2(b)).

The input sketches accepted by the system possibly can contain internal edges or be nonconnected (see Figure 2(b)) because the proposed method is not based on sketches represented by closed curves, as usually CBIR systems do.

Table 2 summarizes the results concerning the first 10 correct retrieved images for each input sketch.

Table 1. Database composition

Object type	Number of items contained in the database
Guitars	46
Cars	27
Horses	25
Watches	24
Pipes	15
Crucifixes	15
Saxophones	14
Faces	14
Other Animals	13
Mushrooms	12
Bottles	11
Pistols	10
Tennis Rackets	10
Vases	10
Fruits	7
Airplanes	7
Balls	6
Fishes	5
Glasses and Cups	4
Other Objects	19

Unfortunately, a quantitative comparison of our method with other shape-based image retrieval systems is not easy, because at the moment, as far as we know, all the existing shape-based benchmarks are composed of images with only one object and a uniform background (e.g., Jeannin & Bober, 1999). Since our approach is focused on the possibility to localize a given sketch in a real-life, complex images and not on the similarity accuracy of the method, a correct comparison should be performed on images with different and possibly mutual occluding objects.

For a qualitative comparison, we underline that, as mentioned in the Introduction, shape-based image retrieval systems that are able to perform both localization and inexact matching are very uncommon. Moreover, most of them (e.g., those based on a deformable template matching approach) either need a partial manual segmentation of the database images or perform quite slow exhaustive iterations.

Table 2. Correct retrieved items in the first 10 positions

Query	Cardinality of the Corresponding Class	Correct Retrieved Items in the First 10 Positions
Guitar	46	9
Car	27	8
Horse	25	7
Watch	24	7
Pipe	15	6
Crucifix	15	8
Saxophone	14	7
Bottle	11	5
Pistol	10	8
Tennis Racket	10	9
Average	**19.7**	**7.4**

Conclusion

We have presented a CBIR system based on shape. Different from most of the existing shape-based retrieval systems, the proposed method is able to deal with unsegmented images in which the objects of interest are not isolated from the background and the other objects.

The method is based on an alignment approach, which has been proved to be robust in rigid object recognition tasks by a large amount of works in scientific literature but never applied to image retrieval issues until now. We exploit the potentialities of the alignment approaches to deal with segmentation problems. The proposed approach can efficiently deal with all the 2D similarity transformations (translation, rotation, and scaling), while common shape-based retrieval techniques usually are independent on only a subset of such transformations.

Furthermore, our method is able to deal with nonrigid matching issues that are peculiar of a query by a sketch system in which the sketch and the real images do not correspond exactly. The proposed matching strategy uses line segments both to reduce the number of hypotheses to be tested and to take into account the image context when two shapes are compared locally or globally. This makes it possible to reject those alignments in which the sketch randomly overlaps with the lines of the image. A clustering process of the candidate hypotheses makes it possible to merge different transformation hypotheses generated by segments of the image having spatial relationships not exactly corresponding to the spatial relationships of the sketch segments.

We believe that in the next future, reliable CBIR systems (for still images) should be based more and more on shape analysis, especially for general-purpose applications, because other perceptive features such as color or texture do not allow the user to discriminate the desired visual content as well as shape can. Moreover, since most of the digital images available in real-life applications contain complex backgrounds and occluding/touching objects, research in this area should focus its attention on retrieval approaches that are able to deal with localization issues besides having good matching capabilities. We think that alignment techniques, suitably modified for inexact matching treatment, can be a good solution to this type of problem.

Concerning CBIR by shape, some of the most important present or future possible applications can be classified as follows (e.g., Forsyth & Ponce, 2003):

- **Military intelligence:** Typical queries involve finding airplanes, tanks, ships, missiles, or other weapons in large collections of satellite photos.

- **Access to museums:** A museum Web site can be interested in offering a Web visitor the possibility to browse its collection of paints searching for specific images. For instance, the Hermitage museum currently offers such a possibility using the QBIC (Flickner et al., 1995) system (see http://www.hermitagemuseum.org/fcgi-bin/db2www/qbicLayout.mac/qbic? selLang=English).

- **Trademark and copyright enforcement:** There are some organizations such as BayTSP that offer registered users the possibility to search for stolen copies of a picture on the Web. Moreover, trademark registration offices can use a shape-based image retrieval system in order to efficiently scan a database of existing trademarks and search for possible logos similar to a new input shape.

- **Indexing the Web:** While there are a lot of commercial textual-based search engines that specialize in finding textual documents, at the time of this writing, there are no sufficiently accurate CBIR systems that can browse the Web searching for images that contain a generic shape. Another possible Web application of query by shape systems regards the possibility of allowing users to avoid possible offensive images such as those containing naked people (Fleck, Forsyth, & Bregler, 1996).

- **Medical information systems:** The well-known Generalized Hough Transform has been used for detection of cancerous cells in medical diagnosis tasks using suitable templates to model the cell silhouettes (Lee & Street, 2000). Generally speaking, shape-based recognition systems can help doctors to recover medical images from a medical image database.

- **Image collections:** Digital photo archives are diffused in a lot of domain-specific applications, such as artistic, geographic, fashion, design, and historical repositories. In all of these domains, accurate and efficient shape-based image retrieval tools can help the user to find images similar to a template whose shape he or she only needs to know approximately.

The aforementioned (nonexhaustive) set of application examples shows some interesting real domains in which automatic systems for shape-based retrieval either are already used or could be used in a short to medium time.

Acknowledgments

We want to thank students Marco Cascianelli, Emanuele Garuglieri, and Paola Ranaldi for developing the first version of the system's prototype and the anonymous referee for the constructive remarks that led to an improved manuscript version.

References

Ayache, N., & Faugeras, O.D. (1986). HYPER: A new approach for the recognition and positioning of two-dimensional objects. *IEEE Trans. on PAMI, 8*(1), 44–54.

Brand, S., Laaksonen, J., & Oja, E. (2000). Statistical shape features in content-based image retrieval. In *ICPR00* (pp. 1062–1065).

Canny, J. (1986) A computational approach to edge detection. *IEEE Trans. on PAMI, 8*(6), 679–698.

Colombo, C., & Del Bimbo, A. (2002). Image data bases: Search and retrieval of digital imagery. In *Visible image retrieval* (pp. 11–33).

Del Bimbo, A. (1999). *Visual information retrieval*. San Francisco: Morgan Kaufmann.

Del Bimbo, A., & Pala, P. (1997). Visual image retrieval by elastic matching of user sketches. *IEEE Trans. on PAMI, 19*(2), 121–132.

Del Bimbo, A., & Pala, P. (1999). Shape indexing by multi-scale representation. *Image and Vision Computing, 17*, 245–261.

Duda, R. O., Hart, P. E., & Strorck, D.G. (2000). *Pattern classification* (2nd ed.). Wiley Interscience.

Fleck, M. M., Forsyth, D. A., & Bregler, C. (1996). Finding naked people. In *ECCV (2)* (pp. 593–602).

Flickner, M., et al. (1995). Query by image and video content: The QBIC system. *IEEE Computer, 28*(9), 23–32.

Folkers, A., & Samet, H. (2002). Content-based image retrieval using Fourier descriptors on a logo database. In *ICPR02* (pp. 521–524).

Forsyth, D. A., & Ponce, J. (2003). *Computer vision: A modern approach*. Prentice Hall.

Huttenlocher, D. P., & Ullman, S. (1990). Recognizing solid objects by alignment with an image. *International Journal of Computer Vision, 5*(2), 195–212.

Jain, A. K., Duin, R. P. W., & Mao, J. (2000). Statistical pattern recognition: A review. *IEEE Trans. on PAMI, 22*(1), 4–37.

Jain, A.K., Zhong, Y., & Lakshmanan, S. (1996). Object matching using deformable templates. *IEEE Trans. on PAMI, 18*, 267–278.

Jeannin, S., & Bober, M. (1999). *Description of core experiments for MPEG-7 motion/shape* (Tech. Rep. No. ISO/IEC JTC 1/SC 29/WG 11 MPEG99/N2690). Seoul: MPEG-7.

Kimia, B. B. (2002). Image databases: Search and retrieval of digital imagery. In *Shape representation for image retrieval* (pp. 345–372).

Lee, K.-M., & Street, W. N. (2000). Generalized Hough transforms with flexible templates. In *Proceedings of the International Conference on Artificial Intelligence (IC-AI2000)*, Las Vegas, NV (pp. 1133–1139).

Mokhtarian, F. (1995). Silhouette-based isolated object recognition through curvature scale-space. *IEEE Trans. on PAMI, 17*(5), 539–544.

Mokhtarian, F., & Abbasi, S. (2002). Shape similarity retrieval under affine transforms. *Pattern Recognition, 35*(1), 31–41.

Schmid, & Mohr. (1997). Local grey value invariants for image retrieval. *IEEE Trans. on PAMI, 19*(5).

Shapiro, L., & Stockman, G. (2001). *Computer vision*. Prentice Hall.

Smeulders, A. W. M., Worring, M., Santini, S., Gupta, A., & Jain, R. (2000). Content-based image retrieval at the end of the early years. *IEEE Trans. on PAMI, 22*(12), 1349–1380.

Tek, H., & Kimia, B. B. (2001). Boundary smoothing via symmetry transforms. *Journal of Mathematical Imaging and Vision, 14*(3), 211–223.

Ullman, S. (1996). *High-level vision. Object recognition and visual cognition*. Cambridge, MA: The MIT Press.

Xue, Z., Li, S. Z., & Teoh, E. K. (2002). AI-EigenSnake: An affine-invariant deformable contour model for object matching. *Image and Vision Computing, 20*(2), 77–84.

Chapter IV

Statistical Audio-Visual Data Fusion for Video Scene Segmentation

Vyacheslav Parshin, Ecole Centrale de Lyon, France

Liming Chen, Ecole Centrale de Lyon, France

Abstract

Automatic video segmentation into semantic units is important in order to organize an effective content-based access to long video. In this work, we focus on the problem of video segmentation into narrative units called scenes—aggregates of shots unified by a common dramatic event or locale. In this work, we derive a statistical video scene segmentation approach that detects scenes boundaries in one pass, fusing multi-modal audiovisual features in a symmetrical and scalable manner. The approach deals properly with the variability of real-valued features and models their conditional dependence on the context. It also integrates prior information concerning the duration of scenes. Two kinds of features extracted in visual and audio domain are proposed. The results of experimental evaluations carried out on ground truth video are reported. They show that our approach effectively fuses multiple modalities with higher performance compared with an alternative rule-based fusion technique.

Introduction

A constantly growing amount of available digitized video stored at centralized libraries or even on personal computers gives rise to the need for an effective means of navigation that allows a user to locate a video segment of interest. Searching of such a segment sequentially using simple fast-forward or fast-reverse operations provided by most of the existing players is tedious and time-consuming. A content-based access could greatly simplify this task, giving to a user the possibility to browse a video organized as a sequence of semantic units. Such an organization also could facilitate the task of automatic video retrieval, restricting the search by the scope of meaningful semantic segments. Another potential area of application is an automatic generation of video summaries or skims that preserve the semantic organization of the original video.

As the basic building blocks of professional video are shots—sequences of contiguous frames recorded from a single camera—it is natural to divide a video into these units. Unfortunately, the semantic meaning they provide is at too low of a level. Common video of about one or two hours (e.g., a full-length film) usually contains hundreds or thousands of shots—too many to allow for efficient browsing. Moreover, individual shots rarely have complete narrative meaning. Users are more likely to recall whole dramatic events or episodes, which usually consist of several contiguous shots. In this work, we consider the task of automatic segmentation of narrative films, such as most movies, into something more meaningful than shots—high-level narrative units called scenes, or aggregates of shots unified by a common dramatic event or locale. We need shot segmentation at the first preliminary processing step since scenes are generated as groups of shots. Segmentation into scenes can be considered the next level of content generation, yielding a hierarchical semantic structure of video in which shots are preserved to form the lower level. In this work, we are not concerned with the problem of shot segmentation or adopting one of already existing techniques (Boresczky & Rowe, 1996; Lienhart, 1999) but rather focus on the task of video segmentation into scenes.

Sharing a common event or locale, shots of a scene usually are characterized by a similar environment that is perceivable in both the visual and audio domains. So, both the image sequence and the audio track of a given video can be used to distinguish scenes. Since the same scene of a film usually is shot in the same settings by the same cameras that are switched repeatedly, it can be detected from the image track as a group of visually similar shots. The visual similarity is established using low-level visual features such as color histograms or motion vectors (Kender & Yeo, 1998; Rasheed & Shah, 2003; Tavanapong & Zhou, 2004). On the other hand, a scene transition in movie video usually entails abrupt changes of some audio features caused by a switch to other sound sources and sometimes by film editing effects (Cao, Tavanapong, Kim, & Oh, 2003; Chen, Shyu, Liao, & Zhang, 2002; Sundaram & Chang, 2000). Hence, sound analysis provides useful information for scene segmentation as well. Moreover, additional or alternative features can be applied. For example, editing rhythm, which usually is preserved during a montage of a scene, can be used to distinguish scenes as groups of shots of predictable duration (Aigrain, Joly, & Longueville, 1997); classification of shots into exterior or interior ones would allow for their grouping into the appropriate scenes (Mahdi, Ardebilian, & Chen, 1998), and so forth.

In order to provide reliable segmentation, there is a need to properly combine these multiple modalities to compensate for their inaccuracies. The common approach uses a set of rules according to which one source of information usually is chosen as the main one to generate initial scene boundaries, while the others serve for their verification (Aigrain, Joly, & Longueville, 1997; Cao, Tavanapong, Kim, & Oh, 2003) or further decomposition into scenes (Mahdi, Ardebilian, & Chen, 2000). Rules-based techniques, however, are convenient for a small number of features, generally do not take into account fine interaction between them, and are hardly extensible. Another frequent drawback of the existing methods is binarization of real-valued features that often leads to losses of information.

In this work, we derive a statistical scene segmentation approach that allows us to fuse multiple information sources in a symmetrical and flexible manner and is easily extensible to new ones. Two features are developed and used as such sources: video coherence that reveals possible scene changes through comparison of visual similarity of shots and audio dissimilarity reflecting changes in the audio environment. For the moment, we fuse these two types of information, but our approach easily can be extended to include additional data. In contrast to the common rule-based segmentation techniques, our approach takes into account a various confidence level of scene boundary evidence provided by each feature.

In our earlier work (Parshin, Paradzinets, & Chen, 2005) we already proposed a simpler approach (referenced hereafter as maximum likelihood ratio method) for the same scene segmentation task. The advantage of the approach proposed in this work (referenced hereafter as sequential segmentation method) is that it is based on less restrictive assumptions about observable feature vectors, allowing for their conditional dependence from the context, and takes into consideration a nonuniform statistical distribution of scene duration.

We have evaluated the proposed technique using a database of ground-truth video, including four full-length films. The evaluation results showed a superior segmentation performance of our sequential segmentation technique with respect to the previous maximum likelihood ratio, one that in its turn outperforms a conventional rule-based, multi-modal algorithm.

The remainder of this chapter has the following organization. First, the background and related work section briefly describes prior work on the scene segmentation problem and introduces the basic ideas that facilitate distinguishing video scenes in the visual and audio domain; coupling of multi-modal evidence about scenes is discussed as well. In the next section, we derive our sequential segmentation approach and make the underlying assumptions. Then, in the Feature Extraction section, we derive our video coherence measure used to distinguish scenes in the visual domain and provide details on our audio dissimilarity feature. In the Experiments section, we report the results of the experimental evaluation of the proposed scene segmentation approach using ground-truth video data. Final remarks then conclude this work.

Background and Related Work

Video Scene Segmentation Using Visual Keys

The common approach to video scene segmentation in the visual domain exploits the visual similarity between shots, which stems from specific editing rules applied during film montage (Bordell & Thompson, 1997). According to these rules, video scenes usually are shot by a small number of cameras that are switched repeatedly. The background and often the foreground objects shot by one camera are mostly static, and hence, the corresponding shots are visually similar to each other. In the classical graph-based approach (Yeung & Yeo, 1996), these shots are clustered into equivalence classes and are labeled accordingly. As a result, the shot sequence of a given video is transformed into a chain of labels that identifies the cameras. Within a scene, this sequence usually consists of the repetitive labels. When a transition to another scene occurs, the camera set changes. This moment is detected at a cut edge of a scene transition graph built for the video. For example, a transition from a scene shot by cameras A and B to a scene taken from cameras C and D could be represented by a chain $ABABCDCD$ in which the scene boundary would be pronounced before the first C. An analogous approach was proposed by Rui, Huang, and Mehrotra (1999) in which shots first were clustered into groups that then were merged into scenes. Tavanapong and Zhou (2004) in their ShotWeave segmentation technique use additional rules to detect specific establishment and reestablishment shots that provide a wide view over the scene setting at the beginning and the end of a scene. They also suggest using only specific regions of video frames to determine more robustly the intershot similarity.

To overcome the difficulties resulting from a discrete nature of the segmentation techniques based on shot clustering, such as their rigidity and the need to choose a clustering threshold, continuous analogues have been proposed. Kender and Yeo (1998) reduce video scene segmentation to searching of maxima or minima on a curve describing the behavior of a continuous-valued parameter called video coherence. This parameter is calculated at each shot change moment as an integral measure of similarity between two adjacent groups of shots based on a short-memory model that takes into consideration the limitation and preferences of the human visual and memory systems. Rasheed and Shah (2003) propose to construct a weighted undirected shot similarity graph and detect scene boundaries by splitting this graph into subgraphs in order to maximize the intra-subgraph similarities and to minimize the inter-subgraph similarities.

In this work, we propose a continuous generalization of the discrete clustering-based technique, which is analogous to the approach of Kender and Yeo (1998) in the sense that it yields a continuous measure of video coherence. This measure then is used in our multimodal segmentation approach as a visual feature providing a flexible confidence level of the presence or absence of a scene boundary at each point under examination; the lower this measure is, the more possible is the presence of a scene boundary (see Figure 2). In contrast to the video coherence of Kender and Yeo (1998), which is a total sum of intershot similarities, our measure integrates only the similarity of the shot pairs that possibly are taken from the same camera.

Video Scene Segmentation in the Audio Domain

As the physical setting of a video scene usually remains fixed or changes gradually (when, for instance, the cameras follow moving personages), the sources of the ambient sound rest stable or change their properties smoothly and slowly. A scene change results in a shift of the locale, and hence, the majority of the sound sources also changes. This change can be detected as the moment of a drastic change of audio parameters characterizing the sound sources.

Since short-term acoustic parameters often are not capable of properly representing the sound environment (Chen, Shyu, Liao, & Zhang, 2002), these parameters often are combined within a long-term window. The resulting characteristics are evaluated within two adjacent time windows that adjoin a point of potential scene boundary (usually shot breaks) or its immediate vicinity (as sound change sometimes shifted by a couple of seconds during montage to create an effect of interscene connectivity) and then compared. A scene boundary is claimed if the difference is large enough. Sundaram and Chang (2000) model the behavior of various short-term acoustic parameters, such as cepstral flux, zero crossing rate, and so forth, with correlation functions that characterize the long-term properties of the sound environment. A scene change is detected when the decay rate of the correlation functions, the total for the all acoustic parameters, reaches a local maximum, as it means low correlation between these parameters caused by the change of the sound sources. Cao, Tavanapong, Kim, and Oh (2003) approximate long-term statistical properties of short-term acoustic parameters using normal distribution. At a potential scene boundary, these properties are compared by applying a weighted Kullback-Leibler divergence distance.

In this work, we adopt Kullback-Leibler distance as an audio dissimilarity feature providing the evidence of the presence or absence of a scene boundary in the audio domain. This distance represents the divergence between distributions of shot-term spectral parameters that are estimated using the continuous wavelet transform.

Multi-Modal Data Fusion

The common approach to segmentation of narrative video into scenes is based only on visual keys extracted from the image stream. In order to combine information extracted from the audio and image streams into one more reliable decision, a set of simple rules is usually applied. The audio stream can be used as an auxiliary data source to confirm or reject scene boundaries detected from the image sequence. For example, Cao, Tavanapong, Kim, and Oh (2003) first segment video into scenes in the visual domain and then apply sound analysis to remove a boundary of suspiciously short scenes, if it is not accompanied by a high value of audio dissimilarity. Jiang, Zhang, and Lin (2000) propose first to segment the video in the audio domain and to find potential scene boundaries at shot breaks accompanied by a change in the sound environment; these boundaries then are kept in the final decision if they are confirmed by low visual similarity between preceding and succeeding shots. Sundaram and Chang (2000) first segment video into scenes independently in the video and audio domains and then align visual and audio scene boundaries as follows. For visual and audio scene boundaries lying within a time ambiguity window, only the visual scene boundary is

Figure 1. Scene duration pdf

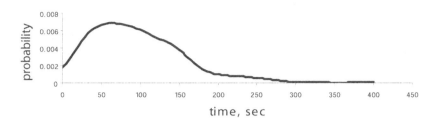

claimed to be the actual scene boundary; the rest of the boundaries are treated as the actual scene boundaries.

Rule-based approaches suffer from rigidity of the logic governing the feature fusion. Generally, each feature provides evidence about the presence or absence of a scene boundary with a different level of confidence, depending on its value. Making intermediate decisions, rule-based techniques ignore this difference for one or several features. Moreover, these techniques require the proper choice of thresholds, which usually are more numerous, the more rules that are applied. In this work, we derive a segmentation approach that fuses multiple evidences in a statistical manner, dealing properly with the variability of each feature. This approach is easily extensible to new features, in contrast to rule-based techniques that often become too complicated and cumbersome when many features are treated. We also take into consideration a nonuniform distribution of scene durations (see Figure 1) by including it as prior information.

Sequential Segmentation Approach

In this section, we derive our segmentation approach, which makes decisions about the presence or absence of a scene boundary at each candidate point. As scenes are considered as groups of shots, their boundaries occur at the moments of shot transitions. Therefore, in this work, these transitions are chosen as candidate points of scene boundaries. It is assumed that evidence about a scene boundary at an arbitrary candidate point is provided by a locally observable audiovisual feature vector. Further in this section, we first do some assumptions about observable features in the following subsection. Then an estimate of the posterior probability of scene boundary is derived, and the final segmentation algorithm is given in the next two subsections.

Conditional Dependence Assumptions about Observable Features

Let's consider an observable audiovisual feature vector D_i measured at a scene boundary candidate point i independently from the rest of vectors. In the general case, this vector is conditioned on the fact of presence or absence of a scene boundary not only at this point but at the neighboring points as well. Indeed, in the visual domain, the corresponding feature usually represents visual similarity between two groups of shots adjoining to the point under examination. If a scene boundary appears exactly between these groups, then the similarity measure usually has a local extremum. But if a scene boundary lies inside one of these groups, then the similarity measure takes an intermediate value that is closer to the extremum, the closer the scene boundary is (see Figure 2). Similar considerations also hold true for the audio data (see Figure 2).

For the purpose of simplification, we assume that local features are conditionally dependent on the distance to the closest scene boundary and are independent of the position of the rest of the scene boundaries. As the visual feature used in this work is a similarity measure applied to the whole shots, it is reasonable to assume the conditional dependence of this feature on the distance expressed in the number of shots. Let's denote a time-ordered sequence of scene boundaries as $B = \{b_1, b_2, ..., b_n\}$, in which each boundary is represented by the order number of the corresponding candidate point. As the scene boundary closest to an arbitrary candidate point i is one of two successive boundaries b_{k-1} and b_k surrounding this point so as $b_{k-1} \leq i < b_k$, the likelihood of video feature v_i measured at point i given partitioning into scenes B can be written as:

$$P(v_i \mid B) = P(v_i \mid b_{k-1}, b_k) = P(v_i \mid \Delta_i), \tag{1}$$

in which Δ_i is the distance from point i to its closest scene boundary b_c defined as:

$$\Delta_i = i - b_c, \tag{2}$$

$$b_c = \begin{cases} b_{k-1}, \text{if } i - b_{k-1} \leq b_k - i \\ b_k \text{ otherwise.} \end{cases} \tag{3}$$

The audio feature is defined in this work as a change in acoustic parameters measured within two contiguous windows of the fixed temporal duration. Therefore, we assume conditional dependence of this feature on the time distance to the closest scene boundary. Denoting the time of i-th candidate point as t_i, the temporal distance from point i to its closest scene boundary—as τ_i, we write the likelihood of audio feature a_i measured at point i as:

$$P(a_i \mid B) = P(a_i \mid b_{k-1}, b_k) = P(a_i \mid \tau_i), \tag{4}$$

in which:

$$\tau_i = t_i - t_c, \tag{5}$$

$$t_c = \begin{cases} t_{b_{k-1}}, \text{if } t_i - t_{b_{k-1}} \le t_{b_k} - t_i \\ t_{b_k} \text{ otherwise.} \end{cases} \tag{6}$$

In this work, we calculate likelihood values $P(v_i \mid \Delta_i)$ and $P(a_i \mid \tau_i)$ using the corresponding probability density functions (pdf) considered to be stationary (i.e., independent of time index i). It is assumed that observable features are dependent on the closest scene boundary only if the distance to it is quite small (i.e., lower than some threshold that is on the order of the length of the time windows used to calculate these features). This assumption facilitates the learning of parameters of pdf estimates based on a set of learning data.

Taking into account expression (1) and (4), the likelihood of the total feature vector $D_i = \{v_i, a_i\}$ given partitioning into scenes B can be reduced to:

$$P(D_i \mid B) = P(D_i \mid b_{k-1}, b_k). \tag{7}$$

In this work, we assume conditional independence of the components of D_i given B:

$$P(D_i \mid B) = P(v_i \mid B)P(a_i \mid B) = P(v_i \mid b_{k-1}, b_k)P(a_i \mid b_{k-1}, b_k). \tag{8}$$

If more observable data are available, expression (8) can include additional feature vector components that provide an easy extensibility of our segmentation approach.

Segmentation Principles

Statistical analysis of scene duration shows that it has nonuniform distribution, as most scenes last from half a minute to two to three minutes (see Figure 1). In order to take into account the information about scene duration, we include a prior of a scene boundary that depends on the time elapsed from the previous scene boundary and does not depend on the earlier ones, much as it is done in the case of variable duration hidden Markov models (Rabiner, 1989). Furthermore, the posterior probability of a scene boundary b_k at point i is assumed to be conditionally dependent solely on local feature vector D_i given the position b_{k-1} of the previous scene boundary. This assumption agrees with the intuition that evidence of the presence or absence of a scene boundary at an arbitrary point is determined by the feature vector measured at the same point. Indeed, this feature vector reflects the degree of change in the visual and audio environment of a scene, and the larger this change is, the higher is the probability of a scene change. Using Bayes rule, the posterior probability of k-th scene boundary at point i given b_{k-1} is written as:

$$P(b_k = i \mid D_i, b_{k-1}) = \frac{P(D_i \mid b_{k-1}, b_k = i)P(b_k = i \mid b_{k-1})}{P(D_i \mid b_{k-1}, b_k = i)P(b_k = i \mid b_{k-1}) + P(D_i \mid b_{k-1}, b_k \neq i)P(b_k \neq i \mid b_{k-1})}$$

$$= 1 \Big/ \left[1 + \frac{P(D_i \mid b_{k-1}, b_k \neq i)P(b_k \neq i \mid b_{k-1})}{P(D_i \mid b_{k-1}, b_k = i)P(b_k = i \mid b_{k-1})} \right], \ \forall i > b_{k-1}. \tag{9}$$

In expression (9), it is further assumed that the next scene boundary b_{k+1} takes place a long time after boundary b_k, so that the likelihood of D_i given $b_k < i$ always is conditioned on b_k when computed according to expressions (1) through (6). We denote this assumption as $b_{k+1} = +\infty$. It also is supposed that scene boundary duration is limited in time by a threshold value S. Then a possible position of k-th scene boundary is limited by a value m_k defined as:

$$m_k = \max\{l \mid t_l - t_{b_{k-1}} \leq S\}. \tag{10}$$

Under these assumptions, expression (9) is continued as:

$$P(b_k = i \mid D_i, b_{k-1}, b_{k+1} = +\infty) =$$

$$= 1 \Big/ \left[1 + \frac{\displaystyle\sum_{l=b_{k-1}+1}^{i-1} P(D_i \mid b_k = l, b_{k+1} = +\infty)P(b_k = l \mid b_{k-1}) + \sum_{l=i+1}^{m_k} P(D_i \mid b_{k-1}, b_k = l)P(b_k = l \mid b_{k-1})}{P(D_i \mid b_k = i)P(b_k = i \mid b_{k-1})} \right]$$

$$\tag{11}$$

It is assumed that the prior probability $P(b_k \mid b_{k-1})$ of scene boundary b_k is determined by the duration of the scene, which ends up at this boundary and is calculated using pdf of scene duration p_s as:

$$P(b_k \mid b_{k-1}) = \alpha p_s(t_{b_k} - t_{b_{k-1}}). \tag{12}$$

Normalizing coefficient α can be omitted when this expression is substituted in equality (11), as only the ratio of probability values is taken into account. In this work, we use a nonparametrical estimate of pdf p_s with Gaussian kernel (Duda & Hart, 1973) and limit its range of definition by lower and upper boundaries.

We propose to segment an input video into scenes sequentially, choosing each next scene boundary based on the position of the previous one. So, the video can be segmented in real

Figure 2. Audio dissimilarity (upper curve), video coherence (middle curve), and scene boundary posterior probability in sequential segmentation approach (partially overlapping curves in the bottom) vs. frame number; vertical dashed lines delimit scenes

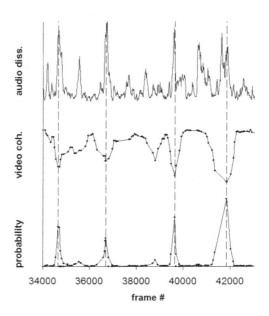

time with a time delay of the order of the maximal scene duration S. Knowing the position of scene boundary b_{k-1}, we select the next boundary b_k using the posterior probability estimated at each candidate point i, $i > b_{k-1}$, on time length S according to expression (11). In this chapter, the boundary b_k is placed at the point of the maximal probability, as such a decision criterion has appeared to work well in experimental evaluations. This criterion is based on a relative comparison of the evidence of a scene boundary at each point under consideration provided by the feature vector measured at the same point. In this manner, the resulting segmentation procedure resembles the conventional techniques that pronounce scene boundaries at the points of local extremum of some visual or audio similarity curve, expression (11) being considered as a way to fuse multiple data into one cumulative measure. Four posterior probability curves along with audio dissimilarity and video coherence curves obtained for a ground-truth film are depicted in Figure 2. The probability curves are shown partly overlapped; each curve begins at the first candidate point inside a scene, achieves the global maximum at the point of transition to the next scene, and is interrupted at the middle of the next scene (in order not to encumber the figure).

We deliberately include only one local feature vector D_i in expression (11) and exclude surrounding data from consideration. Otherwise, there would be a need to treat properly the strong dependence that usually exists between contiguous observable data. This would complicate the proposed approach and possibly would require more learning data. Experimental tests on a more complicated model that includes the complete set of observable data up to the point under examination, much as the model proposed by Vasconcelos and Lippman (1997) for the task of shot segmentation, suggest that simple neglect of this dependence in such a model degrades considerably the segmentation performance, let alone the increase of the computational complexity. For the same reasons, we do not adopt hidden Markov models that assume conditional independence between observable feature vectors. The problem of dependence between feature vectors is avoided in our model, as the single feature vector D_i in expression (11) usually is placed far enough from boundary b_{k-1} at the most points under examination and, thus, does not depend strongly on the feature vector measured at this boundary.

Final Algorithm

The final segmentation algorithm used in this work is resumed as follows.

Segment an input video into shots and assign candidate points of scene boundaries to be the shot transition moments. Estimate feature vector D_i at each point i.

Place the initial scene boundary b_0 at the beginning of the first scene (which is supposed to be given). Select recursively each subsequent scene boundary b_k based on the position of the previous one b_{k-1} through the following steps:

Calculate the posterior probability of k-th scene boundary at each candidate point i of set $\{b_{k-1} + 1, ..., m_k\}$ according to expression (11) in which m_k is defined by expression (10) and is limited by the last candidate point.

Place the next scene boundary b_k at the point of the highest posterior probability.

If a stopping criterion is fulfilled, exit the algorithm.

The stopping criterion is used mostly to keep inside the narrative part of the input video. In this work, we suppose that the position of the last scene boundary is given and that the stopping criterion is fulfilled when scene boundary b_k appears to be closer in time to the last scene boundary than a predefined threshold value that is approximately equal to the mean scene duration.

Feature Extraction

In this section, we propose visual and audio features that provide evidence of the presence or absence of a video scene boundary and describe the corresponding likelihood estimates required in our sequential segmentation approach.

Video Coherence

Our video coherence feature is derived as a continuous generalization of the conventional graph-based approach (Yeung & Yeo, 1996). As mentioned in the section describing related work, in this approach, visually similar shots first are clustered into equivalence classes and labeled accordingly. Then, a scene transition graph is built, and scene boundaries are claimed at cut edges of the graph. Let's consider the following shot clustering technique. First, a similarity matrix for an input video is built, each element $Sim(i,j)$ of which is the value of visual similarity between shots i and j. Then, each pair of shots that are similar enough (i.e., their similarity is higher then a threshold T_{cl}) is merged into one cluster until the whole matrix is exhausted. This is almost a conventional clustering procedure, except the radius of the clusters is not limited. In practice, we consider the shots that are far apart in time and, hence, are not likely to belong to one scene as nonsimilar and never combine them into one cluster. So, we need to treat only the elements of the similarity matrix located near the main diagonal, which makes the computational burden approximately linear with respect to the duration of the video.

Let's define for each shot i the following variable:

$$C^0(i) = \max_{a<i, b\geq i} Sim(a,b). \tag{13}$$

If this variable is less than the clustering threshold T_{cl}, then, according to the proposed clustering technique, it means that there are no common clusters that combine at least one shot preceding shot i with shot i or with a shot that follows shot i. In this and only in this case, there would be pronounced a scene boundary (at the transition to shot i) according to the graph-based segmentation method.

Hence, we can reformulate the conventional graph-based procedure of scene segmentation procedure as searching points on the curve C^0 that fall below the threshold value, scene boundaries being claimed in these points. Alternatively, scene boundaries can be pronounced at the points of local minima of the curve.

In real video, visual similarity between shots within the same scene often is not quite high, especially in action films in which there are many dynamic episodes. Because of this, minima of the variable C^0 often are pronounced badly, and it can happen accidentally that a shot of a scene resembles a shot of the previous or the next scene. In this case, the segmentation procedure can miss scene boundaries. Consider, for example, two scenes represented by a shot clusters chain $ABABCDADCD$ in which a real scene boundary occurs before the first shot of cluster C, and because of accidental similarity, one of the shots from the second scene was misclassified as A. Since the shot clusters in this example cannot be divided into two nonintersecting groups, clustering-based segmenting procedure fails to detect the scene boundary.

In order to enhance the robustness of the segmentation procedure, we can try to implicitly exclude isolated misclassified shots from consideration. At first glance, the next maximal value after C^0 could be taken according to expression (13). However, if a single shot is similar to a shot from another scene, it is likely to resemble other shots of the same cluster. In the previous example of a cluster chain, the shot from the second scene, misclassified as cluster A, is similar to two shots of this cluster for the first scene. Hence, exclusion of a single pair of maximally similar shots does not definitely exclude the influence of a single misclassified shot. So, in addition to this pair, we propose not to take into consideration all the maximally similar shots that follow or precede it and to define for each shot i the following variable:

$$C^1(i) = \min\{\max_{a<i,b\geq i, a\neq a_0(i)} Sim(a,b), \max_{a<i,b\geq i, b\neq b_0(i)} Sim(a,b)\},\qquad(14)$$

in which the variables a_0 and b_0 are the shot numbers, for which the expression (13) attains the maximum:

$$\{a_0(i), b_0(i)\} = \arg\max_{a<i,b\geq i} Sim(a,b) \cdot\qquad(15)$$

By recursion, we can derive variables to exclude the influence of the second misclassified shot, the third one, and so forth:

$$C^n(i) = \min\{\max_{a<i,b\geq i, a\notin\{a_0(i),...,a_{n-1}(i)\}} Sim(a,b), \max_{a<i,b\geq i, b\notin\{b_0(i),...,b_{n-1}(i)\}} Sim(a,b)\},\qquad(16)$$

$$a_n(i) = \arg\max_{a<i,b\geq i, a\notin\{a_0(i),...,a_{n-1}(i)\}} Sim(a,b),\qquad(17)$$

$$b_n(i) = \arg\max_{b\geq i, b\notin\{b_0(i),...,b_{n-1}(i)\}, a<i} Sim(a,b).\qquad(18)$$

The variable C^k has sharp local minima at scene boundaries only if they correspond to k misclassified shots. Otherwise, these minima are not well-pronounced. Generally, as the same pair of maximally similar shots can correspond to several contiguous shots, the previously defined variables C can remain constant during a period of time. If this constant region corresponds to a local minimum, the scene boundary position cannot be located precisely. In order to use all the variables C together and to reduce the probability of wide local minima, an integral variable is defined:

$$C_{int}(i) = \frac{1}{N}\sum_{k=0}^{N-1} C^k(i),\qquad(19)$$

in which N denotes the number of terms C determined by expression (13) through (18). By analogy with Kender and Yeo (1998), we refer to variable $C_{int}(i)$ as video coherence and consider it a visual feature that provides evidence of the presence or absence of a scene boundary at the beginning of shot i.

The similarity $Sim(a,b)$ between shots a and b involved in expression (13) through (18) can be calculated in various manners. In our experimental evaluations that will be described next, it is calculated as normalized color histogram intersection for the pair of maximally similar key frames representing the shots. The histogram is defined in HSV-color space quantized at 18 hue, 4 saturation, and 3 value points, and included additional 16 shades of gray. Video coherence feature includes three terms; that is, in expression (19), N is equal to 3 and is calculated for two contiguous groups of five shots that adjoin the point under consideration.

In our sequential segmentation approach, we consider the video coherence feature as a random value generated by a stationary process. In order to evaluate the likelihood of this variable mentioned in expression (1), we use a nonparametrical estimate of the corresponding pdf based on a Gaussian kernel and obtained for a set of presegmented ground-truth data. It is calculated separately for each possible value of the distance to the closest scene boundary Δ. We assume that this distance is limited by a range $[-n_1, n_2]$ in which n1 and n2 are natural numbers of the order of value N in expression (19). If it happens that $\Delta < -n_1$, we set $\Delta = -n_1$, and if $\Delta > n_2$, we set $\Delta = n_2$.

Audio Dissimilarity

In order to calculate the short-term acoustic feature vector for a sound segment, we divide the spectrum obtained from Continuous Wavelet Transform (CWT) into windows by application of triangular weight functions W_i with central frequencies f_i in Mel scale as it is done in the case of Mel Frequency Cepstrum Coefficients calculation (see Figure 3). Unlike the FFT, which provides uniform time resolution, the CWT provides high time resolution and low frequency resolution for high frequencies, and low time resolution with high frequency resolution for low frequencies. In that respect, it is similar to the human ear, which exhibits similar time-frequency resolution characteristics (Tzanetakis, Essl, & Cook, 2001).

Then energy values E_i in each spectral window are computed, and finally, the matrix of spectral bands ratios is obtained as:

$$K_{ij} = \log(E_i / E_j).$$
(20)

Values from the top-right or bottom-left corner of the matrix K are taken as our acoustic features.

The mentioned acoustic feature vector (matrix) is not affected by main volume change, unlike spectral coefficients. At the same time, it allows us to detect changes in acoustic environment.

The procedure of audio dissimilarity curve calculation is done by moving two neighboring windows (with size 8 and step 0.5 seconds in our experiments) along the audio stream and by obtaining the distance between the distributions of the corresponding acoustic features. Various measures may be used as a distance or dissimilarity for the task of acoustic segmentation: Bayesian Information Criterion (Chen & Gopalakrishnan, 1998), Second-Order

Figure 3. Triangular weight functions with central frequencies in Mel scale

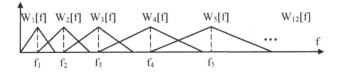

Statistics (Bimbot, Magrin-Chagnolleau, & Mathan, 1995), Kullback-Leibler (KL) distance applied directly to distribution of spectral variables (Harb & Chen, 2003).

The KL-measure is a distance between two random distributions (Cover & Thomas, 2003). In the case of Gaussian distribution of random variables, the symmetric KL distance is defined as:

$$KL(X_1, X_2) = \left(\frac{\sigma_1^2}{\sigma_2^2} + \frac{\sigma_2^2}{\sigma_1^2} \right) + (\mu_1 - \mu_2)^2 \left(\frac{1}{\sigma_1^2} + \frac{1}{\sigma_2^2} \right), \tag{21}$$

in which μ and σ are the mean value and the variance of compared distributions.

Instead of multi-dimensional KL applied to a feature vector of spectral bands ratios, a sum of KL distances applied to each element of the vector is used in this work as audio dissimilarity measure:

$$D = \sum_{ij} KL(K1_{ij}, K2_{ij}), \tag{22}$$

in which K1 and K2—feature matrices for the neighboring windows.

As an observable feature of a scene boundary in the audio domain, in this work, we extract the maximal value of audio dissimilarity in a time window of about four seconds centered in the corresponding candidate point in order to tolerate small misalignments between the audio and image streams of video. The likelihood of this feature included in expression (4) is calculated from the joint probability as:

$$P(a|\tau) = \frac{P(a,\tau)}{P(\tau)}, \tag{23}$$

in which, as earlier, a stands for the feature value, τ—for the time distance to the closest scene boundary. We approximate the joint probability with a nonparametric estimate of pdf using a Gaussian kernel on a set of learning data. Just as for the visual feature, we limit the range of τ by a value having the order of duration of the neighboring time windows used to calculate the audio dissimilarity.

Experiments

In this section, we report the results of experiments that are designed to test the proposed video scene segmentation approach. For the lack of common benchmark data, a database of four ground-truth movies (drama: *A Beautiful Mind*; mystery: *Murder in the Mirror*; French comedy: *Si J'Etais Lui*; and romance: *When Harry Met Sally*) was prepared and manually segmented into semantic scenes. The performance comparisons were made inside time intervals that had the total duration of about 22,000 seconds and included 234 manually labeled scene boundaries.

The segmentation into shots at the preprocessing stage was carried out automatically using a twist-threshold method (Zhang, Qi, & Zhang, 2001) based on color histogram similarity measure. To reduce the computational complexity of the segmentation algorithm, likelihood values of audio and visual features in expressions (1) and (4) were calculated using linear interpolation between tabled values. In the feature domains (fixed through all experiments described next), where estimates of the corresponding pdf became unstable due to the lack of learning data, the likelihoods were extrapolated as constant functions. Only a small portion of data fell into these domains, and experimental evaluations demonstrated that the choice of their boundaries was not crucial for segmentation performance.

Segmentation performance was measured by using the value of precision p and recall r defined as:

$$r = \frac{n_c}{n_c + n_m}, \quad p = \frac{n_c}{n_c + n_f},$$

(24)

in which n_c, n_m, and n_f are the number of correctly detected scene boundaries, the number of missed boundaries, and the number of false alarms, respectively. Detected scene boundary was considered correct if it coincided with a manual scene boundary within an ambiguity of five seconds. Otherwise, it was considered a false alarm. A manual scene boundary was considered missed if it did not coincide with any of the automatically detected boundaries within the same ambiguity of five seconds. As a unified performance score, F1 measure was used:

$$F1 = \frac{2rp}{r + p}.$$

(25)

Segmentation performance of the proposed sequential segmentation algorithm relative to various films entered into our database is compared in Table 1. Feature likelihoods and scene duration pdf were estimated on the learning set including all four films. The highest integral performance F1 for the film *Murder in the Mirror* was caused mainly by the most stable behavior of the video coherence curve, as the scenes were shot by relatively slow-moving or static cameras. In contrast, the outsider film *Si J'Etais Lui* was characterized by intensive camera movements. A reason for a relatively low performance for the film *A Beautiful Mind* was a less accurate shot segmentation for gradual shot breaks, which sometimes merged shots that were contiguous to a scene boundary.

In order to evaluate the generalization capability of the segmentation approach learned on a set of presegmented data, the cross-validation tests were carried out. The learning set included three films, and the test set consisted of the resting fourth. The overall results for all four films are given in Table 2. Three trials were made: the first one did not use cross-validation at all, serving as a reference; the second used a separate set to learn only the pdf estimates for the audio and visual features, while the scene duration pdf was estimated on a common set including all four films; the third trial supposed separate learning and test sets for all the pdf estimates. As it follows from Table 2, our segmentation approach does not

Table 1. Performance of the sequential segmentation algorithm for various films

Film	Precision, %	Recall, %	F1, %
A Beautiful Mind	67.7	67.7	67.7
Murder in the Mirror	88.9	66.7	76.2
Si J'Etais Lui	66.7	63.2	64.9
When Harry Met Sally	69.8	71.2	70.5
Total for four films	72.4	67.1	69.6

suffer much from parameters over-fitting, providing quite a general model for video scene segmentation. The perceptible sensitivity to the estimate of scene duration pdf suggests the importance of taking into account of prior information about scene duration. The results given next in this section assume the same learning and test set, which includes all four films of the ground truth.

The capability of our sequential segmentation approach to fuse audiovisual features is shown in Table 3, in which the first row presents the segmentation performance when only the visual feature was used, the second row gives the performance only for the audio feature, and the third for both features. As it follows from the table, fusing the visual and audio features enhances both recall and precision.

Table 2. Performance of the sequential segmentation algorithm in cross-validation tests

Using Cross-Validation	Precision, %	Recall, %	F1, %
Non	72.4	67.1	69.6
For the feature pdf only	69.9	67.5	68.7
Total for the feature pdf and the scene duration pdf	67.6	65.0	66.2

Table 3. Performance of the sequential segmentation algorithm for audio-visual feature fusion

Feature Used	Precision, %	Recall, %	F1, %
Visual	61.7	64.1	62.9
Audio	39.9	48.7	43.8
Visual + Audio	72.4	67.1	69.6

In order to compare our segmentation approach with related multi-modal techniques, we considered the following rule-based scene segmentation algorithm. First, an input video was segmented solely in the visual domain. Strong scene boundaries then were claimed as the actual scene boundaries, while weak scene boundaries were kept only if they were confirmed by a high level of the audio dissimilarity that had to be above threshold A. This is a scheme of audiovisual data fusion somewhat analogous to that of Cao, Tavanapong, Kim, and Oh (2003). We also refused from the use of scene duration distribution and adopted a segmentation technique that searched scene boundaries at local minima of the video coherence curve. A local minimum was claimed as a scene boundary if it was a global minimum of enough depth in a surrounding time window and had the absolute value below some threshold $T1$. A scene boundary was considered weak if the corresponding video coherence value was above a second threshold $T2$, $T1 > T2$; otherwise, it was marked as a strong boundary. Thresholds A, $T1$, and $T2$ were selected in order to maximize the overall performance measure F1.

The performance of this rule-based algorithm is given in the first row of Table 4, where it can be compared with the performance of our earlier maximum likelihood ratio approach (Parshin, Paradzinets, & Chen, 2005) and the sequential segmentation one derived in this work. The maximum likelihood ratio algorithm uses the same audio feature as the others; as the visual feature, it used the video coherence C^0 given by expression (13) since it is less dependent on the context and, hence, is more suitable for this algorithm. To compare the efficiency of audiovisual data fusion provided by our rule-based algorithm, we also include the test results for a segmentation algorithm, referenced as "local minima of video coherence," which works solely in the visual domain. This algorithm detects scene boundaries at local minima on the video coherence curve in the same way as our rule-based algorithm with the difference that it uses only one threshold value that maximizes performance measure F1. A comparison of the results given in Table 4 allows us to conclude that the sequential segmentation approach has the best performance measured by both precision and recall.

As for computational time required by our sequential segmentation algorithm, it is quite fast, given that audiovisual features are precomputed and take less than a second on our Intel Pentium 4 1.8GHz computer for one film. This is because the computational complexity is approximately linear with respect to the film length due to a limited time search for

Table 4. Performance of different segmentation approaches

Segmentation approach	Precision, %	Recall, %	F1, %
Rule-based	61.0	63.9	62.4
Local minima of video coherence	54.1	64.3	58.8
Maximum likelihood ratio	63.2	63.2	63.2
Sequential segmentation	72.4	67.1	69.6

each scene boundary. The main computational burden for a raw video file stems from its decoding and feature extraction, which, however, can be done in real time without much optimization for MPEG 4 video format.

Conclusion

A statistical video scene segmentation approach is proposed that combines multiple mid-level features in a symmetrical and flexible manner. In contrast to its rule-based counterparts, it deals properly with real-valued observable features by taking into account the variability of scene boundary evidence provided by these features. This approach also models the duration of scenes, including it as prior information. Two kinds of features are proposed to be used in scene segmentation: video coherence and audio dissimilarity extracted in the visual and the audio domain, respectively. In contrast to the video coherence measure obtained using a conventional short-term memory model, the measure proposed in this work compares only the shots that probably are taken by one camera. Currently, our approach fuses only two types of observable features, but it easily can be extended to include new data. The results of experimental tests carried out on ground truth video showed enhancement of the segmentation performance when multiple modalities are fused. Superior performance also was demonstrated with respect to a rule-based segmentation algorithm.

As our future work, we expect to extend the proposed approach to new features. Useful information, for example, could be provided by automatic person tracking since the same scene usually includes the same personages. New features may appear to be strongly conditionally dependent on each other. So there would be a need to propose more complicated fusion framework. Also, we are going to apply our approach to other types of video (e.g., sports broadcasting, news programs, or documentary video).

References

Aigrain, P., Joly, P., Longueville, V. (1997). Medium knowledge-based macrosegmentation of video into sequences. In *Intelligent multimedia information retrieval* (pp. 159-173). Cambridge, MA: AAAI Press and MIT Press.

Bimbot, F., Magrin-Chagnolleau, I., & Mathan, L. (1995). Second order statistical measures for text-independent speaker identification. *Speech Communication, 17*(1-2), 177–192.

Boresczky, S., & Rowe, L. A. (1996). A comparison of video shot boundary detection techniques. In *Proceedings of the SPIE Conference on Storage & Retrieval for Image and Video Databases IV* (pp. 170–179).

Borwell, D., & Thompson, K. (1997). *Film art: An introduction* (5th ed.). New York: McGraw-Hill.

Cao, Y., Tavanapong, W., Kim, K., & Oh, J. (2003). Audio assisted scene segmentation for story browsing. In *Proceedings of the International Conference on Image and Video Retrieval* (pp. 446–455).

Chen, S. C., Shyu, M. L., Liao, W., & Zhang, C. (2002). Scene change detection by audio and video clues. In *Proceedings of the IEEE ICME* (pp. 365–368).

Chen, S. S., & Gopalakrishnan, P. S. (1998). Speaker environment and channel change detection and clustering via the Bayesian Information Criterion. In *Proceedings of the DARPA Speech Recognition Workshop.*

Cover, T., & Thomas, J. (2003). *Elements of information theory.* John Wiley & Sons.

Duda, R. O., & Hart, P. E. (1973). *Pattern classification and scene analysis.* New York: Wiley.

Harb, H., & Chen, L. (2003). A query by example music retrieval algorithm. In *Proceedings of the 4th European Workshop on Image Analysis for Multimedia Interactive Services (WIAMIS03),* London (pp. 122–128).

Jiang, H., Zhang, H., & Lin, T. (2000). Video segmentation with the support of audio segmentation and classification. In *Proceedings of the IEEE International Conference on Multimedia and Expo (ICME2000).*

Kender, J. R., & Yeo, B. L. (1998). Video scene segmentation via continuous video coherence. In *Proceedings of the IEEE CVPR* (pp. 367–373).

Lienhart, R. (1999). Comparison of automatic shot boundary detection algorithms. In *Proceedings of the SPIE Conference on Storage and Retrieval for Image and Video Databases VII* (pp. 290–301).

Mahdi, W., Ardebilian, M., & Chen, L. (1998). Improving the spatio-temporel clues by the use of rhythm. In *Proceedings of the 2nd European Conference on Research and Advanced Technology for Digital Libraries (ECDL'98),* Heraklion, Crete, Greece (pp. 169–181).

Mahdi, W., Ardebilian, M., & Chen, L. (2000). Automatic scene segmentation based on exterior and interior shots classification for video browsing. In *Proceedings of the IS&T/SPIE's 12th International Symposium on Electronic Imaging.*

Parshin, V., Paradzinets, A., & Chen, L. (2005) Multimodal data fusion for video scene segmentation. In *Proceedings of the 8ᵗʰ International Conference on Visual Information Systems (VIS2005),* Amsterdam, The Netherlands (pp. 279–289).

Rabiner, L. R. (1989). A tutorial on hidden markov models and selected applications in speech recognition. *Proceedings of the IEEE, 77*(2), 257–286.

Rasheed, Z., & Shah, M. (2003). A graph theoretic approach for scene detection in produced videos. In *Proceedings of the Multimedia Information Retrieval Workshop*, Toronto, Canada.

Rui, Y., Huang, T. S., & Mehrotra, S. (1999). Constructing table-of-content for videos. *ACM Multimedia Syst., 7*(5), 359-368.

Sundaram, H., & Chang, S. F. (2000). Determining computable scenes in films and their structures using audio-visual memory models. In *Proceedings of the ACM Multimedia* (pp. 95–104).

Tzanetakis, G., Essl, G., & Cook, P. (2001). Audio analysis using the discrete wavelet transform. In *Proceedings of the WSES International Conference on Acoustics and Music: Theory and Applications (AMTA 2001)*, Skiathos, Greece.

Vasconcelos, N., & Lippman, A. (1997). A Bayesian video modeling framework for shot segmentation and content characterization. In *Proceedings of the IEEE Workshop on Content-Based Access of Image and Video Libraries*.

Yeung, M. M., & Yeo, B. L. (1996). Time-constrained clustering for segmentation of video into story units. In *Proceedings of the International Conference on Pattern Recognition* (Vol. C, pp. 375–380).

Zhang, D., Qi, W., & Zhang, H.J. (2001). A new shot boundary detection algorithm. In *Proceedings of the IEEE Pacific Rim Conference on Multimedia* (pp. 63–70).

Section III

Image and Video Annotation

Chapter V

A Novel Framework for Image Categorization and Automatic Annotation

Feng Xu, Tsinghua University, Beijing, China

Yu-Jin Zhang, Tsinghua University, Beijing, China

Abstract

Image classification and automatic annotation could be treated as effective solutions to enable keyword-based semantic image retrieval. Traditionally, they are investigated in different models separately. In this chapter, we propose a novel framework that unites image classification and automatic annotation by learning semantic concepts of image categories. In order to choose representative features, a feature selection strategy is proposed, and visual keywords are constructed, including discrete method and continuous method. Based on the selected features, the integrated patch (IP) model is proposed to describe the image category. As a generative model, the IP model describes the appearance of the combination of the visual keywords, considering the diversity of the object. The parameters are estimated by EM algorithm. The experimental results on Corel image dataset and Getty Image Archive demonstrate that the proposed feature selection and image description model are effective in image categorization and automatic image annotation, respectively.

Introduction

Although content-based image retrieval (CBIR) has been studied well over decades, it is still a challenging problem to search images from a large-scale image database because of the well-acknowledged semantic gap between low-level features and high-level semantic concepts. An alternative solution is to use keyword-based approaches, which usually associate images with keywords either by manually labeling or automatically extracting surrounding text. Although such a solution is adopted widely by most existing commercial image search engines, it is not perfect. First, manual annotation, though precise, is expensive and difficult to extend to large-scale databases. Second, automatically extracted surrounding text might be incomplete and ambiguous in describing images, and even more, surrounding text may not be available in some applications.

To overcome these problems, automated image classification and annotation are considered two promising approaches to understanding and describing the content of images. Besides obtaining text annotation, a successful image categorization significantly will enhance the performance of the content-based image retrieval system by filtering out images from ir-relevant classes during matching.

In this chapter, we propose a novel framework for image classification and automatic an-notation. First, feature selection strategy is explicitly introduced, including discrete feature method and continuous feature method. In both methods, the salient patches are detected and quantized for every image category. According to some rules, the informative salient patches are selected. The clusters of the selected salient patches are called the visual keyword dictionary. Second, for each selected patch, a 64-dimensional feature is extracted. Finally, considering the diversity of the same object in different images, the integrated patch (IP) model is proposed. The parameters of the model are estimated by EM algorithm. For a new test image, its posterior probability in each class is calculated. If the label with the largest probability is assigned to it, the classification result is achieved; if multi-words with the larg-est probability are assigned to it, the annotation result is achieved. The proposed framework, including the feature selection and the probabilistic model can be considered generative and have the potential to be implemented on larger-scale image databases.

On the surface, the proposed IP model appears to be similar to some existing models with the GM-mixture model and EM algorithm. However, the topological structure of the IP model is quite different from others. Compared with most of the annotation methods, the IP model regards the annotation problem as a visual categorization with a semantic concept. Compared with some visual categorization methods, the IP model adopts the feature selection to learn a semantic concept to avoid over-complexity. Therefore, through the IP model, the visual categorization and the automatic image annotation can be implemented effectively.

The main contributions of this chapter can be highlighted as follows:

- First, novel feature selection algorithms are proposed. Taking into account visual quantization, both the discrete and continuous feature methods are capable of select-ing the most informative features based on the detected salient patches.

- Second, a generative model, the IP model, is proposed to describe image concept. Compared with some discriminative models, such as SVM, the proposed model is capable of treating larger-scale image databases. Compared with some object recognition models, this model requires less computation.

- Third, in the proposed method, the image categorization and automatic annotation are considered in the same framework—classification according to the image concept. Since the image annotation can be regarded as a multi-classification problem, the categorization and annotation can be implemented in the same way.

The rest of the chapter is organized as follows. First, some background information and related works are introduced. Second, the framework is presented. Third, the feature selection strategy and the IP model are proposed. Then the experimental results are shown to evaluate the performance of our technique. Finally the concluding remarks and future works are presented.

Related Works

Many good results have been reported in two-class image classification tasks such as city vs. landscape (Vailaya, Jain, & Zhang, 1998), indoor vs. outdoor (Szummer & Picard, 1998). Recently, many approaches for general object recognition have been proposed and demonstrated to be promising to solve multiple-class image classification tasks. Fergus, Perona, and Zisserman (2003) proposed a constellation model, which is learned in a Bayesian manner, to recognize several classes of objects. The model can be learned from unlabeled and unsegmented cluttered scenes in a scale-invariant manner and is capable of recognizing six object classes. This classification scheme is improved further by Li, Fergus, and Perona (2003) to classify more categories with less training samples. A good application of this scheme is filtering Google images (Fergus, Perona, & Zisserman, 2004). Taking into account shape, appearance, occlusion, and relative scale, the constellation model describes well an object in multiple semantic aspects with low-level features and demonstrates promising potentials in image understanding. However, its computational cost is too expensive in both learning and recognition, and it is difficult to extend the algorithm to large-scale image databases. Csurka, Dance, Fan, Willamowski, and Bray (2004) proposed bags of key points of objects as features. Based on that, the visual vocabulary is constructed by k-means clustering algorithm. Both Naïve Bayes and SVM classifiers are applied to categorization. After training, labels are propagated to more categories. But the discriminative binary classifier is inefficient and difficult to extend to large-scale image databases. Another method for image classification based on discovered knowledge from annotated images using WordNet has been proposed by Benitez and Chang (2003). The novelty of this work is the automated class discovery and the classifier combination using the extracted knowledge.

As another way to support keyword-based image retrieval, automatic image annotation has been an active research topic in recent years. A typical annotation method is implemented by classification (Fan, Gao, & Luo, 2004), in which multi-level annotation is used. Images are segmented, and salient objects are detected by region classification. In this method, semantic concepts correspond perfectly to salient objects. However, the salient object detection depends on a syntax classification tree and is difficult to be scaled up. The generative model is successful in image annotation. Mori, Takahashi, and Oka (1999) proposed a Co-Occurrence Model in which they looked at the co-occurrence of words with image regions created using a regular grid. Duygulu, Barnard, Freitas, and de Forsyth (2002) proposed describing images using a vocabulary of blobs. First, regions are created using a segmentation algorithm like normalized cuts. For each region, features are computed and then blobs are generated by clustering the image features for these regions across images. Each image is generated by using a certain number of these blobs. Their Translation Model applies one of the classical statistical machine translation models to translate from the set of keywords of an image to the set of blobs forming the image. Jeon, Lavrenko, and Manmatha (2003) instead assumed that this could be viewed as analogous to the cross-lingual retrieval problem and used a cross-media relevance model (CMRM) to perform both image annotation and ranked retrieval. Lavrenko, Manmatha, and Jeon (2003) proposed continuous-space relevance model (CRM), which assumes that every image is divided into regions, and each region is described by a continuous-valued feature vector. Given a training set of images with annotations, a joint probabilistic model of image features and words is computed. Then the probability of generating a word given the image regions can be predicted. Compared with the CMRM, the CRM directly models continuous features, does not rely on clustering, and consequently does not suffer from the granularity issues. Feng, Manmatha, and Lavrenko (2004) also proposed a probabilistic generative model that uses a Bernoulli process to generate words and kernel density estimate to generate image features. The results show that it outperforms CRM and other models. Blei and Jordan (2003) extended the latent dirichlet allocation (LDA) model and proposed a correspondence LDA model that relates words and images. This model assumes that a dirichlet distribution can be used to generate a mixture of latent factors. This mixture of latent factors then is used to generate words and regions. EM is used to estimate this model.

For the automatic image categorization and annotation, it is useful to have access to high-level information about objects contained in the images to manage image collections. The high-level information must be learned from low-level features. As low-level features are usually noisy and uninformative, feature selection is of great importance and needs to be conducted before the modeling object. Vasconcelos and Vasconcelos (2004) exploited recent connections between theoretic feature selection and minimum Bayesian error solutions to derive feature selection algorithms that are optimal in a discriminant feature sense without compromising scalability. However, they do not provide feature selection from image content, which makes the semantic concept not clear. Although there are many feature selection approaches, an effective feature selection method based on the image content is necessary.

Recent progress in object recognition and image annotation has shown that local salient features are more informative than global features in describing image content (Csurka et al., 2004; Fan et al., 2004; Fergus et al., 2003). In the image classification, features are required to be common for the same class and discriminative for different classes. In the object semantic concept learning, the model should emphasize the object in an image category. Therefore, it is essential to select the common features and meanwhile reduce noisy features contributed by various backgrounds. However, to the best of our knowledge, there are no related works explicitly eliminating noises. Thus, a robust feature selection strategy based on the image category is crucial and worthy of investigation. On the other hand, it is difficult to extend discriminative image classification algorithms to large-scale databases. Take the SVM classifier, for example. If there is some change in the training dataset, the classifier should be trained again for all the categories. In such situations, generative methods with low complexity are more effective.

Feature Selection Strategy

In most existing classification methods, all the local salient features are used to train the classifier, such as works by Csurka et al. (2004) and Fergus et al. (2003). However, some salient features may be noises and are contributed by only a few images in an image category. To model the image category, the common features are the most important, because the common features are considered to be capable of expressing the object away from the different backgrounds. Thus, it is essential to conduct feature selection before modeling image categories. In this section, we describe the proposed feature selection strategy in detail. Features used in the classification model are generated from two stages: salient patch selection and feature extraction.

For image categorization and annotation, two types of feature selection strategies are proposed. First, salient feature selection strategy based on visual keywords is proposed. Since the visual keywords are quantized by k-means clustering, this strategy is called discrete feature method. Second, feature selection based on density is proposed. Since the salient patches are clustered according to continuous density, this strategy is called continuous feature method.

Salient Patch Selection Based on Discrete Feature

Salient patch selection consists of three steps: salient patch detection, visual keywords construction, and noise exclusion (Xu & Zhang, 2005).

Salient Patch Detection

In this step, salient patches are detected by the local salient feature detector proposed by Kadir and Brady (2001). This detector finds regions that are salient over both location and scale. For each point in an input image, a number of intensity histograms are calculated in circular regions of different radiuses (scales). The entropy of each histogram then is calculated, and the local maximum is selected as a candidate region. The regions with the highest saliency over the image provide the features. In practice, this method gives stable identification of features over a variety of sizes and copes well with intraclass variability. Only intensity information is used to detect and represent features.

Once the regions are identified, they are cropped from the image and rescaled to the size of a small pixel patch. Because a high dimensional Gaussian is difficult to manage, principal component analysis (PCA) is performed on the patches from all images. Then, each patch is represented by a vector of the coordinates within the first 15 principal components.

Figure 1. Salient patch histogram

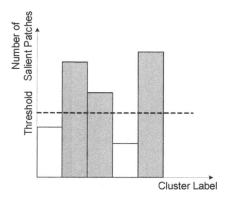

Visual Keyword Construction

The 15-dimensional feature vectors are used to construct the visual keyword dictionary. The vector quantization is performed on the vectors of all the images within one category, conducted by k-means clustering. Clusters of the vectors are the visual keywords for the category. Cluster histogram of salient patches shows its distribution in which each bin corresponds to a visual keyword. Those visual keywords with a large number of salient patches or over a predetermined threshold, which are regarded as the most important features for the image category, are selected. The visual keywords with a small number of salient patches are considered noises from different backgrounds. An example of cluster histogram is illustrated in Figure 1, in which each salient patch cluster in the image is denoted by a bin in the histogram. According to the number of salient patches, the shadowed bins are selected. For the image category, all the salient patch clusters from each image are accumulated to form the histogram, and the salient patch clusters with larger numbers are selected as the informative features.

Noise Exclusion

Two types of noises can be excluded. First, the most similar noise can be excluded by the region of dominance (ROD) (Liu & Collins, 2000). In a histogram of salient patches, for each bin, ROD is defined in the feature space as the maximal distance between the current bin and the detected local maxima. If the maximal distance is smaller than a preset threshold, the current local maximum is regarded as similar to one of the detected local maxima and will not be preserved as a visual keyword.

Second, the most non-common noises can be excluded by the salient entropy. The salient entropy is defined as follows:

$$H(n) = -\sum_{m=1}^{M} p_m(n) \log(p_m(n)) \tag{1}$$

Figure 2. Feature filter

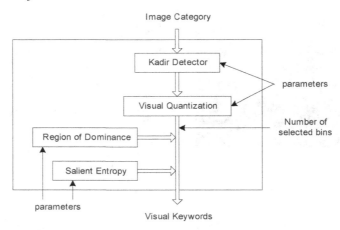

where *n* denotes the index of the cluster, *m* denotes the index of the image, and *M* denotes the total number of images in one category. $p_m(n)$ is the ratio between the number of salient patches in the *n*-th cluster of the *m*-th image and the total number of salient patches in this cluster of all the images.

Salient entropy reflects the distribution of a certain salient patch cluster in each image within the same category. If this distribution is uniform, the selected feature is more common. So, the visual keywords with larger entropies are preserved. According to the entropy measure, those visual keywords contributed by a few images are excluded, despite the large number of salient patches in the histogram.

This three-step feature selection strategy can be modeled as a feature filter shown in Figure 2. Through this feature filter, the most important and common features are reserved while the most possible noisy patches on the background are removed. Then, for each reserved patch, more detailed low-level features can be extracted to further model the object.

Some ship images from Corel image dataset are shown in Figure 3 to demonstrate the feature selection results. The first row shows the salient patches without noise reduction, while the second row shows the salient patches after feature filtering. For illustration, five visual key-

Figure 3. Ship image examples with salient patches in corel dataset

words are constructed, and the different color represents the different visual keyword. After feature filtering, three visual keywords for the ship category are preserved. For each color, the circles in the second row denote the preserved visual keywords. From these examples, it can be found that the preserved salient patches correspond to the visual keywords that are the most informative and relevant to the semantic concept of the image category, while the noises on the background or irrelevant to the category concept are removed. For example, the patches marked in yellow denote the hull, especially where the windows are located. The hull is a necessary part of a ship. But the circles on the sky are removed.

Salient Patch Selection Based on Continuous Feature

The continuous feature selection includes salient patch detection, density estimation, and patch ranking according to pointwise mutual information (Xu & Zhang, 2006). First, image data are transformed into scale-invariant coordinates relative to local features by SIFT (Scale Invariant Feature Transform by Lowe, 1999). SIFT is built by selecting key locations at maxima and minima of a difference of Gaussian function applied in scale space. Maxima and minima of this scale-space function are determined by comparing each pixel in the pyramid to its neighbors. Then the detected salient patches also are performed by PCA. Each patch is represented by a vector of the coordinates within the first 10 principal components. The orientation of the gradient is regarded as the weight. The visual keyword dictionary is constructed by DENCLUE (DENsity-based CLUstEring) algorithm (Hinneburg & Keim, 1998). It is assumed that $f(x)$ is the influence function of a data object. The density function is defined as the sum of the influence functions of all data points. Given N data objects described by a set of feature vectors $\mathbf{D} = \{\mathbf{x}_1, ..., \mathbf{x}_N\}$, the density function is defined as:

$$f(\mathbf{x}) = \sum_{i=1}^{N} f^{\mathbf{x}_i}(\mathbf{x}) \tag{2}$$

with Gaussian influence function:

$$f^{\mathbf{x}_i}(\mathbf{x}) = e^{-\frac{d(\mathbf{x},\mathbf{x}_i)^2}{2\sigma^2}} \tag{3}$$

The corresponding density function is:

$$f(\mathbf{x}) = \sum_{i=1}^{N} e^{-\frac{d(\mathbf{x},\mathbf{x}_i)^2}{2\sigma^2}} \tag{4}$$

After estimating the density, a Parzen estimation algorithm (Duda, Hart, & Stork, 2001) is applied to obtain the clusters. Assume that the current partition with density function $f(\mathbf{x}_i \mid C_j)$ for each data $\mathbf{x}_i$ in each cluster $C_j, j = 1, ..., J$. The objective function is:

$$f(\mathbf{x}_i | C_j) = \max_j f(\mathbf{x}_i | C_j) \tag{5}$$

So, three steps are conducted in this clustering algorithm:

- **Step 1:** Initializing clusters for the data set.
- **Step 2:** For each data $\mathbf{x}_i$, conditional densities in each cluster are calculated. Then label $\mathbf{x}_i$ according to equation (5).
- **Step 3:** If some labels of $\mathbf{x}_i$ are changed, go to step 2.

The distribution function can be regarded as a probability density function after normalization. Thus, the pointwise mutual information between the salient patch $\mathbf{x}_i$ and the cluster C_j is:

$$I(\mathbf{x}_i, C_j) = \log \frac{f(\mathbf{x}_i | C_j)}{f(\mathbf{x}_i)} \tag{6}$$

where $f(\mathbf{x}_i)$ is the estimated probability of the salient patch in the image category. Then the top M salient patches with higher pointwise mutual information are selected to describe the content and concept of the images.

The salient patches selected by density and pointwise mutual information represent not only the relation to the image category but also the feature distribution. These selected salient patches can represent the image category well. Figure 4 shows two pairs of image examples in the beach category, in which the left image is shown with all the detected patches, and the right image is shown with the selected patches.

Feature Extraction

For each selected salient patch, a 64-dimensional feature vector is extracted, in which 44 elements are banded autocolor correlogram, 14 elements are color texture moment, and six elements are color moment.

Color correlogram is proposed by Huang, Kumar, Mitra, Zhu, and Zabih. (1997). A color correlogram expresses how the spatial correlation of pairs of colors changes with distance.

Figure 4. Beach image examples with salient patches in Corel dataset

The probability that a pixel at a certain distance away from the given pixel is of a certain color is calculated as the color correlogram. To reduce feature dimension, we calculate banded autocolor correlogram at distances 1, 3, 5, and 7.

Color texture moment is proposed by Yu, Zhang, and Feng (2002). The local Fourier transform is adopted as a texture representation scheme to derive eight characteristic maps for describing different aspects of co-occurrence relations of image pixels in each channel of the color space. Then, the first and second moments of these maps are calculated as a representation of the natural color image pixel distribution.

To compensate for the lack of the global color feature, the color moment is computed. As a widely used feature in CBIR (Deng, Manjunath, Kenney, Moore, & Shin, 2001; Vailaya, Zhang, Yang, Liu, & Jain, 2002), the color moment computes an index containing only the dominant color features instead of storing the complete color distribution. The way to implement this approach is to store the first two moments of each color channel in an image as the index. For example, for an HSV image, we need to store only six floating-point numbers.

Analysis of the Feature Selection

The correctness of the proposed feature selection strategy can be verified by Fisher Discriminat Analysis (FDA) (Duda et al., 2001). Assume that the salient feature set in class $c\{c = 1, ..., M\}$ is $X^c = \{x_j^c \mid j = 1, ... N\}$. We define the feature selection criteria function for image category c as the ratio between scatters of intraclass and interclass:

$$R^c = \frac{\sum_{i=1}^{N_c} \sum_{j=1}^{N_c} d(x_i^c, x_j^c)}{\sum_{m=1}^{M} \sum_{n=1}^{M} \sum_{k=1}^{N_m} \sum_{l=1}^{N_n} d(x_k^m, x_l^n)} \qquad (7)$$

where $d(x_k^m, x_l^n)$ is a measure between two salient features in two different categories m and n, and $d(x_i^c, x_j^c)$ is the same measure between two salient features in the same category. The measure can be any distance, discriminative information or other measures that can present the similarity between two salient features. The ratio between the intraclass measure and the interclass measure is the objective to optimize; that is:

$$\arg \min_{\Gamma}(R^c) \qquad (8)$$

where Γ defined as the indicator function of X^c, i.e.,

$$\Gamma = \{\delta_j^c \mid j = 1, ... N, \delta_j^c \in \{0,1\}\} \qquad (9)$$

$$\delta_j^c = \begin{cases} 1 & \text{if } x_j^c \text{ is selected} \\ 0 & \text{if } x_j^c \text{ is removed} \end{cases} \qquad (10)$$

Minimizing R^c will result in the optimal feature set $\{x_j^c\}$.

In order to obtain the most common features within one class, the salient features are clustered in the same category. However, this will lead to some extremely similar visual keywords in different classes, that is $\{x_k^{c_1} \simeq x_l^{c_2}\}$ so that $\{d(x_k^{c_1}, x_l^{c_2}) \simeq 0\}$. These visual keywords cannot distinguish different categories and are regarded as uninformative feature points. If those uninformative features can be removed, the sum of interclass distances (the denominator in equation (7)) will be almost invariable. However, the sum of intraclass distances (the numerator in equation (7)) will decrease significantly, so that R^c will decrease. Until the minimal R^c is achieved, the reversed feature points are regarded as the important features for the classification. Also, since the selected points are based on the analysis of the image content, they are regarded as the informative features for the category concept. Therefore, it can be concluded that the proposed feature selection not only achieves the classification according to the FDA, but also emphasizes the semantic concept of the category by the selected features.

Image Category Description Model

In order to classify the images at the object level, we model each image category based on the selected salient patches. The model could learn the object concept and then classify the new test images.

In the previous feature selection, the detector proposed by Kadir and Brady (2001) is used, which depends on the local intensity information. Then the 64-dimensional feature is extracted so that more color and texture information is included. Considering the diversity of the newly included features, we use the finite mixture model. Each category is modeled as a combination of all the salient patch clusters, and the appearance of the visual keywords is defined as Gaussian density distribution.

Suppose we are given M images in *ONE* category. Let K denote the optimal number of mixture components. In an image, there are N salient patch clusters that correspond to the selected visual keywords for the image category. Let $\mathbf{X}_{mn}$ denote the feature vector for cluster n in image m.

For an image category I, the model can be defined as follows:

$$p(I|\Theta) = \prod_{m=1}^{M} p(I_m|\Theta) = \prod_{m=1}^{M} \left[\sum_{k=1}^{K} \prod_{n=1}^{N} p(\mathbf{X}_{mn}, c_k | \Theta) \right] \tag{11}$$

where $p(I_m|\Theta)$ is the probability of the m-th image. $p(\mathbf{X}_{mn}, c_k|\Theta)$ is the joint probability of the n-th patch in the m-th image and the k-th mixture component. For each component, there are N independent means and N covariance matrices corresponding to N clusters. Θ is the set of the parameters for these mixture components, c_k is the weight of the k-th component in the mixture model. Therefore, $\Theta = \{\boldsymbol{\mu}_{kn}, \boldsymbol{\Sigma}_{kn}, k = 1, \dots, K, n = 1, \dots, N\}$, $c_k = \{c_k, k = 1, \dots, K\}$. $\mathbf{X}_{mn}$ is the 64-dimensional feature vector.

It should be mentioned that the visual properties of a certain type of objects may look various at different lighting and capturing conditions. For example, the ship consists of various appearances, especially various colors, such as "red ship pattern," "white ship pattern," and "yellow ship pattern," which have very different properties. Thus, the data distribution for a certain type of objects is approximated by using multiple mixture components in order to accommodate the variability of the same type of objects (i.e., presence/absence of distinctive parts, variability on overall shape, changing of visual properties due to the object patterns and viewpoints, etc.).

The key of the proposed model is the multiplication of N cluster probabilities, each of which corresponds to a visual keyword selected by the algorithm in the previous section. It is assumed that the different visual keyword is independently identical distribution (i.i.d.), and can be defined as a Gaussian density distribution. The constellation model proposed by Fergus et al. (2003) is an object description model for general object recognition. A probabilistic representation is used for all aspects of the object: shape, appearance, occlusion, and relative scale. This model is quite effective in recognition based on Bayesian manner. However, the performance depends on the hypothesis of the parts, which is searched over all possible hypotheses and which makes the model too inefficient to be applied to classification in large-scale image databases. In our model, only appearance is used for each visual keyword, and the order of each visual keyword has been determined in the feature selection stage. Therefore, it is more efficient and can be extended to large-scale image databases. The optimal model structure and parameters $(\hat{c}_k, \hat{\Theta})$ for an image category are determined by:

$$(\hat{c}_k, \hat{\Theta}) = \arg\max_{c_k,\Theta}\{L(\Theta|I)\} = \arg\max_{c_k,\Theta}\{\log p(I|\Theta)\} \tag{12}$$

The likelihood function is:

$$L(\Theta|I) = \log p(I|\Theta) = \sum_{m=1}^{M}\log\left[\sum_{k=1}^{K}p(I_m,c_k|\Theta)\right] \geq \sum_{m=1}^{M}\left[\sum_{k=1}^{K}q_m(c_k)\log\frac{p(I_m,c_k|\Theta)}{q_m(c_k)}\right] \triangleq B(\Theta;c_k)$$

$$\tag{13}$$

where $B(\Theta;c_k)$ is the lower bound, and the inequality is deduced by Jesen Inequality. $q_m(c_k)$ is an arbitrary distribution of c_k, which is satisfied the constraint:

$$\sum_{k=1}^{K}p(c_k) = 1 \tag{14}$$

The maximum likelihood estimation (MLE) can be achieved by using the EM algorithm (Dempster, Laird, & Rubin, 1977). In E-step, the posterior distribution of c_k is computed:

$$q_m(c_k) = p\left(c_k|I_m,\Theta\right) = \frac{p(I_m|c_k,\Theta)p(c_k)}{\sum_{k=1}^{K}p(I_m|c_k,\Theta)p(c_k)} = \frac{\left[\prod_{n=1}^{N}p(\mathbf{X}_{mn}|c_k,\Theta)\right]p(c_k)}{\sum_{k=1}^{K}\left[\prod_{n=1}^{N}p(\mathbf{X}_{mn}|c_k,\Theta)\right]p(c_k)} \tag{15}$$

In M-step, let the partial differential of $B(\Theta; c_k)$ for μ_{kn}, Σ_{kn} and c_k equals to zero, respectively, and then we get the iterative solution for each parameter:

$$c_k^{new} = \frac{1}{M} \sum_{m=1}^{M} p(c_k | I_m, \Theta) \qquad (16)$$

$$\mu_{kn}^{new} = \frac{\sum_{m=1}^{M} \mathbf{X}_{mn} p(c_k | I_m, \Theta)}{\sum_{m=1}^{M} p(c_k | I_m, \Theta)} \qquad (17)$$

$$\Sigma_{kn}^{new} = \frac{\sum_{m=1}^{M} p(c_k | I_m, \Theta)(\mathbf{X}_{mn} - \mathbf{X}_{kn}^{new})(\mu_{mn} - \mu_{kn}^{new})^T}{\sum_{m=1}^{M} p(c_k | I_m, \Theta)} \qquad (18)$$

Because the number of salient patch is various in different images, the representative patch must be selected for each visual keyword in order to keep the probability multiplication of visual keywords in the same number. The selection of the representative patch can be implemented in two ways. First, the patch nearest to the center of the visual keyword is selected. Second, the average patch in the same visual keyword is selected.

This model can be explained at three levels. At the first level, the probabilities of images are multiplied based on the same category concept, because the images are assumed as i.i.d. Then, at the second level, a finite mixture model is applied for each image category. At the third level, the probability multiplication of the visual keywords is computed for each image.

As the parameter estimation results, there are K components in one category, which are modeled as Gaussian functions. And for each component, there are N visual keywords, which are modeled as independent Gaussian functions with mean μ_{kn} and covariance Σ_{kn}. So the total number of parameters is $2KN+K$.

For a test image, salient patches are detected by the detector and labeled as the nearest visual keyword. Then the 64-dimensional feature is extracted. Thus, for each image, the posterior probability is computed as follows:

$$p(c_k | I_m, \Theta) \propto p(I_m | c_k, \Theta) p(c_k) = \left[\prod_{n=1}^{N} p(\mathbf{X}_{mn} | c_k, \mu_{kn}, \Sigma_{kn}) \right] p(c_k) \qquad (19)$$

The largest posterior probability is taken as the prediction, and the image is labeled as the corresponding category.

The posterior probability of the test image is also the multiplication of the salient patch probabilities. In fact, the similarity measure in this model is the combination of the similar salient patches. If the salient patches in each visual keyword between two images are correspondingly similar, then these two images are considered similar. Therefore, we call this model the Integrated Patch model.

Experimental Results

Experiment Setting

The experiments are shown as image categorization and annotation, respectively. First, image categorization is conducted on Corel image dataset, in which all the image categories are labeled with the object semantic concepts. Overall, we select 3,500 images from 35 categories. In each image category, 80 images are chosen randomly to train the model, and the other 20 images are used as test images. Second, image annotation is conducted on an extensive dataset. This extensive dataset is built by downloading medium-resolution thumbnails from the Getty Image Archive Web site (http://creative.gettyimages.com), in which 400 images are used for training and 60 images are annotated automatically. After the model has been trained, k-NN classifier is used to annotate the test images.

Our algorithm evaluation focuses on: (1) feature selection for each image category to evaluate the object semantic concept and (2) training model for each image category and test the classification and annotation performance.

The benchmark metric of the experiments is precision α and recall β, defined as:

$$\alpha = \frac{\phi}{\phi + \varepsilon}, \qquad \beta = \frac{\phi}{\phi + \eta} \qquad (20)$$

where ϕ is the number of true positive samples that are classified correctly to their corresponding semantic category, ε is the number of true negative samples that are irrelevant to the corresponding semantic category and are classified incorrectly, and η is the number of false positive samples that are related to the corresponding semantic concept but are misclassified.

In our algorithm, several parameters need to be tuned in order to achieve better performance.

First, the total number of clusters is important in visual keyword construction. Too few visual keywords may lead to low discriminative results between category models, while too many visual keywords may lead to low distribution entropy of salient patches. According to equation (7), we give the curve of the ratio in Figure 5, in which the horizontal axis denotes the number of clusters in an image category, and the vertical axis denotes the ratio.

Although only a few points are calculated and illustrated, it is clear to see the trend that the number of clusters has a value to minimize the ratio. In the experiment, we set this number as 50, around the minimum of the ratio curve. Thus, there are 50 visual keywords for each image category.

Second, the number of selected visual keywords is significant for the description of image category. In the experiments, we set the selected number of both visual keywords and salient entropy. The intersection clusters between them are used as the final visual keywords. Thus, the number of selected visual keywords is determined adaptively in various image categories.

Figure 5. Curve of ratio between intraclass and interclass scatters

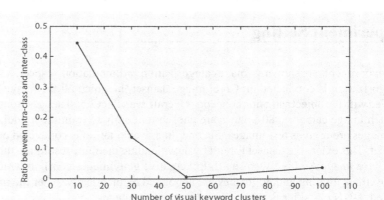

Third, the number of mixture components in the model is another important parameter. Since the color and texture information is used to describe the various patterns of the object, the number of components should be determined according to the number of color and texture patterns in the image category. In our experiments, the component number is variable in the range from 2 to 6.

In the test, the salient patches are selected by the distance from the centers of visual keywords. For each test image, the 100 salient patches nearest to some visual keywords are preserved to represent the image.

Performance Evaluation

Classification Performance and Comparison

The classification precisions for 35 image categories in Corel image dataset are shown in Figure 6.

Since SVM (support vector machine) (Vapnik, 1998) is a well-known classifier to produce state-of-the-art results in high-dimensional problems, we compare the performance of the proposed classification system with the SVM classifier. For the feature selection based on the discrete feature method, the results between average precisions without feature selection and with feature selection as well as average precisions of the proposed model and SVM are both compared. For feature selection based on continuous feature method, the results between average precisions without feature selection and with feature selection are compared.

In Table 1, the average classification precisions by different combinations of feature selection methods and classifiers are shown.

From this table, it can be concluded that the proposed feature selection strategies apparently improve the effectiveness of the classification. For the same SVM classifier, both the discrete feature selection and the continuous feature selection improve the precisions about 10%. With the feature selection, the most informative features are selected, and the combination of salient patches is much stronger than pixels in describing objects, so different categories can be classified more accurately. Each visual keyword corresponds to a certain semantic concept. Only if all the corresponding keywords are similar enough can the images be considered in the same category. Thus, the semantic concept is enhanced by the combination of the salient patches. For the proposed model, although the precision is slightly lower than that by SVM classifier, the proposed classification method has advantages in scalability because it is independent in modeling each image category. Nevertheless, SVM is a discriminative binary classifier, and its efficiency will decrease with the growth of the number of image categories.

Annotation Performance and Comparison

For image annotation, the selected N visual keywords are combined without the Gaussian mixture model, and k-NN classifier is used. Here, k equals 5. In the five nearest neighbors, the top 10 most probable words are assigned to each test image. The average annotation precision and recall are shown in Table 2.

Although the results are produced from different image datasets, we still compare the average precision and recall of the proposed model with that of the MBRM (Feng et al., 2004) and the CRM (Lavrenko et al., 2003) (the results of MBRM and CRM in Table 2 are from Yavlinsky, Schofield, & Ruger, 2005). MBRM and CRM both annotate images in Corel dataset, while our experiment is conducted on Getty Image Archive. Since the images in Getty Image Archive are more diverse than images in Corel dataset, the comparison can prove the effectiveness of our algorithm.

Figure 6. Classification precisions

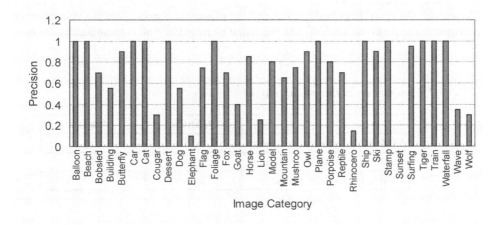

Table 1. Comparison of the average classification precision

	Without Feature Selection	With Feature Selection
Discrete feature and GMM	N/A	0.73
Discrete feature and SVM	0.64	0.75
Continuous feature and SVM	0.57	0.69

Some annotated image examples are illustrated in Figure 7, in which each image is annotated by four text words. The annotated words describe the image content and can be used as the keywords for image retrieval.

Visual Keyword Dictionary

The object semantic concept can be decomposed into the concepts of its components. For example, *ship* consists of *backstay, fore, stern, bottom, hull*, and *baluster*. Every component is characterized with some low-level features, especially some local features. Therefore, objects can be represented by the combination of some local features that are called visual keywords.

In the proposed method, the visual keyword dictionary is constructed with the feature selection. Every image is represented as a group of selected visual keywords. At the semantic concept level, images can be searched according to the similarity of the visual keywords. Whether the semantic concept of two images is similar or not is determined by their corresponding visual keywords. Thus, the gap between the object semantic concept and the low-level features can be narrowed.

In Figure 8, an example image in the *ship* category with six visual keywords marked by colored circles is illustrated. For each visual keyword, six patches extracted from the training images are shown. The visual keyword in red corresponds to the *backstay*; the visual keyword in blue corresponds to the *bottom in water*; the visual keyword in yellow corresponds to the *windows*; the visual keyword in green corresponds to the *fore* or *stern*; the visual keyword in purple corresponds to the *hull*; and the visual keyword in pink corresponds to the *baluster*.

Table 2. Comparison of the annotation performance

	Precision	Recall
IP Model and Getty Image Archive	0.35	0.34
MBRM and Corel Image Dataset	0.24	0.25
CRM and Corel Image Dataset	0.16	0.19

For each image, the object, *ship*, is represented as the combination of the salient patches in various visual keywords. For each image category, there are dozens of visual keywords to describe the object concept.

Discussion

From the experimental results, it can be concluded as follows:

- The proposed feature selection strategy is effective in image classification and automatic annotation. Through the feature selection, images are represented by the informative visual keywords. The combination of visual keywords is much stronger than simple low-level features in describing objects. Moreover, this feature selection makes the unsupervised object modeling possible.

- For the proposed IP model, the scalability is another advantage except for the better performance. The IP model is independent in learning for each image category so that it can be extended to even larger-scale image databases.

- The proposed IP model is more appropriate for object concept than for scene concept. For some image categories with more concrete concept, the classification precision is quite high, such as *balloon*, *car*, *ship*, *tiger*, *train*, and so forth. There are also some image categories not appropriate to be modeled by this classification method, such as *sunset*. It may be better to model the scene concept by global features. In practice, it is possible to detect such cases in the feature selection stage by examining the clustering result and salient entropies. We will investigate this issue further in the future.

Figure 7. Annotated image examples in Getty Image Archive

| day, tree, | cloud, lake, | beach, day, | autumn, forest, |
| outdoor, house | mountain, forest | summer, sand | leaf, tree |

Figure 8. Examples of visual keywords for ship category

Visual Keyword *backstay* (top white circle)

Visual Keyword *bottom in water* (bottom black circle)

Visual Keyword *windows* (top black circle)

Visual Keyword *fore or stern* (right gray circle)

Visual Keyword *hull* (bottom white circle)

Visual Keyword *baluster* (left gray circle)

Conclusion and Future Work

We have presented a novel framework to automatically classify and annotate images based on the semantic concept of the image category. The feature selection strategy is effective in generating visual keywords to describe image categories. Through the feature selection, objects are emphasized, while noises are reduced. In the description model, the IP model is proposed to represent objects. Both the categorization experiment on Corel image database and automatic image annotation experiment on Getty Image Archive demonstrate that the proposed model is effective and efficient for image categories with concrete semantic concepts. As the IP model is category-independent, it potentially can be extended to large-scale image databases.

However, there are also some limitations in our algorithm. First, the discriminative features among image categories are not well-leveraged. The discriminative information possibly can improve the classification performance. Second, EM algorithm can lead to local extremum, and therefore, it is worthy to investigate adaptive parameter estimation algorithms in the future.

Acknowledgment

This work has been supported by the Grant SRFDP-20050003013.

References

Benitez, A. B., & Chang, S.-F. (2003). Image classification using multimedia knowledge networks. In *Proceedings of the IEEE International Conference on Image Processing, Barcelona*, Spain (pp. 613–616).

Blei, D. M., & Jordan, M. I. (2003). Modeling annotated data. In *Proceedings of the 26th International ACM SIGIR Conference*, Toronto, Canada (pp. 127–134).

Csurka, G., Dance, C. R., Fan, L., Willamowski, J., & Bray, C. (2004). Visual categorization with bags of keypoints. *In Proceedings of the 8th European Conference on Computer Vision,* Prague, Czech Republic (pp. 11–14).

Dempster, A., Laird, N., & Rubin, D. (1977). Maximum likelihood from incomplete data via the em algorithm. *JRSS B, 39*, 1–38.

Deng, Y., Manjunath, B. S., Kenney, C., Moore, M. S., & Shin H. (2001). An efficient color representation for image retrieval. *IEEE Transactions on Image Processing, 10*(1), 140–147.

Duda, R. O., Hart, P. E., & Stork, D. G. (2001). *Pattern classification* (2nd ed.). John Wiley & Sons.

Duygulu, P., Barnard, K., Freitas, N., & de Forsyth, R. (2002). Object recognition as machine translation: Learning a lexicon for a fixed image vocabulary. In *Proceedings of the Seventh European Conference on Computer Vision,* Copenhagen, Denmark (pp. 97–112).

Fan, J., Gao, Y., & Luo, H. (2004). Multi-level annotation of natural scene using dominant image components and semantic concepts. In *Proceedings of the 12th Annual ACM International Conference on Multimedia,* New York (pp. 540–547).

Feng, S., Manmatha, R., & Lavrenko, V. (2004). Multiple Bernoulli relevance models for image and video annotation. In *Proceedings of the IEEE Conference on Computer Vision and Pattern Recognition,* Washington, DC (pp. 1002–1009).

Fergus, R., Perona, P., & Zisserman, A. (2003). Object class recognition by unsupervised scale-invariant learning. In *Proceedings of the IEEE Computer Society Conference on Computer Vision and Pattern Recognition,* Madison, WI (pp. 264–271).

Fergus, R., Perona, P., & Zisserman, A. (2004). A visual category filter for Google images. In *Proceedings of the 8th European Conference on Computer Vision,* Prague, Czech Republic (pp. 242–256).

Hinneburg, A., & Keim, D. A. (1998). An efficient approach to clustering in large multimedia databases with noise. In *Proceedings of the International Conference on Knowledge Discovery and Data Mining,* New York (pp. 58–65).

Huang, J., Kumar, S. R., Mitra, M., Zhu, W.-J., & Zabih, R. (1997). Image indexing using color correlograms. In *Proceedings of the IEEE Computer Society Conference on Computer Vision and Pattern Recognition,* San Juan, Puerto Rico (pp. 762–768).

Jeon, J., Lavrenko, V., & Manmatha, R. (2003). Automatic image annotation and retrieval using cross-media relevance models. In *Proceedings of the 26th International ACM SIGIR Conference,* Toronto, Canada (pp. 119–126).

Kadir, T., & Brady, M. (2001). Scale, saliency and image description. *International Journal of Computer Vision, 45*(2), 83–105.

Lavrenko, V., Manmatha, R., & Jeon, J. (2003). A model for learning the semantics of pictures. In *Proceedings of Advances in Neutral Information Processing Systems,* Whistler, British Columbia, Canada.

Li, F.-F., Fergus, R., & Perona, P. (2003). A Bayesian approach to unsupervised one-shot learning of object categories. In *Proceedings of the IEEE International Conference on Computer Vision,* Nice, France (pp. 1134–1141).

Liu, Y., & Collins, R. T. (2000). A computational model for repeated pattern perception using frieze and wallpaper groups. In *Proceedings of the IEEE Computer Society Conference on Computer Vision and Pattern Recognition,* SC (pp. 537–544).

Lowe, D. G. (1999). Object recognition from local scale-invariant features. In *Proceedings of the IEEE International Conference on Computer Vision,* Kerkyra, Corfu, Greece (pp. 1150–1157).

Mori, Y., Takahashi, H., & Oka, R. (1999). Image-to-word transformation based on dividing and vector quantizing images with words. In *Proceedings of the 1st International Workshop on Multimedia Intelligent Storage and Retrieval Management,* Orlando, FL.

Szummer, M., & Picard, R. (1998). Indoor-outdoor image classification. In *Proceedings of the IEEE International Workshop on Content-Based Access of Image and Video Databases*, Bombay, India (pp. 42–51).

Vailaya, A., Jain, A., & Zhang, H. J. (1998). On image classification: City vs. landscape. *Pattern Recognition, 31*(12), 1921–1935.

Vailaya, A., Zhang, H.-J., Yang, C., Liu, F.-I., & Jain, A. K. (2002). Automatic image orientation detection. *IEEE Transactions on Image Processing, 11*(7), 746–754.

Vapnik, V. (1998). *Statistical learning theory*. Wiley.

Vasconcelos, N., & Vasconcelos M. (2004). Scalable discriminant feature selection for image retrieval and recognition. In *Proceedings of the IEEE Computer Society Conference on Computer Vision and Pattern Recognition,* Washington, DC (pp. 770–775).

Xu, F., & Zhang, Y.-J. (2005). Salient feature selection for visual concept learning. In *Proceedings of the Pacific-Rim Conference on Multimedia*, Jeju Island, Korea (pp. 617–628).

Xu, F., & Zhang, Y.-J. (2006). Feature selection for image categorization. In *Proceedings of Asian Conference on Computer Vision*, Hyderabad, India (pp. 653–662).

Yavlinsky, A., Schofield, E., & Ruger, S. (2005). Automated image annotation using global features and robust nonparametric density estimation. In *Proceedings of the 4th International Conference on Image and Video Retrieval*, Singapore (pp. 507–516).

Yu, H., Li, M., Zhang, H.-J., & Feng, J. (2002). Color texture moments for content-based image retrieval. In *Proceedings of the International Conference on Image Processing*, Rochester, NY (pp. 929–932).

Chapter VI

Automatic and Semi-Automatic Techniques for Image Annotation

Biren Shah, University of Louisiana at Lafayette, USA

Ryan Benton, University of Louisiana at Lafayette, USA

Zonghuan Wu, University of Louisiana at Lafayette, USA

Vijay Raghavan, University of Louisiana at Lafayette, USA

Abstract

When retrieving images, users may find it easier to express the desired semantic content with keywords than with visual features. Accurate keyword retrieval can occur only when images are described completely and accurately. This can be achieved either through laborious manual effort or automated approaches. Current methods for automatically extracting semantic information from images can be classified into (a) text-based methods that use metadata such as ontological descriptions and/or text associated with images to assign and/or refine annotations, and (b) image-based methods that focus on extracting semantic information directly from image content. The focus of this chapter is to create an awareness and understanding of research and advances in this field by introducing them to basic concepts and theories and then by classifying, summarizing, and describing works from the published literature. It also will identify unsolved problems and offer suggestions for future research directions.

Introduction

A picture is worth a thousand words. As human beings, we are able to tell a story from a picture based upon what we see and our background knowledge. Can a computer program discover semantic concepts from images, build models from them, and recognize them based on these models? The short answer is yes. However, there is no perfect solution, and there are several challenges that need to be addressed.

Automatic image annotation is important to both video and image retrieval and computer object recognition.[1] It potentially can be applied to many areas, including biomedicine, commerce, military, journalism, education, digital libraries, advertising, and Web searching. Decades of research has shown that it is extremely difficult to design a set of generic computer algorithms that can learn concepts from images to automatically translate their contents to linguistic terms.

The mission of this chapter is to create an awareness and understanding of research and advances in the field of automatic image annotation by introducing the basic concepts and theories and then by classifying, summarizing, and describing published works. The chapter also will identify unsolved problems and offer suggestions for future research directions.

The chapter will address the following questions:

- What is a semantic gap?
- Why is there a need for automatic techniques for image annotation?
- What are the various automatic annotation methods in the literature?
- How does one assess the quality of annotations?

The rest of the chapter is organized as follows. The next section provides background and related work. The text-based and image-based methods for automatic annotation are discussed in the third and fourth sections, respectively. In the fifth section, we provide several metrics for evaluating the quality of annotations. Finally, we conclude the chapter with a general discussion on future and emerging trends.

Background and Related Work

"Given a new image, find similar images in the current database that best describe its content. Given a large image database, find images that have tigers." These are examples of queries encountered by many current image indexing and retrieval systems. Effective techniques are needed to model and search the content of large digital image archives.

One solution is to use content-based image retrieval (CBIR) systems (Flickner et al., 1995; Shah, Raghavan, Dhatric, & Zhao, 2006) that compute relevance based on the visual similarity of low-level image features such as color, texture, shape, and so forth. However, visual

Figure 1. Sample Images described with keywords

keywords: sun, sky, sea, waves

keywords: horse, people, house

keywords: bird, shrub

keywords: festival, people, car

similarity is not equivalent to semantic similarity; this lack of equivalency forms the semantic gap. For example, a CBIR system, when presented an image of a beach scene, may return images from various other classes that also contain large areas of sky.

Image semantics can be represented more accurately by keywords than by low-level visual features. For example, Figure 1 shows sample images, each of which is described by a set of keywords reflecting the image's semantic concepts. An added benefit of using keyword approaches is that traditional keyword-based search methodologies can be applied to image retrieval. The implication of adopting a keyword approach is that the problem focus shifts from improving content-based retrieval to annotating images with meaningful concepts. It also introduces several complications. First, each image must be described adequately. For instance, assigning only one keyword (e.g., snow) to an image is often not adequate. The picture could be a scene of a mountain, a plain covered by snow, or a scene of a house. Second, manually describing each image is a time-consuming affair and is costly. Furthermore, for nonstatic image collections, describing the image collection is not a one-time affair. Thus, one or more personnel must be retained to annotate new images. Third, for manual descriptions, the keywords used are, to an extent, subjective and can be influenced by the person's mood, health, and other factors. Hence, similar images, which should have a large number of overlapping keywords, may, due to user input, have few keywords in common. Thus, there is a need to investigate and develop techniques accurately and completely for automatic image annotation with semantic concepts.

The current methods for extracting semantic information from images can be classified broadly into two main categories[2]:

- **Text-based methods:** In text-based methods, the text associated with the image objects is analyzed, and the system extracts those that appear to be relevant. Most text-based

Figure 2. Image annotation classification

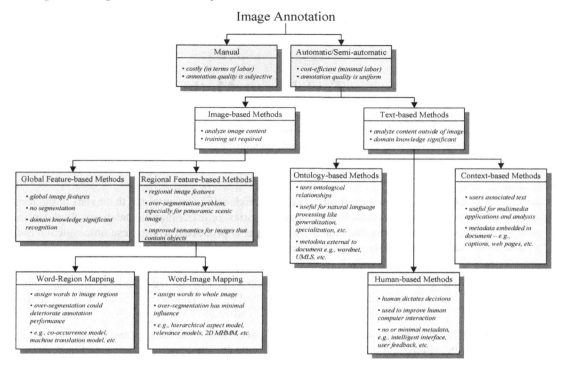

methods require some form of domain knowledge (e.g., the presence of high-quality textual information associated with the image objects). To relax this requirement, more sophisticated methods based on image content analysis have been proposed.

- **Image-based methods:** Many studies focus on extracting semantic information directly from image content. This converts the annotation problem into the problem of semantic classification of images. Image-based annotation methods can be classified into the following two types:

  - **Regional feature-based methods:** Most image annotation methods rely on regional features, which, in turn, rely on semantically meaningful segments or regions. Image segmentation and/or partitioning techniques are employed to detect objects in an image. Next, models are generated by calculating the probability of co-occurrence of words with regional image features. These models are used to annotate new images.

  - **Global feature-based methods:** Unlike regional feature-based methods, global feature-based methods make use of global image features and are very attractive when creating a low-cost framework for feature extraction and image classification. The tradeoff compared to regional feature-based methods is that global features typically do not capture detailed information about individual objects. Thus, some types of annotations may not be possible using global-based features.

Since the semantics of an image usually are well-represented by semantically meaningful regions, the bulk of the image-based annotation efforts have been via region-based methods.

Text-Based Methods

Text-based methods use additional knowledge about images that exist in associated text, ontologies, and so forth, for annotation. Generally speaking, text-based annotation can be divided into three main categories:

- **Ontology-based methods:** These methods utilize an ontology that exists in the form of lists of keywords, hierarchical categories, or lexicographic databases. Verbose and unrestricted text associated with images is processed using these ontological terms, rules, and/or relations to generate the annotation.

- **Context-based methods:** Images may exist in the context of a multimedia document. For example, an image on a Web page may be surrounded by text, or an image frame in a video clip may have text captions and/or audio narrative. Such contextual content from the multimedia document can be utilized to annotate images.

- **Human-based methods:** These methods assist humans to assign annotations by providing convenient user interfaces and by utilizing supportive methods that are designed to alleviate the amount of required manual operations.

In the following subsections, we will discuss these three types of text-based methods.

Ontology-Based Methods

An ontology offers a convenient means to characterize a given domain by conceptualizing and specifying knowledge in terms of entities, attributes, and relationships. It allows one to distinguish among various types of objects and to describe object relationships, dependencies, and properties. Ontology constructs the relational structure of concepts that one can use to describe and reason about aspects of the world. "They can play both a passive representational or taxonomic role, in terms of prior ontological commitments to certain objects and categories, and an active inferential role, which defines how the relevant properties of the world may be identified" (Town, 2004, p. 25).

The World Wide Web Consortium (W3C) project referred to as the Semantic Web (http://www.w3.org/2001/sw/BestPractices/MM/image_annotation.html) aims at allowing people to find, share, and combine information more easily by relying on machine-readable information and metadata. This initiative has resulted in the emergence of many ontologies, general-purpose and domain-specific standards, and tools such as the Web Ontology Language (http://www.w3.org/TR/owl-features/) and the resource description framework (http://www.w3.org/RDF/).

This has boosted ontology-based image annotation efforts; it is one of the most researched areas in text-based annotation.

Several studies, such as Hu, Dasmahapatra, Lewis, and Shadbolt (2003), Hollink, Schreiber, Wielemaker, and Wielinga (2003), and Soo, Lee, Li, Chen, and Chen (2003), demonstrate the usage of specialized ontologies in medicine, art, and history in order to perform domain-specific annotations. Alternately, general-purpose ontologies also are used widely for annotation. For example, Benitez and Chang (2002) proposed automatic methods for extracting semantic concepts by disambiguating, using a lexicographical database, WordNet (Miller, 1995), the senses of words found in a collection of clustered annotated images. The three-step method is as follows: (1) tokenizing and chunking the textual annotations and tagging the words with their POS (e.g., noun and verb); (2) semantic concepts are extracted by disambiguating the senses of the content words using WordNet and image clusters. The intuition is that images belonging to the same cluster likely share the same semantics. This step consists of two tasks: (a) the different senses of words annotating the images in a cluster are ranked based on their relevance, and (b) the ranks for the same word and the same image that are obtained from the various clusters to which the image belongs are combined into a single rank. The sense of each word is considered the highest-ranked sense. Finally, (3) utilize the relations and additional concepts from WordNet that relate to the detected word senses. All semantic relationships and the intermediate senses are added to the extracted semantic knowledge. Hence, the constructed knowledge will not be restricted to the detected senses but includes the intermediate senses among them.

Context-Based Methods

Many studies have been conducted on context-based methods, especially with respect to the World Wide Web. Good context-based examples of automatic Web-based image annotation and retrieval systems include search engines such as Google image search (http://images.google.com). The engines analyze the text on the Web page adjacent to an image, its caption, and other factors in order to determine the image content.

The weight chainNet model by Shen, Ooi, and Tan (2000) is one of the representative works for Web image annotation and retrieval. It is based on the observation that an image on a Web page is related semantically to its surrounding text. The weight chainNet model is based on using lexical chains (LCs) that represent the semantics of an image from its nearby text. At the annotation stage, an image is associated with several LCs, such as image title LC and Web page title LC; each LC is a sequence of semantically related words. Since certain HTML components such as the caption of an image may provide more semantic information than others, weights are applied to reflect the same. At the retrieval stage, for a given user query, a free sentence that describes the image content is represented as a Query LC. A formula called the list space model is used for computing semantic similarity between the Query LC and the LCs associated with an image. Finally, the image similarity is defined as the sum of all similarities between the Query LC and each of the LCs belonging to an image.

Human-Based Methods

Unlike the previous methods, human-based methods are utilized for images without associated text (i.e., there is minimal or no usable context). Human indexers make the ultimate decision to the annotations. Usually, intelligent user interfaces are provided in order to minimize the manual operation, to provide a few initial options for users to choose from, and/or to learn from users' histories, which helps to make future annotations more accurate and efficient. Lieberman, Rosenzweig, and Singh (2001) introduced one such interface system, Aria, through which users can create annotations for images when editing e-mail messages.

Researchers also have explored the use of multiple types of text-based methods in order to achieve annotation and retrieval. For example, in the Show&Tell (Srihari & Zhang, 2000) annotation system, human annotators select objects in an image via a mouse click and then record a speech that corresponds to each of the selected objects. After speech recognition and natural language processing are applied to synchronize speech with the mouse input, domain-specific ontologies are used to generate annotations for indexing and retrieval.

One major constraint for most text-based methods is the requirement that high-quality textual information be associated with the image objects. In many situations, this requirement is not satisfied. In response, annotation methods based upon image content have been proposed, which is the focus of the next section.

Image-Based Methods

Unlike text-based methods, image-based methods do not expect images to appear within a particular context; hence, they can be domain-independent. An image-based method is composed of two distinct parts. The first part is to derive one or more mappings (models) between content features (e.g., color, texture, etc.) and keywords and/or concepts, using a set of previously annotated images (also called training set). The second part is to use the mappings to efficiently and accurately annotate new images and/or regions.

The annotated data for the images in the training set are generated in one of two ways:

1.   It can be sourced directly from any of the text-based methods.
2.   One or more personnel are assigned to manually annotate the images.

When enough annotated images are not available, one can adopt techniques, such as boot-strapping (Feng, Shi, & Chua, 2004) or active learning (Jin, Chai, & Si, 2004), which require only a small set of annotation examples to kick start the learning process. However, for the purposes of this section, it will be assumed that those utilizing image-based methods have acquired a sufficient training dataset.

As discussed in the second section, image-based methods make use of either regional or global image features. The following subsection discusses regional feature-based annotation, whereas the second subsection presents some global feature-based methods.

Regional Feature-Based Methods

Starting from a training set of annotated images, regional feature-based methods attempt to discover the statistical links between visual features and words followed by estimating the correlations between words and regional image features, hence elegantly posing annotation as a statistical inference model.

Given a training set of images T, one or more of the following steps are employed in tackling this problem:

- **Segmentation:** General-purpose segmentation algorithms (e.g., Blobworld [Carson, Thomas, Belongie, Hellerstein, & Malik, 1999], Normalized cuts [Shi & Malik, 2000], etc.) are used to divide images into regions, followed by the extraction and quantification of features from the regions. Regions are described by a set of low-level features like color, texture, shape, and so forth. Image segmentation greatly influences the quality of the automatic image annotations. While several segmentation methods exist in the literature, none guarantee the creation of semantically meaningful regions; nevertheless, the regions produced often suffice. Alternatively, one could partition an image into a fixed set of rectangular regions, which avoids the computational costs of segmentation. Since segmentation is not the focus of this chapter, we leave the choice of either using a segmentation algorithm or fixed partitioning to the reader.

- **Clustering:** Regions are grouped into a set of similar regions; each group (cluster) is called a blob. The assumption here is that similar regions belong to the same cluster. Since most region-based methods rely on clustering, the annotation performance is influenced strongly by the quality of clustering. Each region typically is represented by high-dimensional features. This poses a problem for well-known clustering algorithms such as K-means (Hartigan & Wong, 1979), which assume that all features are equally important. As the number of dimensions increase, data become sparse, and meaningful clusters are difficult to define. This can be addressed by more sophisticated clustering algorithms (Jin, Shi, & Chua, 2004; Wang, Liu, & Khan, 2004) that account for feature relevance based on the assigned weights. In the rest of the discussion, we will use the terms *region* and *blob* interchangeably.

Let there be m blobs $B = \{b_1, b_2, ..., b_m\}$, which are annotated with words selected from a set of n words $W = \{w_1, w_2, ..., w_n\}$ that form the vocabulary of all the images in the training set T. Each image $I \in T$ is represented as $I = B_I \cup W_I$, in which $B_I = \{b_1, b_2, ..., b_{p_I}\}$ represents the p_I blobs corresponding to regions of the image, and $W_I = \{w_1, w_2, ..., w_{q_I}\}$ corresponds to the q_I words (annotations) in the image.

- **Correlation:** The last step is to analyze the correlation among words and regions in order to discover hidden semantics. This usually is also the most difficult task, since image datasets usually do not provide explicit correspondence between the two. The regional feature-based methods can be classified further as word-region mapping and word-image mapping. These are explained in more detail in the following subsections[3]:

Word-Region Mapping

The word-region process maps frequent words to every region (i.e., words are assigned to regions instead of whole images). However, with the exception of a small vocabulary, it is very difficult to find such learning data. With the increase in the size of the data, it becomes increasingly difficult to assign words to individual regions of an image.

To overcome this problem, one can use multiple instance learning (MIL) (Dietterich, Lathrop, & Lozano-Perez, 1997), in which the task is to learn a concept, given positive and negative bags of instances. Each bag may contain many instances, but a bag is labeled positive, even if only one of the instances in it falls within the concept. A bag is labeled negative only if all the instances in it are negative. In regional feature-based image annotation, each region is an instance, and the set of regions from the same image is a bag.

Alternatively, one could use expectation maximization (EM) algorithm (Moon 1996), which is an iterative optimization method for finding maximum likelihood estimates of unknown parameters in probabilistic models; the unknown parameters are referred to as *unobserved latent variables*.

Two word-region mapping models—the co-occurrence model (Mori, Takahashi, & Oka, 1999), which uses MIL; and the machine translation model (Duygulu, Barnard, Freitas, & Forsyth, 2002) based on the EM algorithm—are presented; however, the reader should be aware that this is not exhaustive.

Co-Occurrence Model

The co-occurrence model (Mori et al., 1999) for automatic image annotation collects co-occurrence counts between words and regional features and uses them to annotate new images. The idea is to reduce noise (i.e., unsuitable correlation) by accumulating similar partial patterns from many annotated images. MIL is used to achieve this goal as illustrated by the following example:

Suppose an image $I_1 \in T$ that is annotated with two words—*sky* and *mountain*—is segmented into two regions: R_{11} (containing *sky*) and R_{12} (containing *mountain*). According to MIL, each region of an image is assigned with words that are inherited from the whole image (i.e., both R_{11} and R_{12} are annotated with *sky* and *mountain*). Note that the word *mountain* is an inappropriate annotation for R_{11}. However, if there is another image $I_2 \in T$ that is annotated with two words—*sky* and *river*—that also is divided into two regions, R_{21} (containing *sky*) and R_{22} (containing *river*), then after accumulating the words from these two images, the *sky* pattern will have two *sky*, one *mountain*, and one *river*. In such a way, MIL assumes that the rate of inappropriate words gradually decreases by accumulating similar patterns.

The likelihood $P(w_i|b_j)$ for each word w_i and blob b_j is estimated by accumulating their frequency of occurrences as follows:

$$P(w_i \mid b_j) = \frac{P(b_j \mid w_i)P(w_i)}{\sum\limits_{k=1}^{n} P(b_j \mid w_k)P(w_k)} \quad (by\ Bayes\ theorem)$$

$$\approx \frac{\left(\dfrac{m_{ji}}{n_i}\right)\left(\dfrac{n_i}{N}\right)}{\sum\limits_{k=1}^{n}\left(\dfrac{m_{jk}}{n_k}\right)\left(\dfrac{n_k}{N}\right)}$$

$$= \frac{m_{ji}}{\sum\limits_{k=1}^{n} m_{jk}},$$

$$\tag{1}$$

in which n is the number of words in vocabulary, m_{ji} is the number of occurrences of word w_i in blob b_j, n_i is the number of occurrences of word w_i in all blobs, and N is the sum of the number of occurrences of all words in all blobs.

Once the statistical model is trained using the vocabulary from the images in the training set, a new image is annotated first by classifying its segments to find the closest blobs and then finding the corresponding words for each blob by choosing the words with the highest probability.

Machine Translation Model

The machine translation model (Duygulu et al., 2002) is a substantial improvement over the co-occurrence model. The model views image annotation as the task of translating from a vocabulary of blobs to a vocabulary of words. It uses the EM algorithm, which integrates correspondence learning into image annotation.

The translation model uses the training set to construct a probability table linking blobs with words. This table is the conditional probability $P(w_i \mid b_j)$ of a word w_i given a blob b_j (i.e., the probability of translating blob b_j into word w_i). These probabilities are obtained by maximizing the likelihood $l(T)$ of annotated images from the training set T:

$$l(T) = \prod_{I \in T} \prod_{i=1}^{q_I} \sum_{j=1}^{p_I} P(w_i \mid b_j) v_{b_j}, \tag{2}$$

in which v_{b_j} is the number of blobs b_j that appears in image I. The EM algorithm is applied to find the optimal solution, which suggests the following iterative strategy:

- Compute the expectation (estimate) of the probability table to predict correspondences.

- Use the correspondences to refine the estimate of the probability table (i.e., find the new maximum).

Once the probability table is obtained, then the process of annotating an image is the same as the co-occurrence technique. However, due to the EM algorithm and the high-dimensionality of the image feature vector, it is time-consuming to train and apply the machine translation model.

One difficulty with word-region mapping arises from the skewed distribution of word frequency. If an image blob co-occurs more frequently with the word *sky* than with other words, it will be more likely for the blob to be associated with *sky*. The term *frequency of annotated words* follows the Zipfian law (i.e., a small number of words appears very often in image annotations, and most words are used only by a few images). This inaccurate co-occurrence statistic allows common annotated words to be associated with many irrelevant image blobs and thus degrades the annotation quality. A solution would be to raise the number of blobs that is associated with rare words by scaling the conditional probability $P(w_i|b_j)$ of different words with different factors (Kang, Jin, & Chai, 2004). In other words, to reduce the distribution imbalance of the number of blobs associated with every word, one could ensure that each word is associated with at least one blob with high probability.

Word-Image Mapping

Assigning words to image regions can give rise to many errors due to the over-segmentation of images. Thus, unlike word-region mapping, word-image mapping uses regional features to map frequent words to every image (i.e., words are assigned to whole images and not to image regions).

Three word-image mapping statistical models are presented: (a) the hierarchical aspect model (Barnard & Forsyth, 2001); (b) the relevance model (Jeon, Lavrenko, & Manmatha, 2003; Lavrenko, Manmatha, & Jeon, 2003); and (c) the two-dimensional multi-resolution hidden Markov model (2D MHMM) (Li, Gray, & Olshen, 2000); however, the reader should be aware that this is not an exhaustive treatment. With the exception of 2D MHMM, these models take advantage of the joint distribution of words and whole images (represented as a union of all blobs in an image).

Hierarchical Aspect Model

The hierarchical aspect model (Barnard & Forsyth, 2001) for automatic image annotation is inspired by the generative hierarchical model (Hofmann, 1998), a hierarchical combination of the asymmetric clustering model that maps images into clusters, and the symmetric clustering model that models the joint distribution of images and its features. With its hierarchical structure, the model is well-suited for information retrieval tasks such as database browsing and searching for images based on text and/or image features.

The hierarchical model is trained using the EM algorithm. The training data are modeled as being generated by a fixed hierarchy of nodes with the leaves of the hierarchy corresponding to clusters. Each node in the tree has some probability of generating a word or a blob. The images belonging to a given cluster are modeled as being generated by the nodes along the path from the leaf corresponding to the cluster up to the root node, with each node being weighted on an image and cluster basis. Conceptually, an image belongs to a specific cluster, but given finite data, it is only possible to model the probability that an image belongs to a cluster. Taking all clusters into consideration, an image is modeled by the sum over the clusters weighted by the probability that the image is in the cluster.

The probability that a test image $J \notin T$ containing blobs B_J emits a proposed word w_i from the vocabulary is computed as follows:

$$
\begin{aligned}
P(w_i \mid J) \approx P(w_i \mid B_J) &= P(w_i \mid b_1, b_2, \ldots b_{p_J}) \\
&= \sum_c P(w_i \mid c, b_1, b_2, \ldots b_{p_J}) P(c \mid b_1, b_2, \ldots b_{p_J}) \\
&\approx \sum_c P(w_i \mid c, b_1, b_2, \ldots b_{p_J}) P(b_1, b_2, \ldots b_{p_J} \mid c) P(c) \\
&= \sum_c P(w_i \mid c, b_1, b_2, \ldots b_{p_J}) \prod_{j=1}^{p_J} P(b_j \mid c) P(c) \\
&= \sum_c \left(\sum_l P(w_i \mid c, l) P(l \mid c, b_1, b_2, \ldots b_{p_J}) \right) \prod_{j=1}^{p_J} \left(\sum_l P(b_j \mid l, c) P(l \mid c) \right) P(c),
\end{aligned}
$$

$$(3)$$

in which c indexes clusters and l indexes levels of the hierarchical structure. Similar to the co-occurrence model, the words from the vocabulary with the highest probability are used to annotate new images.

Relevance Model

Relevance-based language models (Lavrenko & Croft, 2001) were introduced to facilitate query expansion and relaxation. These models have been used successfully for both ad-hoc retrieval and cross-language retrieval. Cross-media relevance model (CMRM) (Jeon et al., 2003), which is extended from language models, takes advantage of the joint distribution of words and images.

The model assumes for each image J that there exists some underlying probability distribution $P(\cdot \mid J)$, which is referred to as the language model of J. The language model can be thought of as an urn that contains all possible blobs that could appear in image J as well as all words that could appear in the annotation of J. Since $P(\cdot \mid J)$ itself is unknown, the probability that an image $J \notin T$ containing blobs B_J emits a proposed word w_i from the vocabulary is approximated as follows:

$$
\begin{aligned}
P(w_i \mid J) \approx P(w_i \mid B_J) &= P(w_i \mid b_1, b_2, \ldots b_{p_J}) \\
&= \sum_{I \in T} P(I) P(w_i, b_1, b_2, \ldots, b_{p_J} \mid I) \\
&= \sum_{I \in T} \left\{ P(I) P(w_i \mid I) \prod_{k=1}^{p_J} P(b_k \mid I) \right\}
\end{aligned}
$$

$$(4)$$

The prior probabilities $P(I)$ can be kept uniform over all images in T. Since the images I in the training set contain both words and blobs, maximum likelihood estimates could be used to estimate the probabilities in equation (4).

The words from the vocabulary with the highest probability then are used to annotate new images. Empirical studies (Jeon et al., 2003) have shown that CMRM significantly outperforms the co-occurrence model and machine translation model for automatic image annotation.

All of the models discussed (co-occurrence, machine translation, hierarchical, and CMRM) are discrete and cannot take advantage of the continuous image features. Additionally, they rely heavily on clustering of feature vectors into blobs. Thus, the quality of these annotation models are sensitive to clustering errors and are dependent on the a priori selection of the right cluster granularity; too many clusters result in extreme sparseness of the space, while too few will confuse discriminating various objects in the images. A continuous-space relevance model (CRM) (Lavrenko et al., 2003) has been proposed that models continuous features and is not dependent on clustering; consequently, CRM does not suffer from granularity issues.

According to CRM, an image $J \notin T$ containing z_J distinct regions $\{r_1, r_2, ..., r_{z_J}\}$, produced from the corresponding set of real (continuous) vectors $\{g_1, g_2, ..., g_{z_J}\}$ is then annotated with word w_i from the vocabulary as follows:

$$P(w_i \mid J) \approx P(w_i \mid r_1, r_2, ..., r_{z_J})$$

$$= \sum_{I \in T} \left\{ P_T(I) P_V(w_i \mid I) \prod_{l=1}^{z_J} \int P_R(r_l \mid g_l) P_G(g_l \mid I) dg_l \right\},$$

(5)

in which $P_T(I)$ is the probability of selecting an image $I \subset T$, $P_V(\cdot \mid I)$ is the Multinomial distribution that is assumed to have generated the annotation word w_i, $P_G(\cdot \mid I)$ is the density function responsible for generating the feature vectors $\{g_1, g_2, ..., g_{z_J}\}$, and $P_R(r \mid g)$ is a global probability distribution responsible for mapping real-valued generator vectors g to actual image regions r. Details about estimating the parameters P_T, P_V, P_G, and P_R are beyond the scope of this chapter.

Empirical results (Lavrenko et al., 2003) have demonstrated that CRM model (i) outperforms the co-occurrence-based model by almost an order of magnitude, (ii) is more than 2.5 times better than the machine translation model, and (iii) is at least 1.6 times as good as an annotation method that uses the CMRM.

Two-Dimensional Multi-Resolution Hidden Markov Model (2D MHMM)

Another approach, as described by Li and Wang (2003), is different from several other models previously discussed. This approach assumes that each image has been assigned to a category in which each category corresponds to a concept and contains a set of keywords. A keyword can belong to more than one category.

Categorized images are used to train a dictionary of statistical models, with each model representing a concept. In their approach, the system extracts block-based features from each training image at several resolutions with region-level color and texture information extracted at each resolution. A 2D MHMM (Li et al., 2000), which is a type of statistical model, is generated for each category; each model determines the probability of an image belonging to the category, based upon the various regional textures and colors extracted from the image. In addition, for each keyword, a count of the number of categories to which the keyword belongs is stored in a global index. This concludes the model-generating section.

To annotate a new image, the system extracts the image's content information at multiple resolutions and uses each model to determine the probability of the image belonging to the model's category. The top k categories are selected, and a count of the number of times a keyword appears in the k categories is performed. This count, along with the information in the global index, is used to determine how surprising a word is. A threshold is used to retain only the surprising words. This process has a tendency to favor rare words at the expense of occasionally discarding more common keywords.

One of the important advantages of training a statistical model for each unique concept is that this strategy favors the incremental addition of new images into the training set (i.e., if images representing new concepts or new images in existing concepts are added into the training set, then only the statistical model for the involved concepts need to be trained or retrained). As a result, the system naturally has good scalability without invoking an extra mechanism to address the issue. The relative effectiveness of using 2D MHMM, however, is questionable, since it has been independently evaluated and not compared against any of the existing annotation models.

Global Feature-Based Methods

Unlike region feature-based methods, global feature-based methods assume that correct annotation requires the consideration of the entire image rather than individual sections. Thus, the annotation process utilizes a virtual representation of the image from which a global feature set is derived. Since annotations are based upon global features, global feature-based methods are not categorized based on the derived mapping. Rather, the categorization is based upon the amount of user interaction. Information about both automated and semi-automated approaches will be presented in the next two subsections.

Automated Image Annotation

For automated image annotation, the primary processes are displayed in Figures 4a and 4b. Like the region-based approaches, the automated approach is defined by the feature extraction and model generation steps. Thus, modifying either of the two steps results in a new annotation method. Therefore, one could use almost any combination of feature generation and machine learning techniques to construct an automated annotation system. Consequently, providing an exhaustive coverage is not feasible. Rather, two recent examples of the global feature annotation process will be provided.

Figure 3. Comparative analysis of regional feature-based methods

	Co-occurrence Model	*Machine Translation Model*	*Hierarchical Aspect Model*	*Relevance Model*	*2D MHMM*
Mapping	word-region	word-region	word-image	word-image	word-image
Learning Method	MIL	EM	EM	language modeling and EM	markov chain
Specialty	image and region retrieval	image and region retrieval	image browsing and searching	ranked image retrieval	image classification
Scalability	scalable	scalable	scalable up to a few thousand images	scalable	scalable and incremental
Overall Performance	good	better	good	best	better

Figure 4. (a) Generating annotation model; (b) annotating an image

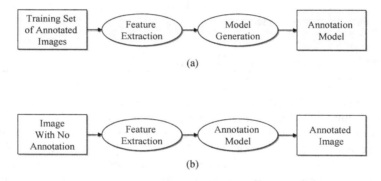

Content-Based Soft Annotation (CBSA)

The CBSA system (Chang, Goh, Sychay, & Wu, 2003) generates a set of color and texture features for each image. Following that, the CBSA generates a model using Bayesian point machines (Herbrich, Graepel, & Campbell, 2001) for each keyword; each model is designed to determine the probability based upon the content features of the keyword being applicable to the image. Once a model has been created for each keyword, the model generation process is completed. When a new image is presented, the system generates a vector of the form $< k_1, p_1, k_2, p_2, \ldots, k_n, p_n >$ in which k_x is keyword x, p_x is the probability that k_x accurately describes the image, and n is the number of keywords (and models).

Support Vector Machines: SVM-Based Annotation

In Feng et al. (2004), a similar approach to the CBSA is utilized. The primary differences are the types of model generators and the choice of descriptors. To generate the annotation model, probabilistic SVM (Vapnik, 1995) are utilized. For feature generation, Feng et al. (2004) extract both color and texture features, as does CBSA. However, in CBSA, the colors of the images are binned into 12 coarse colors. Then, within each bin, the color mean, spread, elongation, and variance are calculated, as well as a histogram of the bin's colors. The textures are generated via wavelet decomposition of the entire image.

For Feng et al. (2004), a global color histogram is used as the color feature. For the texture features, the image is segmented into a series of blocks, each of which is decomposed into 16 frequency bands using discrete cosine transform and wavelet transforms. The mean and variance of each band then is computed, using adaptive pursuit.

Semi-Automated Image Annotation

For automated annotation, the goal is to avoid any human interaction within the annotation process. However, the derived annotation models are often imperfect, resulting in missing assignments and spurious annotations to new images. To resolve this issue, several approaches (Wenyin et al., 2001; Yang, Dong, & Fotouhi, 2005) have been devised to update annotations during retrieval.

Initially, images are annotated either manually or by one of the previously discussed methods. At this point, both the images and their annotations are loaded into a retrieval system. The retrieval system must be capable of performing both keyword- and content-based queries; this permits images that are missing relevant keywords to be retrieved and annotated. The user then executes a retrieval query. Based on the results, the user either ends the search or provides feedback on one or more images. The feedback can be either positive (an image is relevant) or negative (the image is irrelevant). In the latter case, the system performs the following steps:

1. If the initial query is keyword-based, for each keyword in the query
   a. For relevant images
      i. If keyword is not present, assign to image with constant weight
      ii. Or else increase the keyword weight by fixed amount
   b. For non-relevant
      i. If keyword is present, decrease keyword's weight
2. Retrieve new images similar in content and/or keywords of the positive examples and dissimilar to the negative examples
3. If user is not satisfied with results, return to step 1, or else go to step 4
4. Remove from images keywords with a weight below a threshold

5.   If the initial query is content-based

    a.   The user selects the images to be annotated and provides the keywords

Over time, this process will remove spurious keywords and assign any missing keywords.

Evaluation and Quality Assessment

Many evaluation measures are considered to assess the annotation performance of an algorithm. Furthermore, since facilitating the retrieval of images is one of the motivations for automated annotation, retrieval performance measures could be used to assess the annotation quality. Selecting an appropriate measure, however, is dependent on the application type and context.

Let T' denote the test set. Let $J \in T'$ be a test image, W_J be its real (manual) annotation set, and W_J^a be its automatic annotation set. We will now describe the several standard measures that are used for analyzing the performance from both the retrieval and the annotations perspective.

Assessment from the Retrieval Perspective

To assess the performance from the retrieval perspective, auto-annotated test images first are retrieved using keywords from the vocabulary. The relevance of the retrieved images is judged by evaluating it against the real (manual) annotations of the images.

To quantify the results, precision and recall are computed for every word in the test set. Precision is the percentage of retrieved images that are relevant. Recall is the percentage of relevant images that are retrieved. For a given query word w_q, precision and recall are defined as:

$$precision(w_q) = \frac{|\{J \in T' \mid w_q \in W_J^a \wedge w_q \in W_J\}|}{|\{J \in T' \mid w_q \in W_J^a\}|}$$

$$recall(w_q) = \frac{|\{J \in T' \mid w_q \in W_J^a \wedge w_q \in W_J\}|}{|\{J \in T' \mid w_q \in W_J\}|} \tag{7}$$

The $precision(w_q)$ measures the correctness in annotating images with word w_q, and $recall(w_q)$ measures the completeness in annotating images with word w_q. The average precision and recall over different single-word queries are used to measure the overall quality of automatically generated annotations.

The two measures, precision and recall, can be combined into a single measure called the *F*-measure. It is defined as the harmonic mean of precision and recall.

$$F(w_q) = \frac{2 * precision(w_q) * recall(w_q)}{precision(w_q) + recall(w_q)} \qquad (8)$$

In addition to using precision and recall, it is useful to measure the number of single-word queries for which at least one relevant image can be retrieved using the automatic annotations. It is defined as:

$$|\{w_q \mid precision(w_q) > 0 \wedge recall(w_q) > 0\}| \qquad (9)$$

This metric compliments average precision and recall by providing information about how wide the range of words that contribute to the average precision and recall is. This metric is important because a biased model can achieve high precision and recall values only by performing extremely well on a small number of queries with common words.

The precision and recall (and its variations) measure defined previously do not explicitly and/or effectively account for the ordering of the retrieved images. To capture this information, more sophisticated measures such as average un-interpolated precision (Voorhees & Harman, 2000) and R-Norm (Bollmann-Sdorra & Raghavan, 1998) also have been proposed.

Assessment from the Annotations Perspective

To evaluate the performance from the annotations perspective, the accuracy of the auto-annotated test images is measured as the percentage of correctly annotated words. The annotation accuracy for a given test image $J \subset T'$ is defined as:

$$accuracy = \frac{r}{|W_J|} \qquad (10)$$

in which the additional variable r is the number of correctly predicted words in J. The drawback of the accuracy measure is that it does not account for the number of wrong predicted words with respect to the vocabulary size $|W|$.

Normalized score (*NS*), which directly extends from accuracy, penalizes for wrong predictions. The measure is defined as:

$$NS = \frac{r}{|W_J|} - \frac{r'}{|W| - |W_J|} \qquad (11)$$

in which the additional variable r' denotes the number of wrong predicted words in J. Predicting *all* and *only* the right words produces a value of 1, predicting *all but* the right words produces a value of -1, and predicting *all* the words in the vocabulary produces a value of 0.

Conclusion

In this chapter, we presented an overview of the automatic image annotation problem and several approaches that attempt to resolve this problem. However, despite the variety of proposed solutions, the problem is not yet solved. More research is anticipated along a variety of lines. The following subsections will present some potential lines of investigation; however, the reader is cautioned that this is not exhaustive.

Benchmark Datasets

Currently, the published results for annotation utilize different datasets, models, and features. Thus, directly comparing the performance of different approaches in terms of accuracy and efficiency is difficult. Even when comparisons are performed, researchers typically compare the performance of their approaches against one other approach. In the future, it is expected that more comparative work will be done to determine the impact upon automated annotation by the choice of features, the choice of models, and the quality of the data.

To facilitate comparison studies, annotated datasets must be developed and made publicly available. These common datasets are needed for three reasons. First, an annotation method's performance is dependent upon the annotations associated with images. If different image sets and annotations are utilized, then conflicting conclusions may arise. Second, the cost of obtaining annotated data can be prohibitive. This represents a barrier to researchers entering the field, resulting in prevention of or delays in the introduction of new perspectives. Third, common datasets facilitate the development of standardized measures of data quality and annotation performance; this, consequently, results in a better understanding of the annotation process.

In recent years, some common datasets have arisen. The Corel (www.corel.com) dataset has been the most widely used dataset for evaluation. However, to this author's knowledge, the Corel dataset is not readily available, since the company no longer sells it. Another example is ImageCLEF (http://ir.shef.ac.uk/imageclef) (Clough, Mueller, & Sanderson, 2005), which is organized similarly to Text REtrieval Collection (http://trec.nist.gov). The goal of ImageCLEF is to evaluate image retrieval with respect to multilingual document collections; this effort includes image retrieval and annotation. The 2005 and 2006 tracks provide both medical and general photography annotation sets. In a related effort, Li and Wang (2003) and Wang, Li, and Wiederhold (2001) provide several annotated datasets for comparison work, ranging from 1,000 to 60,000 images (http://wang.ist.psu.edu/docs/related). As practitioners in the community will start utilizing these resources, more data sources eventually will be developed.

Natural Language Processing

Although image annotation techniques largely can be automated, they suffer from the same problems encountered by text-based information retrieval systems. Because of widespread synonymy and polysemy in natural language, the annotation precision may be very low, and the resulting recall inadequate. Synonymy occurs when different users describe the same concept using different terms; this may be a result of different contexts or linguistic habits. Polysemy occurs when the same term takes on a varying referential significance, depending on the context. In addition, linguistic barriers and the nonuniform textual descriptions for common image attributes also limit the effectiveness of annotation systems. Domain knowledge, combined with natural language processing, could be utilized to address some of these problems. However, to the best of our knowledge, the field is quite immature, while the challenges are many.

Hybrid Methods

The chapter focused on text-based and image-based methods. Text-based methods, although highly specialized in domain- (context-) specific image annotations, are usually semi-automatic. On the flip side, image-based methods are automatic; however, they are domain-independent and could deliver arbitrarily poor annotation performance when applied to fields such as biomedicine, art, and so forth. An interesting yet challenging problem would be to combine the characteristics of these two classes of methods into what we call hybrid methods, which (potentially) could exploit the synergistic relationship between them in an attempt toward solving the well-known semantic problem of visual information retrieval.

References

Barnard, K., & Forsyth, D. (2001). Learning the semantics of words and pictures. In *Proceedings of the Conference on computer vision*, Vancouver, Canada (pp. 408–415).

Benitez, A., & Chang, S. F. (2002). Semantic knowledge construction from annotated image collection. In *Proceedings of the IEEE Conference on Multimedia*, Lausanne, Switzerland (pp. 205–208).

Bollmann-Sdorra, P., & Raghavan, V. (1998). On the necessity of term dependence in a query space for weighted retrieval. *Journal of the American Society for Information Science, 49*(13), 1161–1168.

Carson, C., Thomas, M., Belongie, S., Hellerstein, J., & Malik, J. (1999). Blobworld: A system for region-based image indexing and retrieval. *Proceedings of the Conference on Visual Information Systems* (pp. 509–516). Amsterdam, Netherlands.

Chang, E., Goh, K., Sychay, G., & Wu, G. (2003). CBSA: Content-based soft annotation for multimodal image retrieval using Bayes point machines. *IEEE Transactions on*

Circuits and Systems for Video Technology: Special Issue on Conceptual and Dynamical Aspects of Multimedia Content Description, 13(1), 26–38.

Clough, P., Mueller, H., & Sanderson, M. (2005). The CLEF 2004 cross language image retrieval track. In *Proceedings of the Workshop on Cross-Language Evaluation Forum,* Heidelberg, Germany (pp. 597–613).

Dietterich, T., Lathrop, R., & Lozano-Perez, T. (1997). Solving the multiple instance learning with axis-parallel rectangles. *Artificial Intelligence, 89*(1-2), 31–71.

Duygulu, P., Barnard, K., Freitas, N., & Forsyth. D. (2002). Object recognition as machine translation: Learning a lexicon for a fixed image vocabulary. In *Proceedings of the European Conference on Computer Vision,* Copenhagen, Denmark (pp. 97–112).

Feng, H., Shi, R., & Chua, T. (2004). A bootstrapping framework for annotating and retrieving WWW images. In *Proceedings of the ACM Conference on Multimedia,* New York (pp. 960–967).

Flickner, M., et al. (1995). Query by image and video content: The QBIC system. *IEEE Computer, 28*(9), 23–32.

Hartigan, J., & Wong, M. (1979). A K-means clustering algorithm. *Applied Statistics, 28*(1), 100–108.

Herbrich, R., Graepel, T., & Campbell, C. (2001). Bayes point machines. *Journal of Machine Learning Research, 1,* 245–279.

Hofmann, T. (1998). Learning and representing topic: A hierarchical mixture model for word occurrence in document databases. In *Proceedings of the Workshop on Learning from Text and Web,* Pittsburgh, PA.

Hollink, L., Schreiber, A., Wielemaker, J., & Wielinga, B. (2003). Semantic annotation of image collections. In *Proceedings of the Workshop on Knowledge Markup and Semantic Annotation,* Sanibel Island, FL.

Hu, B., Dasmahapatra, S., Lewis, P., & Shadbolt, N. (2003). Ontology-based medical image annotation with description logics. In *Proceedings of the IEEE Conference on Tools with Artificial Intelligence,* Sacramento, CA (p. 77).

Jeon, J., Lavrenko, V., & Manmatha. R. (2003). Automatic image annotation and retrieval using cross-media relevance models. In *Proceedings of the ACM Conference on SIGIR,* Toronto, Canada (pp. 119–126).

Jin, R., Chai, J., & Si, L. (2004). Effective automatic image annotation via a coherent language model and active learning. In *Proceedings of the ACM Conference on Multimedia,* New York (pp. 892–899).

Jin, W., Shi, R., & Chua, T. (2004). A semi-naïve Bayesian method incorporating clustering with pair-wise constraints for auto image annotation. In *Proceedings of the ACM Conference on Multimedia,* New York (pp. 336–339).

Kang, F., Jin, R., & Chai, J. (2004). Regularizing translation models for better automatic image annotation. In *Proceedings of the Conference on Information and Knowledge Management,* Washington, DC (pp. 350–359).

Lavrenko, V., & Croft, W. (2001). Relevance-based language models. In *Proceedings of the ACM Conference on SIGIR,* New Orleans, LA (pp. 120–127).

Lavrenko, V., Manmatha, R., & Jeon, J. (2003). A model for learning the semantics of pictures. In *Proceedings of Advances in Neutral Information Processing*, Vancouver, Canada.

Li, J., Gray, R., & Olshen, R. A. (2000). Multiresolution image classification by hierarchical modeling with two dimensional hidden markov models. *IEEE Transactions on Information Theory, 34*(5), 1826–1841.

Li, J., & Wang, J. (2003). Automatic linguistic indexing of pictures by a statistical modeling approach. *IEEE Transactions on Pattern Analysis and Machine Intelligence, 25*(9), 1075–1088.

Lieberman, H., Rosenzweig, E., & Singh, P. (2001). Aria: An agent for annotating and retrieving images. *IEEE Computer, 34*(7), 57–61.

Miller, G. (1995). WordNet: A lexical database for English. *Communications of the ACM, 38*(11), 39–41.

Moon, T. (1996). The expectation-maximization algorithm. *IEEE Signal Processing, 13*(6), 47–60.

Mori, Y., Takahashi, H., & Oka, R. (1999). Image-to-word transformation based on dividing and vector quantizing images with words. In *Proceedings of the Workshop on Multimedia Intelligent Storage and Retrieval Management*, Orlando, FL.

Shah, B., Raghavan, V., Dhatric, P., & Zhao, X. (2006). A cluster-based approach for efficient content-based image retrieval using a similarity-preserving space transformation method. *Journal of the American Society for Information Science and Technology, 57*(12), 1694-1707.

Shen, H., Ooi, B., & Tan, K. (2000). *Giving meanings to WWW images*. In *Proceedings of the ACM Conference on Multimedia*, Marina del Rey, CA (pp. 39–47).

Shi, J., & Malik, J. (2000). Normalized cuts and image segmentation. *IEEE Transactions on Pattern Analysis and Machine Intelligence, 22*(8), 888–905.

Soo, V., Lee, C., Li, C., Chen, S., & Chen, C. (2003). Automatic semantic annotation and retrieval based on sharable ontology and case-based learning techniques. In *Proceedings of the Joint Conference on Digital Libraries*, Houston, TX (61–72).

Srihari, R., & Zhang, Z. (2000). Show&Tell: A semi-automated image annotation system. *IEEE Multimedia, 7*(3), 61–71.

Town, C. (2004). *Ontology based visual information processing* [doctoral dissertation]. Cambridge, UK: University of Cambridge.

Vapnik, V. (1995). *The nature of statistical learning theory*. New York: Springer-Verlag.

Voorhees, E., & Harman, D. (2000). Overview of the sixth text retrieval conference. *Information Processing and Management, 36*(1), 3–35.

Wang, J., Li, L., & Wiederhold, G. (2001). SIMPLIcity: Semantics-sensitive integrated matching for picture libraries. *IEEE Transactions on Pattern Analysis and Machine Intelligence, 23*(9), 947–963.

Wang, L., Liu, L., & Khan, L. (2004). Automatic image annotation and retrieval using subspace clustering algorithm. In *Proceedings of the Workshop on Multimedia Databases*, Washington, DC (pp. 100–108).

Wenyin, L., et al. (2001). Semi-automatic image annotation. In *Proceedings of the Conference on Human-Computer Interaction*, Tokyo, Japan (pp. 326–333).

Yang, C., Dong, M., & Fotouhi, F. (2005). Semantic feedback for interactive image retrieval. In *Proceedings of the ACM Conference on Multimedia*, Singapore (pp. 415–418).

Endnotes

[1]   A video database may be searched using images, as video clips can be segmented into shots, which are represented by a set of keyframe images. Video annotation is thus reduced to annotation of keyframe images.

[2]   Please refer to Figure 2 for a precise and clear distinction between the various choices.

[3]   The reader is encouraged to reference Figure 3 for a comparative analysis of the various regional feature-based annotation models discussed in this chapter.

Chapter VII

Adaptive Metadata Generation for Integration of Visual and Semantic Information

Hideyasu Sasaki, Ritsumeikan University, Japan

Yasushi Kiyoki, Keio University, Japan

Abstract

The principal concern of this chapter is to provide those in the visual information retrieval community with a methodology that allows them to integrate the results of content analysis of visual information (i.e., the content descriptors) and their text-based representation in order to attain the semantically precise results of keyword-based image retrieval operations. The main visual objects of our discussion are images that do not have any semantic representations therein. Those images demand textual annotation of precise semantics, which is to be based on the results of automatic content analysis but not on the results of time-consuming manual annotation processes. We first outline the technical background and literature review on a variety of annotation techniques for visual information retrieval. We then describe our proposed method and its implemented system for generating metadata or textual indexes to visual objects by using content analysis techniques by bridging the gaps between content descriptors and textual information.

Introduction

Images contain valuable information within themselves for content-based multimedia analysis, though image processing of visual features does not generate any textual indexes as metadata that allow efficient and intuitive keyword-based retrieval. Those in the visual information retrieval community demand content-based metadata indexing that realizes the following four agendas: (1) provide digital images with automatically generated metadata; (2) generate metadata of human cognizable information (i.e., textual labels or indexes); (3) include external information that is not available from image processing of data or objects into metadata generation; and (4) accept external knowledge of the expertise in certain domains in metadata generation (e.g., metadata generation of brain disease images for expert physicians).

Some method of automatic indexing (i.e., metadata generation) is indispensable to keyword-based image retrieval that is more precise and speedier than content-based image retrieval (CBIR) in networked multimedia databases (Moreno, van Thong, Logan, & Jones, 2002). The content descriptors obtained from images are usable to automatically generate metadata. Content analysis employed in CBIR is to provide a desirable solution to that automatic metadata generation (Smeulders, Worring, Santini, Gupta, & Jain, 2000; Rui, Huang, & Chang, 1999; Yoshitaka & Ichikawa, 1999). Digital images per se, however, do not contain any seeds of textual labels or indexes for metadata generation. They are different from documents that contain direct sources for textual indexing.

Metadata generation using content analysis has four issues. First, object images as queries are not always proper for the structural similarity computation by content analysis, because images often are distorted or noisy. Second, content descriptors are binary indexes that are not proper for keyword-based retrieval that demands textual indexes attached to images. Image processing approach as content analysis automatically extracts visual features out of images and dynamically generates binary indexes as content descriptors of images that lack human cognizable meanings. Third, the direct use of the results of content analysis is not satisfactory for metadata generation, because those obtained descriptors often are conflicting in classification of object images. In addition, image processing focuses on internal information stored in images, though external knowledge is indispensable to metadata generation that refers to categorization of metadata. We need a solution for content-based metadata indexing that is based on image processing techniques as automatic content analysis and that is also accessible to textual information and external knowledge on structural properties of visual objects for textual indexing.

In this chapter, we provide those in the visual information retrieval community with the fundamental theory and implementation on content-based metadata indexing that allows semantic-based image acquisition that bridges the gaps between the results of content analysis of visual objects and their textual information in categorization. The main object of our discussion is to present our automatic adaptive metadata generation method and its implemented system using content analysis of sample images by exemplary textual indexing to digital images as a prototype system for content-based metadata indexing with a variety of experimental results. Instead of costly human-created metadata, our method and its implemented system rank sample images by distance computation on their structural similarity to query images and automatically generate metadata as textual indexes that represent geometric structural properties of the most similar sample images to the query images, as outlined in Figure 1. Here, we simply define *metadata* as "data about other data and objects" (Rusch-Feja, 1998) in human cognizable forms such as textual indexes as well as HTML tags and so forth. In the case of ranking conflict, the proposed method generates new metadata that are adaptive to selected domains and identify query images within, at most, two geometric types of visual objects. The objective indexing to sample images provides performance evaluation with objective measure on appropriate indexing. We outline several experimental results that the proposed prototype system has generated appropriate metadata to digital images of geometric figures with 61.0% recall ratio and 20.0% fallout ratio overall.

The goal of this chapter is to design a method and its implemented system for metadata generation that mitigates the four issues already discussed by using content analysis and databases of sample images and their sample indexes. After the introduction, Section Two discusses the state of arts in the area of automatic annotation to visual information. In Section Three, we present our method and its implemented system, which consists of three modules (see Figure 1).The first module screens out improper object images as queries for metadata generation by using CBIR that computes structural similarity between sample images and object images as queries. We have realized automatic selection of proper threshold values in the screening module (Sasaki & Kiyoki, 2006). The second module generates metadata by selecting sample indexes attached to sample images that are structurally similar to object images as queries. The selection of sample indexes as metadata candidates is optimized by thresholding the results of content analysis of the sample images and the object images as queries (Sasaki & Kiyoki, 2004). The third module detects improper metadata and regenerates proper metadata by identifying common and distinctive indexes in wrongly selected metadata (Sasaki & Kiyoki, 2005a, 2005b). In Section Four, we provide mathematical formulation to each process in the modules of our method and its implemented system. In Section Five, we present its cost-efficient implementation using a typical package solution of CBIR with databases. In Section Six, we evaluate the performance of our system using CBIR and databases of query and sample images that are basic geometric figures. We have

Figure 1. Architecture of the automatic adaptive metadata generation

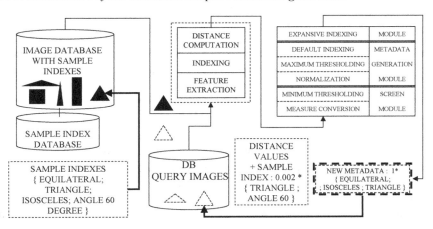

Figure 2. Metadata generation bridges computer vision and databases

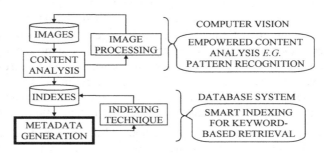

used various types of geometric figures for the experiments of this research. Our system has generated proper metadata to the images of basic figures at 61.0% recall ratio and 20.0% fallout ratio overall. This result has been bettered off by 23.5% on recall ratio and 37.0% on fallout ratio than just using content analysis for metadata generation. In Section Seven, we discuss the experimental results. Our method and its implemented system show their feasibility with well-selected sample images and indexes that represent typical structural properties of object images in specific domains. Our method has its extensibility to specific object domains with the inclusion of computer vision techniques. In Section Eight, we conclude the chapter with future perspectives in this area. Our method is to integrate the contributions in computer vision and database researches (here, content analysis and indexing technique) as an automatic solution for precise and speedy image retrieval in networked multimedia databases (see Figure 2).

Background

A content-based approach for semantic image retrieval demands metadata indexing that allows precise keyword-based retrieval for a large number of visual objects stored in databases. Metadata indexing is classified in two main approaches regarding indexing techniques: keyword-based indexing and content-based indexing. Those individual indexing approaches correspond to two types of analysis objects: textual information and visual features (Sheth & Klas, 1998). Table 1 gives an overview of taxonomy of metadata indexing in this research field.

Keyword-based indexing was developed as an automatic solution to document indexing. Human-created indexes are costly in selecting and indexing proper labels (Marshall, 1998). Documents contain textual labels that are easy to automatically extract (Kobayashi & Takeda, 2000; Sebastiani, 2002). Keyword-based indexing is classified in two subclass approaches: textual closed caption (Brown, Foote, Jones, Jones, & Young, 1995) and associated text incorporation (Srihari, Zhang, & Rao, 2000). This classification is about associative/sec-

Table 1. Taxonomy of metadata indexing approaches

Approach	Primary Analysis Object	Primary Indexing Technique	Associative Indexing Technique	Index Format	Application Domains
Textual Closed Caption	Textual Information in Documents	Keyword-Based Indexing	Human Language Technology, Ontology, etc.	Textual Labels	General Domains
Associated Text Incorporation	Textual Information Associated with Images	Keyword-Based Indexing	Term Categorization	Textual Labels	General Domains
Image Processing	Visual Features in Images	Content-Based Indexing	Visual Feature Extraction	Binary Indexes	General Domains
Categorization Modeling for Content Analysis	Visual Features in Images	Content-Based Indexing	Model-Based Categorization	Textual Labels	Model-Specified Types of Specific Domains
Adaptive Metadata Generation (Our Approach)	Visual Features in Images	Content-Based Indexing	Exemplary Textual Labeling Based on Structural Distance Computation	Textual Labels	Specific Domains

ondary indexing techniques and corresponds to various attributes on primary analysis objects: textual information in documents and textual information associated with images, respectively. Typical studies on textual closed captions include automatic content annotation and indexing to documents based on human language technology (HLT) (Bontcheva, Maynard, Cunningham, & Saggion, 2002), multilingual cataloging of document databases by vocabulary indexes (Larson, Gey, & Chen, 2002), unfamiliar metadata searching by entry vocabulary modules (Gey et al., 2002), HTML tags-based WWW document organization (Doan, Beigbeder, Girardot, & Jaillon, 1998), and content table creation by HTML tags (Dolin et al., 2001). Typical studies on associated text incorporation include HTML source code analysis by reference information (e.g., tags or metadata to images) (Tsymbalenko & Munson, 2001), term linkage with images in PDF documents such as maps and descriptions (Smith, Mahoney, & Crane, 2002), categorical term analysis for metadata indexing to images (Duygulu, Barnard, de Freitas, & Forsyth, 2002; Yee, Swearingen, Li, & Hearst, 2003), semi-automatic textual indexing to images for classification by using lexical databases (Kiyoki, Kitagawa, & Hayama, 1994), and similar approaches are found in audio indexing (Magrin-Chagnolleau, 2000) and speech indexing (Amato, Gennaro, & Savino,

2001). Keyword-based indexing realizes summarization of documents and image/document commingling objects with textual indexes. The latter approach, nonetheless, is not a direct solution for indexing images. Images per se contain visual information that is different from the textual but peripheral information associated with those images.

Content-based indexing was introduced to the multimedia research field as a solution for direct indexing to images in the form of content descriptors with image processing techniques as content analysis that often are employed in CBIR systems (Smeulders et al., 2000; Rui et al., 1999; Blei & Jordan, 2003). In CBIR systems, image processing techniques are used to automatically generate binary indexes to images by feature extraction and to classify those generated indexes by computation of structural distance on visual features in general domains. Content-based indexing is classified in several subclass approaches: image processing, categorization modeling for content analysis, and so forth. This classification is about associative/secondary indexing techniques and corresponds to individual types of application domains: general domains, model-specified types of specific domains, and so forth. Typical studies on image processing include multi-spectral image indexing (Barros, French, Martin, Kelly, & White, 1994) in pattern analysis with a number of applications of CBIR: QBIC (Flickner et al., 1995; Ashley et al., 1995), WebSEEK (Smith & Chang, 1997), MARS (Ortega, Rui, Chakrabarti, Mehrotra, & Huang, 1997), and so forth. Similar content-based approaches also are found in speech analysis (Wilcox & Boreczky, 1998).

Image processing automatically generates binary but not textual indexes (i.e., content descriptors) to images. More advanced solutions come from combining keyword-based indexing and content-based indexing (Moreno et al., 2002). The categorization modeling approach realizes textual indexing to images. Its typical studies are found in the field of medical image indexing and retrieval. For example, semantic (i.e., knowledge) modeling for textual indexing to brain tumor images is employed in a brain disease image retrieval system (Chu, Ieong, & Taira, 1994). A number of studies has employed such modeling as methods that discover direct linkages between the results of content analysis of visual objects and the prospective metadata that often describe sophisticated categories. For example, a small black bulb of extracted features is associated with a symptom of fatal stroke in neoplastic regions of human brains. However, that direct linkage involves not only its costly modeling and implementation but also the limitation of its applications to model-specified types of specific and narrower domains that have hierarchical categories on image classification (e.g., a metadata categorization in a field of brain tumor images). In addition, the direct linkage is not intuitive in human recognition mechanisms in which geometric or simple objects contained in images are perceived and associated with their basic structural properties. For example, an image of a tower beside a bridge contains a small horizontal rectangle and a tall vertical rectangle. Those properties are integrated and associated with another image stored in knowledge stock, which represents more sophisticated concepts such as Big Ben in London. It is reasonable and supportive to put an intermediate layer of metadata indexing based on structural properties of extracted features between a layer of content-analysis results of visual objects and another layer of sophisticated metadata association. To remedy that issue and limitation, a desirable solution is the combination of content-based indexing as a primary indexing approach and textual indexing as a secondary indexing technique that is adaptive to specific domains.

Inthe third section we formulate our automatic adaptive metadata generation method and its implemented system.

Adaptive Metadata Generation: Methodology for Integrating Content Descriptors with Textual Information

We present the automatic adaptive metadata generation method and its implemented system with the following two preprocesses for collection of sample images and sample indexes and distance computation and three modules for screening: metadata generation and expansive indexing (see Figure 1). The proposed method allows cost-efficient system implementation by employing two general functions in image processing techniques: feature extraction out of images and structural distance computation of extracted visual features. Our method automatically assigns query images with combinations of textual indexes and numerical indicators as metadata by using results of distance computation (i.e., content descriptors) on structural similarity between query images and the sample images with sample indexes that represent semantics of object images. The sample indexes represent the structural properties of the sample images stored in databases. Numerical indicators represent the proportion of structural properties of the query images in comparison with the sample images. In the case of misevaluation in distance computation, a single query image is calculated as similar to several various types of sample images; then the proposed method generates new adaptive metadata that narrow down the scope of identification of the query image within, at most, two types of structurally similar images. The proposed method with well-defined sets of sample images and textual indexes is adaptive to specific domains without any anterior modeling on metadata categorization. The purpose of focusing on geometric structural properties in metadata indexing is that automatically generated metadata can be used as the objects of text-based analysis, which reflects external and expert wisdom that is accessible in specified domains and generates new metadata that represent more sophisticated concepts.

Collection of Sample Images and Sample Indexes

The purpose of this preprocess to collect images and indexes is to specify each object domain with a set of sample images for metadata generation. This preprocess of manual preprocessing involves four steps. The first step selects sample images that represent typical types of structural properties that are found in the specified domain (e.g., typical shapes of architectures like water pools and baseball parks in urban surroundings). The second step selects a set of well-defined sample indexes as examples that describe structural properties on selected sample images (e.g., { Triangle, 70° ; Isosceles ; … } to a top roof of a certain type of a baseball park. The third step stores sample images into an image database, sample indexes that correspond to the stored sample images into another sample index database, and query images requested for metadata generation to the other image database. Its final step combines sample indexes with sample images.

Example

A sample image of a right angle triangle has a set of sample indexes: { Triangle ; Angle 90° ; Isosceles }. Another sample image of obtuse isosceles triangle has a set of sample indexes: { Obtuse ; Isosceles ; Triangle ; Angle 150° }.

Distance Computation

The purpose of this preprocess is to compare query images with sample images assigned with sample indexes. This preprocess as automatic preprocessing applies visual feature extraction and distance computation functions available in content analysis to query images and the sample images with sample indexes. This preprocess involves three steps: (1) store the sample images and the query images into a CBIR system; (2) fire distance computation functions in the CBIR system; and (3) get results of structural distance computation from the CBIR system. In this preprocess, a CBIR system is to be selected from several CBIR systems. That selected system is to provide query images with sets of values that represent structural distance of the query images from the sample images. Those returned values represent a certain proportion of structural properties that the query images contain against what the sample images represent in the forms of the sample indexes.

Example

As described in Table 2, a query image of a right angle triangle has the same number of structural distance values with right angle triangle and obtuse isosceles triangle as 0.00, which means the selected CBIR system evaluates those images as the least distant images to each other, although that result demands improvement in indexing. Meanwhile, the larger numbers correspond to the more distant images.

Screen Module

This module consists of two processes: measure conversion and minimum thresholding. The purpose of the measure conversion process is to convert the results of distance computation into more intuitive values that represent structural similarity between the query and the sample images. The values on structural distance take 0.00 to the least distant and most similar images, while the larger numbers correspond to the more distant and less similar images. This process converts the current unintuitive measure on distance computation to another intuitive measure on similarity, which evaluates the least distant and most similar images as 1.00 and the largest distant images as 0.00. After the measure conversion, structural distance values of 0.00 are converted to 1.00 when the query images coincide with certain sample images.

The purpose of the minimum thresholding process is to cut off noise information after the measure conversion process. In collection of sample images, prepared images should rep-

Table 2. Sample data processing in default indexing (e.g., query: right angle triangle)

SAMPLE IMAGES	SAMPLE TEXTUAL INDEXES	DISTANCE VALUES	A/F NORMALIZATION
RIGHT ANGLE TRI-ANGLE	TRIANGLE; ANGLE 90°; ISOSCELES	0.00	0.50
OBTUSE ISOSCELES TRIANGLE	TRIANGLE; OBTUSE; ANGLE 150°; ISOS-CELES	0.00	0.50
RHOMBUS	RHOMBUS; QUAD-RILATERAL; QUAD-RANGLE; POLYGON	0.000160	0.00
ELLIPSE	ELLIPSE	0.00256	0.00
ACUTE ISOSCELES TRIANGLE	TRIANGLE; ACUTE; ANGLE 30°; ISOSCE-LES	0.244	0.00
TRAPEZOID	TRAPEZOID; QUAD-RANGLE	0.256	0.00
CIRCLE	CIRCLE	0.310	0.00
EQUILATERAL TRIANGLE	TRIANGLE; EQUILAT-ERAL; ANGLE 60°; ISOSCELES	0.877	0.00
RECTANGLE	RECTANGLE; QUAD-RANGLE; POLYGON	0.974	0.00
SQUARE	SQUARE; RECTAN-GLE; QUADRANGLE; POLYGON; QUADRI-LATERAL	3.603	0.00

resent typical classes of structural difference in object domains. Suppose that a query image is evaluated to be equally similar to more than half of the sample images in a certain result of measure conversion. This case shows that the query image is not proper for metadata generation, because the prepared sample images cannot identify any pattern or class of the query image. Then, those generated metadata show that candidate sample images are not proper representatives of the queries in the perspective of two-party categorization. The minimum thresholding process sets a threshold that is defined as the reciprocal number of the half number of sample images. When the number of sample images is odd, it should be incremented by one. The screen module is to be bettered off for further improvement on noise cut on images by inclusion of image processing techniques such as morphological analysis, including thinning, image segmentation of noisy objects, and so forth.

Metadata Generation Module

This module consists of three processes: normalization, maximum thresholding, and default indexing. The normalization process is to normalize the results after the screen module as they constitute numerical indicators of proportional structural properties that the sample indexes represent. The maximum thresholding process is to emphasize the values that represent meaningful structural properties of the query images in the forms of numerical indicators to the sample indexes. Some values of numerical indicators after the normalization correspond to mutually similar numbers of lower size that represent insignificant structural properties of sample images. This process narrows down the range of those values after the normalization process by cutting off the values of lower size than a certain threshold that is the maximum threshold (e.g., 0.95, 0.40, etc.) and simply converts those values over the maximum threshold to 1.00. In addition, when some values are converted to 1.00, this process evaluates all the other values as noise and converts them to 0.00. The default indexing process is to index query images with combinations of textual indexes and numerical indicators as metadata. This process combines the values after the maximum thresholding process as numerical indicators of proportional structural properties with the sample indexes that represent structural properties and then constitutes newly generated metadata while those metadata appropriately represent structural properties of the query images. Meanwhile, the default indexing process can generate conflicting indexes that represent several different types of sample images to a single query (see Table 2). In that case, that single query image can correspond to different structural types of sample indexes when several normalized values take the same largest size. The default indexing process simply discards those conflicting indexes from metadata candidates and does not generate any proper metadata.

Expansive Indexing Module

The purpose of this module is to generate new metadata that narrow down the scope of identification of a query image within, at most, two types of structurally similar images when that distance computation evaluates the single query image as similar to several different types of sample images. This module involves two steps. The first step extracts common indexes within conflicting indexes of all sample images that default indexing inappropriately evaluates structurally similar to query images. Common indexes constitute a major class that includes a subclass of the discussed sample images and other subclasses of sample images that include the common indexes but are not evaluated as similar to the query images. The second step extracts distinctive indexes from sample indexes in the latter other subclasses of sample images that include the common indexes but are not evaluated as similar to the query images. In the major class, distinctive indexes classify the query images into the former subclass of sample images and differentiate those query images from the latter other subclasses of sample images (see Figure 3).

Figure 3. Exemplary major class and subclass categorization

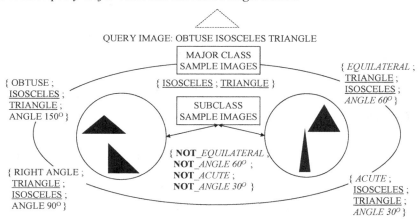

Example

As shown in Table 2, a query image of an obtuse isosceles triangle is evaluated as similar to two sample images: right angle triangle and obtuse isosceles triangle. In the instant case, the two sample images have 0.00 as structural distance values, 1.00 as the values after measure conversion, and 0.50 as normalized values to the single query image. First, this process extracts common indexes from the indexes that tentatively are generated in the default indexing. Those common indexes represent "greatest common measure" of generated indexes that correspond to the same largest size of normalized values to the single query image of the obtuse isosceles triangle as TRIANGLE and ISOSCELES. Other sample images that have normalized values lower than the largest normalized values contain those common indexes. Sample images of an equilateral triangle and an acute isosceles triangle contain those two common indexes: here, TRIANGLE and ISOSCELES within their sample indexes. Those two indexes constitute a group of ISOSCELES TRIANGLE that represents a major class, including right angle triangle and obtuse isosceles triangle with equilateral triangle and acute isosceles triangle. Second, this process finds distinctive factors in addition to common indexes, as outlined in Figure 3. Those two sample images with the lower normalized values also have other sample indexes that are different from the common indexes: ACUTE, EQUILATERAL, ANGLE 30°, and ANGLE 60°. Those various sample indexes work as distinctive factors that constitute a group to represent a subclass within the major class. Distinctive factors are added with a negative sign (e.g., NOT(~)) to constitute distinctive indexes against common indexes: { NOT(~) ANGLE 30°; NOT(~) EQUILATERAL; NOT(~) ACUTE; NOT(~) ANGLE 60°}.

In expansive indexing, a combination of common indexes and distinctive indexes constitutes a set of new metadata that represent a subclass of right angle and obtuse isosceles triangles, which is different from the other subclass of acute isosceles and equilateral triangles within the major class of isosceles triangles.

In Section Four, we provide our method and its implemented system with mathematical formulation.

Formulation of Adaptive Metadata Generation

In this section, we mathematically formulate the processes of the proposed automatic adaptive metadata generation method and its implemented system.

Preprocess: Collection of Sample Images and Sample Indexes

This preprocess is to collect sample images and sample indexes in the following four steps: (1) collections of query images, (2) collections of sample images, (3) collections of sample indexes, and (4) combinations of sample indexes with sample images.

Let define related sets of query and sample images and sample indexes:

Definition 1. (A set of query images)

As i is a continuous natural number, then $\{ I_i \}$ is defined as a set of query images:

$$\{ I_i \} \equiv \{ y \mid y = i , i \in N \} \tag{1}$$

Definition 2. (A set of sample images)

As j is a continuous natural number, then $\{ \hat{I}_j \}$ is defined as a set of sample images:

$$\{ \hat{I}_j \} \equiv \{ x \mid x = j , j \in N \} \tag{2}$$

Definition 3. (A set of sample indexes)

As j and k are continuous natural numbers, respectively, then $\{ \hat{W}_{jk} \}$ is defined as a set of sample indexes:

$$\{ \hat{W}_{jk} \} \equiv \{ z \mid z = \prod_{j \in N} \{ j \times \{ \prod_{k \in N} k \} \} , j \in N , k \in N \} \tag{3}$$

Here, $\prod$ represents production operation (same as follows).

Let define combinations of sample indexes with sample images:

Definition 4. (Combinations of sample indexes with sample images)

$$\{ \hat{I}_j \hat{W}_j \} \equiv \{ \prod_{j \in N} \{ \hat{W}_j \times \hat{I}_j \} \} \tag{4}$$

Preprocess: Distance Computation

Let define a set of structural distance values and a function of distance computation process as follows:

Definition 5. (A set of structural distance values)

$$\{ D_{ij} \} \equiv \{ d \mid d = d_{ij} \in R : \{ i_j \} = \{ \prod_{i \in N} \{ i \times \{ \prod_{j \in N} j \} \} \} \} \tag{5}$$

Definition 6. (A function of distance computation between query images and sample images with Sample Indexes)

$$\phi : \{ \prod_{i \in N} \{ I_i \times \{ \prod_{j \in N} \{ \hat{I}_j \, \hat{W}_j \} \} \} \} \to \{ D_{ij} \} \tag{6}$$

Here, $\to$ represents function operation.

Screen Module: Measure Conversion Process

Let define values after the measure conversion by using the first normalization of reciprocal structural distance values, as follows:

Definition 7. (Values after measure conversion)

$$S_{ij} \equiv \begin{cases} 1 & (when\ D_{ij} = 0) \\ \dfrac{1/D_{ij}}{\sum_{j \in N} 1/D_{ij}} & (otherwise) . \end{cases} \tag{7}$$

Here, $\sum$ represents summation operation. Also, subject to the following computation rules:

$$1/0 \equiv \infty \ ; \ \infty/\infty \equiv 1 \ ; \ 0/0 \equiv 1 \tag{8}$$

Here, the distance computation is converted into an intuitive measure on similarity.

Screen Module: Minimum Thresholding Process

Let define values after the minimum thresholding the results of measure conversion with a certain selected threshold value t :

Definition 8. (Values after the minimum thresholding)

$$\acute{S}_{ij} \equiv \begin{cases} 0 & (when \ ^{\exists}S_{ij} < t = 1 / [\ (\#\{ \ \hat{I}_j \ \} + 1 \ ) / 2] \ ) \\ \\ S_{ij} & (otherwise) \ . \end{cases} \tag{9}$$

Here, [•] represents Gauss' symbol. This thresholding works as an automatic noise cutoff mechanism.

Metadata Generation Module: Normalization Process

Let define values after normalizing the results of the screen module:

Definition 9. (Normalized values)

$$|| \acute{S}_{ij} || \equiv \frac{\acute{S}_{ij}}{\sum_{j \in N} \acute{S}_{ij} \ .} \tag{10}$$

Here, all the values are normalized to represent numerical indicators of proportional structural properties.

Metadata Generation Module: Maximum Thresholding Process

Let define values after the maximum thresholding of the results of normalization with a certain selected threshold value T:

Definition 10. (values after the maximum thresholding)

$$|| \acute{S}_{ij} || \equiv \frac{\acute{S}_{ij}}{\sum_{j \in N} \acute{S}_{ij} \ .} \tag{11}$$

Here, maximum thresholding works as emphasizes the values that represent meaningful structural properties.

Metadata Generation Module: Default Indexing Process

Let define combinations of the normalized values after the maximum thresholding with sample indexes in the default indexing:

Definition 11. (combinations of the values after the maximum thresholding with sample indexes in the default indexing)

$$\| \check{S}_{ij} \| \equiv \begin{cases} 0 & (when \,^{\exists} \| \acute{S}_{ij} \| < T ; 0 < T \leq 1) \\ 1 & (\text{otherwise}) \end{cases} \tag{12}$$

$$\{ L_{ijk} \} \equiv \{ \prod_{i \in N} \{ \prod_{j \in N} \{ \| \check{S}_{ij} \| \} \times \{ \prod_{j \in N} \{ \prod_{k \in N} \{ \hat{W}_{jk} \} \} \} \} \} \tag{13}$$

Let define new metadata generated in the default indexing:

Definition 12. (new metadata generated in the default indexing)

$$\{ M_i \} \equiv \begin{cases} L_{ijk} \text{ for } ^{\exists} j, \text{ such that } \| \check{S}_{ij} \| = 1 & (when \sum_{j \in N} \{ \| \check{S}_{ij} \| \} = 1) \ (for \,^{\exists} i) \\ 0 & (\text{otherwise}) . \end{cases}$$

$$\tag{14}$$

Expansive Indexing Module

Let define common indexes generated in the expansive indexing:

Definition 13. (common indexes)

$$\{ L^c_i \} \equiv \begin{cases} \{ \cap_j \{ L_{ijk} \} \text{ for } ^{\forall} j, \text{ such that } \| \check{S}_{ij} \| = 1 & (when \sum_{j \in N} \{ \| \check{S}_{ij} \| \} \geq \\ 0 & (\text{otherwise}) . \end{cases} \tag{15}$$

Here, $\cap$ represents cap product operation.

Let define distinctive indexes in the expansive indexing:

Definition 14. (distinctive indexes)

$$L^d_i \equiv \begin{cases} \sim\{\{\cup_j\{L_{ijk}\}\}\setminus\{L^c_i\}\} \text{ for } {}^\forall j, \text{ such that } \{\{L_{ijk}\}\cap\{L^c_i\}\}\neq\emptyset\ (when\ \{L^c_i\}\neq\emptyset) \\ 0 \qquad\qquad\qquad\qquad\qquad (otherwise) . \end{cases} \quad (for\ {}^\exists i)$$

(16)

Here $\cup$ represents cup product operation. The $\sim$ over a set represents special operation as each element in the set has an index, NOT ($\sim$), as a negative sign.

Let define a set of newly generated metadata in the expansive indexing by combining common indexes and distinctive indexes with metadata generated in the default indexing:

Definition 15. (a set of new metadata generated in the expansive indexing)

$$\{M'_i\} \equiv \{\{M_i\}\cup\{L^c_i\}\cup\{L^d_i\}\} \quad (17)$$

As shown previously, our proposed automatic adaptive metadata generation provides visual objects as queries with new metadata by bridging content descriptors that represent the structural similarity proportion of the textual indexes of geometric sample figures.

In Section Five, we present a prototype system that is based on the proposed formulation by employing general image processing techniques.

System Implementation

We have implemented our method based on a typical CBIR system that is QBIC enclosed in the DB2 package (Flickner et al., 1995; Ashley et al., 1995). Although the selected system offers feature extraction functions not only on texture but also on color histogram of visual objects, we have focused just on texture in our prototype implementation. The entire system is implemented in a local host machine (CPU 600 MHz; Memory 128 MB). The entire algorithm is programmed in Java and PHP. The images and indexes are transferred from sample and query image databases and a sample index database to a CBIR system (here, QBIC) by booting a server (here, Apache) in the local host machine. In databases, images are transferred via JDBC to the CBIR system. After cataloging the images stored in the CBIR system, each path-linkage locates the images, and the CBIR system defines their data structure. The CBIR system computes and returns structural distance values between query images and sample images. Those returned values are normalized and combined with sample indexes to generate new metadata by our algorithms programmed in PHP. The results are delivered as outputs in HTML format.

In Section Six, we describe several experiments on performance evaluation of our prototype system.

Experiments

The implemented system has generated proper metadata to a total of 4,420 query images of basic figures at 61.0% recall ratio and 20.0% fallout ratio in 9.821 seconds per image in gross average: 8.762 for indexing and 1.059 seconds for distance computation. Then, 884 original images were used repeatedly to generate query images for a series of several experiments. This performance shows that our system has improved metadata generation by 23.5% on recall ratio and 37.0% on fallout ratio rather than just using content analysis.

In the experiments, we first evaluate the bottom line performance of the metadata generation system using just content analysis available in CBIR as a benchmark. As experimental objects, we have prepared 10 types of basic geometric figures for generating a set of sample images (e.g., triangle, rectangle, ellipse, etc.). In addition, we have used more various types of practical figures for the experiments of this research. Query images automatically are generated by translating or affine-transporting, scaling by three times or thickening, or adding sparseness to those 10 types of basic geometric figures. Geometric figures are proper to evaluate the direct identification of object images for metadata generation in the case of just using CBIR, and simple sample indexes work as objective measures in its bottom-line performance evaluation of proper indexing. We then evaluate whether metadata generation is improved by the formulated three modules within a reasonable elapsed time.

As a prototype implementation, the bottom line for its performance evaluation is whether the implemented system automatically generates metadata to requested query images as textual indexes of "objective appropriateness" without any models on categorization in a specific domain and within a reasonable elapsed time. Actual images of scenery pictures (e.g., an image of the rising sun over an ocean) are divided into several parts of geometric figures: ellipse, trapezoid, and so forth. That image could be indexed as Sun, Ocean, or Horizon so that a single metadata selection and description to the image among several candidates of sophisticated metadata inevitably brings subjective factors into performance evaluation (Layne, 1994). The discussed geometric figures provide sample images with indexes of clear definition on individual structural elements of actual images. This objective indexing to sample images provides performance evaluation with objective measure on appropriate indexing.

Performance Evaluation

We have measured how many new metadata follow mathematical definitions of query images in five cases: without the screen module (a benchmark as the bottom-line performance), applying the screen module, the metadata generation module with the maximum threshold 0.40, the same module with the maximum threshold 0.95, and the expansive indexing module. The experiments have been processed to query images in bitmap and jpeg formats. Table 3 describes the results. Query images are provided with their proper metadata when they are indexed with the metadata that follow mathematical definitions of geometric figures represented in the query images. Here, performance is evaluated in recall and fallout ratios.

152   Sasaki & Kiyoki

Table 3. Performance evaluation (*...improved/-...deteriorated)

Image Format/ Recall and Fallout	b/f Screen (benchmark as bottom line)	a/f Screen	Metadata Generation (Threshold 0.40)	Metadata Generation (Threshold 0.95)	Expansive Indexing
BMP Recall Ratio	0.44	0.44	0.44	0.41-	0.61*
BMP Fallout Ratio	0.56	0.36*	0.36	0.12*	0.12*
JPEG Recall Ratio	0.32	0.32	0.32	0.23-	0.61*
JPEG Fallout Ratio	0.68	0.61*	0.60*	0.44*	0.38*

Let define the recall ratio:

Definition 16. (recall ratio)

$$
\text{Recall Ratio} \equiv
\begin{cases}
0 & (\textit{when} \text{ no metadata generated}) \\
\dfrac{\#\text{query images w/ proper metadata}}{\#\text{ all the query images}} & (\text{otherwise}).
\end{cases}
\tag{18}
$$

Let define the fallout ratio:

Definition 17. (fallout ratio)

$$
\text{Fallout Ratio} \equiv
\begin{cases}
0 & (\textit{when} \text{ no metadata generated}) \\
\dfrac{\#\text{query images w/ improper metadata}}{\#\text{ query images w/ metadata}} & (\text{otherwise}).
\end{cases}
\tag{19}
$$

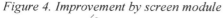

Figure 4. Improvement by screen module

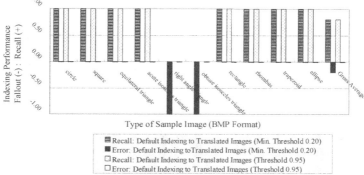

Figure 5. Improvement by maximum thresholding process

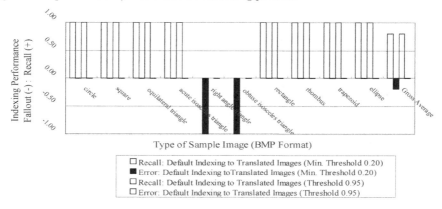

Improvement by Screen Module

The direct use of the results of content analysis does not allow proper metadata generation even in exploiting databases of sample images. The screen module cuts off noise information after the measure conversion process and improves the fallout ratio of metadata generation results with its minimum thresholding. Figure 4 describes its typical improved case of scaled images in jpeg format. Overall, the screen module has improved the average fallout ratio by 13.5% from 62.0% to 48.5%.

Improvement by Maximum Thresholding

Even after the screen module, the normalized values do not always properly represent meaningful structural properties of the query images. The maximum thresholding in the metadata generation module emphasizes the only meaningful values. In the applications of

Figure 6. Improvement by expansive indexing module

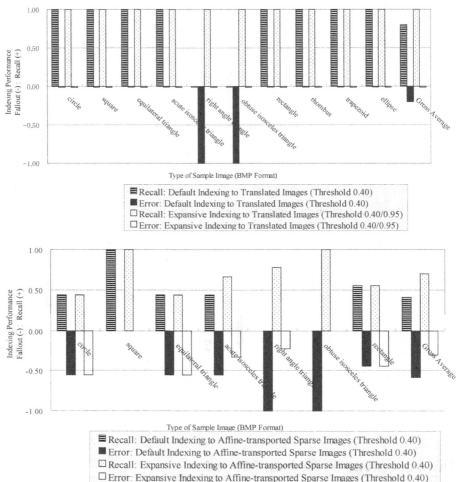

the lower threshold 0.40, the average fallout ratio on the images in jpeg format is improved by only 1.00%. In the applications of the higher threshold 0.95, the average fallout ratios on the images in bitmap and jpeg formats are improved by 24.0% and 17.0%, compared with those ratios after the screen module. Meanwhile, the recall ratios on the same two types of images are deteriorated by 3.00% and 9.00%, respectively.

Figure 5 describes its typical improved case of translated images in bitmap format without any deterioration on its average recall ratio. Overall, the maximum thresholding process has improved the average fallout ratios in the case of threshold 0.40 by 0.50% from 48.5% to 48.0%, and in the case of threshold 0.95 by 20.5% from 48.5% to 28.0%. Meanwhile, the recall ratio of the latter case is deteriorated by 6.00% from 38.0% to 32.0%.

Improvement by Expansive Indexing

Overall, the expansive indexing module has improved the average recall and fallout ratios by 23.0% from 38.0% to 61.0% and 23.0% from 48.0% to 25.0% after the maximum thresholding process of the threshold 0.40 instead of 0.95 in the metadata generation module, respectively. Figure 6 describes its typical improved cases of the translated images and the affine-transported sparse images in bitmap format.

In Section Seven, we discuss the experimental results.

Discussion

In the experiments, we have confirmed the bottom-line performance for metadata generation using just content analysis as a benchmark, which is shown in the result before the application of the screen module as only 38.0% recall ratio and 62.0% fallout ratio overall. First, our system automatically has screened out improper query images for metadata generation and has improved the average fallout ratio by 13.5% overall. Second, our system has improved the overall performance by 23.0% on both the recall and fallout ratios with more practical types of experimental figures. The experimental results assure that our system is applicable in specific domains with well-selected sample images that represent typical structural properties of images in the object domains. Overall, indexing performance has been improved by thresholding adjustment and expansive indexing with common and distinctive indexes. Artificial objects like architectures and drawings often consist of squares and rectangles with clear angles. The experimental results assure that our prototype system has feasibility as metadata generation to digital images of geometric figures.

In Section Eight, we conclude this chapter with our perspective in this research field.

Conclusion

We have presented an automatic metadata generation method and its implemented system using content analysis available in CBIR and databases of sample images with their mathematical formulation and experiments. Our proposed method and its implemented system consist of three modules. The first module screens out improper object images as queries for metadata generation using CBIR by automatically selecting proper threshold values. The second module generates metadata by selecting sample indexes attached to sample images that are structurally similar to object images as queries by optimizing the thresholding processes of the results of content analysis of the sample images and the object images as queries. The third module detects improper metadata and regenerates proper metadata by identifying common and distinctive indexes in wrongly selected metadata. We have realized a cost-efficient implementation of the proposed method using a typical package solution of CBIR with databases. Our system has generated proper metadata to the images of basic figures at 61.0% recall ratio and 20.0% fallout ratio overall. This result has been bettered off by 23.5% on recall ratio and 37.0% on fallout ratio than just using content analysis for metadata generation.

We have implemented the proposed method as an intuitive solution for automatic adaptive metadata generation to digital images by exemplary textual indexing to geometric structural properties of visual objects. Our proposed system has its extensibility to bridge database technology with content analysis in the field of metadata generation for keyword-based retrieval. The incorporation of sophisticated text-based analysis of metadata into our system is to allow the automatically generated metadata to be used as a source to create more sophisticated conceptual metadata, which are able to reflect external and expert wisdom that is accessible in specified domains. Our method with a well-defined set of sample indexes is, therefore, feasible as a general approach of metadata generation by textual indexing to images.

Content-based metadata indexing is an indispensable technique to semantic visual information retrieval. One of the most promising approaches is the combination of multimedia database technology for precise and speedy keyword-based image retrieval and empowered content analysis in computer vision. Our proposed approach is to realize a simple implementation of content-based metadata indexing as a solution for connecting textual information to content descriptors of visual objects to realize precise keyword-based image retrieval.

Acknowledgments

This research is financially supported in part by the Moritani Scholarship Foundation.

References

Amato, G., Gennaro, C., & Savino, P. (2001). Indexing and editing metadata for documentary films on line: The ECHO digital library. In *Proceedings of the International Cultural Heritage Informatics Meeting (ICHIM 01)*, Milan, Italy.

Ashley, J., et al. (1995). The query by image content (QBIC) system. *ACM SIGMOD Record, 36*(7), 475.

Barros, J., French, J., Martin, W., Kelly, P., & White, J. M. (1994). *Indexing multispectral images for content based retrieval* (Tech. Rep. No. CS-94-40). University of Virginia.

Blei, D. M., & Jordan, M. I. (2003). Modeling annotated data. In *Proceedings of the 26th ACM SIGIR*, Toronto, Canada (pp. 127–134).

Bontcheva, K., Maynard, D., Cunningham, H., & Saggion, H. (2002). Using human language technology for automatic annotation and indexing of digital library content. In *Proceedings of the 6th European Conference on Digital Libraries (ECDL 2002)*, Rome, Italy.

Brown, M. G., Foote, J. T., Jones, G. J. F., Jones, K. S., & Young, S. J. (1995). Automatic content-based retrieval of broadcast news. In *Proceedings of the ACM Multimedia* (pp. 35–44).

Chu, W. W., Ieong, I. T., & Taira, R. K. (1994). A semantic modeling approach for image retrieval by content. *VLDB Journal, 3*, 445–477.

Doan, B., Beigbeder, M., Girardot, J., & Jaillon, P. (1998). Using metadata to improve organization and information retrieval on the WWW. In *Proceedings of WebNet98*, Orlando, FL.

Dolin, R. H., et al. (2001). Kaiser Permanente's "metadata-driven" national clinical intranet. In *Proceedings of the MEDINFO 2001*, Amsterdam (pp. 319–323).

Duygulu, P., Barnard, K., de Freitas, J. F. G., & Forsyth, D. A. (2002). Object recognition as machine translation: Learning a lexicon for a fixed image vocabulary. In *Proceedings of the 7th European Conference on Computer Vision,* Copenhagen, Denmark (pp. 97–112).

Flickner, M., et al. (1995). Query by image and video content: The QBIC system. *IEEE Computer, 28*(9), 23–32.

Gey, F., et al. (2002). Advanced search technologies for unfamiliar metadata. In *Proceedings of the 25th Annual International ACM SIGIR Conference on Research and Development in Information Retrieval (SIGIR 2002),* Tampere, Finland (pp. 455–456).

Kiyoki, Y., Kitagawa, Y., & Hayama, T. (1994). A metadatabase system for semantic image search by a mathematical model of meaning. *ACM SIGMOD Record, 23*(4), 34–41.

Kobayashi, M., & Takeda, K. (2000). Information retrieval on the Web. *ACM Computing Surveys, 32*(2), 144–173.

Larson, R. R., Gey, F., & Chen, A. (2002). Harvesting translingual vocabulary mappings for multilingual digital libraries. In *Proceedings of the 2nd ACM/IEEE-CS Joint Conference on Digital libraries (JCDL 2002),* Portland, OR (pp. 185–190).

Layne, S. S. (1994). Some issues in the indexing of images. *Journal of the American Society for Information Science, 45*(8), 583–588.

Magrin-Chagnolleau, I. (2000). Indexing telephone conversations by speakers. In *Proceedings of the 2000 IEEE International Conference on Multimedia and Expo (ICME),* New York (pp. 881–884).

Marshall, C. C. (1998). Making metadata: A study of metadata creation for a mixed physical-digital collection. In *Proceedings of the 3rd ACM International Conference on Digital Libraries (ACM DL),* Pittsburgh, PA (pp. 162–171).

Moreno, P. J., van Thong, J. M., Logan, E., & Jones, G. (2002). From multimedia retrieval to knowledge management. *IEEE Computer, 35*(4), 58–66.

Ortega, M., Rui, Y., Chakrabarti, K., Mehrotra, S., & Huang, T. S. (1997). Supporting similarity queries in MARS. In *Proceedings of the ACM Multimedia* (pp. 403–413).

Rui, Y., Huang, T. S., & Chang, S. F. (1999). Image retrieval: Current techniques, promising directions and open issues. *Journal of Visual Communication and Image Representation, 10*(4), 39–62.

Rusch-Feja, D. (1998). Metadata: Standards for retrieving WWW documents (and other digitized and non-digitized resources). In U. Grothkopf, H. Andernach, S. Stevens-Rayburn, & M. Gomez (Eds.), *Library and information services in astronomy III ASP conference series* (Vol. 153, pp. 157–165).

Sasaki, H., & Kiyoki, Y. (2004). A prototype implementation of metadata generation for image retrieval. In *Proceedings of the 2004 International Symposium on Applications and the Internet—Workshops (SAINT 2004 Workshops)*, Tokyo, Japan (pp. 460–466).

Sasaki, H., & Kiyoki, Y. (2005a). Automatic adaptive metadata generation for image retrieval. In *Proceedings of the 2005 International Symposium on Applications and the Internet—Workshops (SAINT 2005 Workshops)*, Trento, Italy (pp. 426–429).

Sasaki, H., & Kiyoki, Y. (2005b). A prototype implementation of adaptive metadata generation to digital images. *Information Modelling and Knowledge Bases, 16*, 134–151.

Sasaki, H., & Kiyoki, Y. (2006). Theory and implementation on automatic adaptive metadata generation for image retrieval. *Information Modelling and Knowledge Bases, 17*, 68–82.

Sebastiani, F. (2002). Machine learning in automated text categorization. *ACM Computing Surveys, 34*(1), 1–47.

Sheth, A., & Klas, W. (1998). *Multimedia data management: Using metadata to integrate and apply digital media.* New York: McGraw-Hill.

Smeulders, A. W. M., Worring, M., Santini, S., Gupta, A., & Jain, R. (2000). Content-based image retrieval at the end of the early years. *IEEE Transactions on Pattern Analysis and Machine Intelligence, 22*(12), 1349–1380.

Smith, D. A., Mahoney, A., & Crane, G. (2002). Integrating harvesting into digital library content. In *Proceedings of the 2nd ACM/IEEE-CS Joint Conference on Digital Libraries (JCDL 2002)*, Portland, OR (pp. 183–184).

Smith, J. R., & Chang, S. F. (1997). Visually searching the Web for content. *IEEE Multimedia, 4*(3), 12–20.

Srihari, R. K., Zhang, Z., & Rao, A. (2000). Intelligent indexing and semantic retrieval of multimodal documents. *Information Retrieval, 2*(2/3), 245–275.

Tsymbalenko, Y., & Munson, E. V. (2001). Using HTML metadata to find relevant images on the World Wide Web. In *Proceedings of Internet Computing 2001*, Las Vegas, NV (pp. 842–848).

Wilcox, L., & Boreczky, J. S. (1998). Annotation and segmentation for multimedia indexing and retrieval. In *Proceedings of the 31st Annual Hawaii International Conference on System Sciences (HICSS)* (pp. 259–266).

Yee, K. P., Swearingen, K., Li, K., & Hearst, M. (2003). Faceted metadata for image search and browsing. In *Proceedings of the ACM CHI 2003*, Ft. Lauderdale, FL.

Yoshitaka, A., & Ichikawa, T. (1999). A survey on content-based retrieval for multimedia databases. *IEEE Transactions on Knowledge and Data Engineering, 11*(1), 81–93.

Section IV

Human-Computer
Interaction

Chapter VIII

Interaction Models and Relevance Feedback in Image Retrieval

Daniel Heesch, Imperial College London, UK

Stefan Rüger, Imperial College London, UK

Abstract

Human-computer interaction is increasingly recognized as an indispensable component of image retrieval systems. A typical form of interaction is relevance feedback in which users supply relevance information on the retrieved images. This information can subsequently be used to optimize retrieval parameters. The first part of the chapter provides a comprehensive review of existing relevance feedback techniques and also discusses a number of limitations that can be addressed more successfully in a browsing framework. Browsing models form the focus of the second part of this chapter in which we will evaluate the merit of hierarchical structures and networks for interactive image search. This exposition aims to provide enough detail to enable the practitioner to implement many of the techniques and to find numerous pointers to the relevant literature otherwise.

Introduction

Similarity in appearance often is revealing about other and, potentially, much deeper functional and causal commonalities among objects, events, and situations. Things that are similar in some respect are likely to behave in similar ways and to owe their existence to similar causes. It is because of this regularity that similarity is fundamental to many cognitive tasks such as concept learning, object recognition, and inductive inference.

Similarity-based reasoning requires efficient modes of retrieval. It is perhaps only in experimental settings that subjects have direct sensory access to the patterns that they are asked to compare. In most situations, an observed pattern is evaluated by comparing it with patterns stored in memory. The efficiency with which we can classify and recognize objects suggests that the retrieval process itself is based on similarity. According to Steven Wolfram (2004), the use of memory "underlies almost every major aspect of human thinking. Capabilities such as generalization, analogy and intuition immediately seem very closely related to the ability to retrieve data from memory on the basis of similarity." He extends the ambit of similarity-based retrieval to the domain of logical reasoning, which ultimately involves little more than "retrieving patterns of logical argument that we have learned from experience" (p. 627).

The notion of similarity clearly is not without problems. Objects may be similar on account of factors that are merely accidental and that, in fact, shed no light on the relationship that one potentially could unveil. The problem of measuring similarity largely reduces, therefore, to one of defining the set of features that matter. The problem of estimating the relative significance of various features is one of information retrieval in general. However, it is greatly compounded in the case of image retrieval in two significant ways. First, documents readily suggest a representation in terms of their constituent words. Images do not suggest such a natural decomposition into semantic atoms with the effect that image representations, to some extent, are arbitrary. Second, images typically admit to a multitude of different meanings. Each semantic facet has its own set of supporting visual features, and a user may be interested in any one of them.

These challenges traditionally have been studied in the context of the query-by-example paradigm (QBE). In this setting, the primary role of users is to formulate a query; the actual search is taken care of by the computer. This division of roles has its justification in the observation that the search is the computationally most intensive part of the process but is questionable on the grounds that the task of recognizing relevance still is solved best by the human user. The introduction of relevance feedback into QBE systems turns the problem of parameter learning into a supervised learning problem. Feedback on retrieved images can help to find relevant features or better query representations. Although the incorporation of relevance feedback techniques can result in substantial performance gains, it does not overcome the more fundamental limitations of the QBE framework in which they have been formulated. Often, users may not have an information need in the first place and may wish to explore an image collection. Moreover, the presence of an information need does not mean that a query image is readily at hand to describe it. Also, brute force nearest neighbor search is linear in the collection size, and the sublinear performance achieved through hierarchical indexing schemes does not extend to high-dimensional feature spaces with more than 10 dimensions.

Browsing provides an interesting alternative to QBE but, by comparison, has received surprisingly scant attention. Browsing models for image search tends to cast the collection into some structure that can be navigated interactively. Arguably, one of the greatest difficulties of the browsing approach is to identify structures that are conducive to effective search in the sense that they support fast navigation, provide a meaningful neighborhood for choosing a browsing path, and allow users to position themselves in an area of interest.

The first part of this chapter will examine relevance feedback models in the QBE setting. After a brief interlude in which we discuss limitations of the QBE framework, we shift the focus to browsing models that promise to address at least some of these. Each section concludes with a summary table that juxtaposes many of the works that have been discussed.

Query-by-Example Search

Query-by-example systems return a ranked list of images based on similarity to a query image. Relevance feedback in this setting involves users labeling retrieved images, depending on their perceived degree of relevance. Relevance feedback techniques vary along several dimensions, which makes any linear exposition somewhat arbitrary. We structure the survey according to how relevance feedback is used to update system parameters: Query adaptation utilizes relevance information to compute a new query for the next round of retrieval. Metric optimization involves an update of the distance function that is used to compute the visual similarities between the query and database images. Classification involves finding a decision function that optimally separates relevant from nonrelevant images.

Query Adaptation

Query adaptation describes the process in which the representation of an initial query is modified automatically, based on relevance feedback. Query adaptation was among the first relevance feedback techniques developed for text retrieval (Salton & McGill, 1982) and since has been adapted to image retrieval (Rui, Huang, & Mehrotra, 1997; Ishikawa, Subramanya, & Faloutsos, 1998; Porkaew, Chakrabarti, & Mehrotra, 1999; Zhang & Su, 2001; Aggarwal, Ashwin, & Ghosal, 2002; Urban & van Rijsbergen, 2003; Kim & Chung, 2003). The two most important types of query adaptation are query point moving and query expansion. We will be dealing with each in turn.

Query Point Moving

Query point moving is a simple concept and illustrated in Figure 1. Relevant (+) and nonrelevant (o) objects are displayed in a two-dimensional feature space with the query initially being in the bottom-right quadrant (left plot). The images marked by the user correspond to the bold circles. The goal of query point moving is to move the query point toward the relevant images and away from the nonrelevant images. Clearly, if relevant images form

Figure 1. Moving the query point toward positive examples

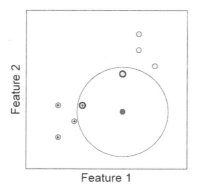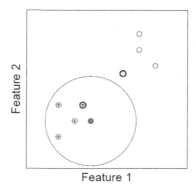

clusters in feature space, this technique should improve retrieval performance in the next step (right plot). Techniques differ in how this movement in feature space is achieved.

In Urban and van Rijsbergen (2003), for example, images are represented in terms of a set of keywords and a color histogram. Given a set of images for which the user has indicated some degree of relevance, the visual part of the query is computed as the weighted average over the relevant images. Meanwhile, Rui et al. (1997) alter the representation of the query using both relevant and nonrelevant images. The method employs Rocchio's (1971) formula originally developed for text retrieval. In particular, given sets R and N of feature vectors of relevant and nonrelevant images, respectively, the learned query vector is computed as:

$$q(t+1) = \alpha q(t) + \beta \left(\frac{1}{|R|} \sum_{x \in R} x \right) - \gamma \left(\frac{1}{|N|} \sum_{x \in N} x \right),$$

in which α, β and γ are parameters to be chosen. For $\alpha = \gamma = 0$, the new query representation is the centroid of the relevant images. The goal of the method is to move the query point closer toward relevant images and further away from nonrelevant images.

Yet another approach is taken by Ishikawa et al. (1998) and Rui and Huang (2000), who find the best query point as that point that minimizes the summed distances to all relevant images. The optimal query representation turns out to be the weighted average of the representations of all relevant images.

Query Expansion

Query point moving suffers from a notable limitation: If relevant images form visually distinct subsets (corresponding to multiple clusters in feature space), the technique easily may fail to move the query into a better position, as relevant images suggest multiple and mutually conflicting directions. The problem arises from the requirement to cover multiple clusters with only one query. A simple modification of the previous approach that alleviates this problem involves replacing the original query point by multiple query points, each located near different subsets of the relevant images. This modification turns query point moving into what more aptly may be described as query expansion (Porkaew et al., 1999; Kim & Chung, 2003; Urban & Jose, 2004a). Differences among techniques are down to details; in particular, to the question of how to choose the precise locations of the query points.

In Porkaew et al. (1999), for example, relevant images are clustered, and the cluster centroids are chosen as new query points. The overall distance of an image to the multi-point query is computed as the weighted average over the distances to each query point, i.e:

$$D(x,Q) = \sum_{q \in Q} w_q d(x,q) \,.$$

The weights are taken to be proportional to the number of relevant images in each cluster. Thus, query points that seem to represent better the user's need have a greater weight in the overall distance computation. One should note, however, that this scheme retains the feature it seeks to overcome by linearly averaging over individual distances. In fact, it can be shown that the overall result is equivalent to query point moving in which the new query point is given by

$$\sum_{q \in Q} w_q q \,.$$

If d is taken to be the Euclidean distance, for example, then the iso-distance lines for a multi-point query remain circles now centered at the new query point.

This is shown on the left plot in Figure 2. The method reduces, therefore, to the solution suggested by Ishikawa et al. (1998) and Rui and Huang (2000). In order to properly account for the cluster structure, it seems more reasonable to treat the multi-point query as a disjunctive query, an approach taken in more recent works by Wu, Faloutsos, Sycara, and Payne (2000), Kim and Chung (2003), and Urban and Jose (2004a). In the second work, for example, the others suggest a remedy in the form of:

$$D(x,Q) = \sum_{q \in Q} d(x,q)^\alpha$$

Figure 2. Multi-point queries: From query point moving toward disjunctive queries

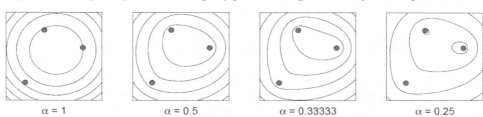

| $\alpha = 1$ | $\alpha = 0.5$ | $\alpha = 0.33333$ | $\alpha = 0.25$ |

in which q is the representation of a cluster and α is a parameter that confers the desired nonconvexity. The iso-distance lines of the model with α ranging from 1 to 1/4 are shown in Figure 2. For $\alpha = 1$, the model reduces to that of Porkaew et al. (1999).

The aforementioned methods have in common that they are concerned with combining distance scores. An alternative approach to multi-point queries is taken by Urban and Jose (2004a), who adapt a rank-based aggregation method, median-rank aggregation (Fagin, Kumar, & Sivakumar, 2003), to multi-point image queries and establish superior performance over the simple score-based method of Porkaew et al. (1999). The rank of an image p is the number of images whose distance scores are smaller or equal to that of p. This particular method involves computing the image ranks with respect to each of the various query points. The median of those ranks becomes the final rank assigned to that image. If an image has a large distance to only a small number of query points, these will have no effect on the final rank of that image. This provides support for more disjunctive queries, but the method is equally robust against unusually small distances.

Distance Metric Optimization

Introduction

The second large group of relevance feedback techniques is concerned with modifying the distance metric that is used to compute the visual similarities between the query and the database images. As noted earlier, one of the problems pertaining to the notion of similarity relates to the question of how to weigh various features. A feature that is good at capturing the fractal characteristics of natural scenes may not be good for distinguishing between yellow and pink roses. How can we infer from relevance feedback which feature is important and which one isn't?

A naïve method of computing the distance between two representations is to concatenate all individual feature vectors into one and measure the distance between two vectors x and y. In hierarchical models, distances are computed between individual features, and the resulting distances are aggregated. In both models, we have to compute the distance between two vectors. In image retrieval, commonly used distance metrics are instances of the general Minkowski metric:

$$D(x,y) = \left[\sum_i |x_i - y_i|^\alpha \right]^{\frac{1}{\alpha}}, \alpha > 0.$$

This reduces to the Euclidean metric for $\alpha = 2$ and to the L_1 metric for $\alpha = 1$. The advantage of the Minkowski metric is that it readily can be parameterized by adding a weight to each componentwise difference. For $\alpha = 2$, we obtain a weighted Euclidean distance:

Figure 3. Changing parameters of the metric

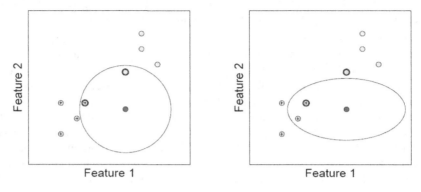

$$D(x,y) = \left[\sum_i w_i (x_i - y_i)^2 \right]^{\frac{1}{2}},$$

(1.1)

in which one typically constrains the weights to sum to one and to be nonnegative. Relevance feedback now can help to adjust these weights so that relevant images tend to be ranked higher in subsequent rounds. The idea is illustrated in Figure 3. Under the weighted Euclidean metric with equal weights, the iso-distance lines in a two-dimensional vector space are circles centered at the query vector. In this example, the one image marked *relevant* is much closer to the query with respect to the second feature. On the assumption that a relevant image has more relevant images in its proximity, we wish to discount distances along the dimension along which the relevant image differs most from the query. Here, this is achieved by decreasing the weight for feature 1 with the effect that the iso-distance lines become ellipsoids with their long axes parallel to that of the least important feature.

In hierarchical models, these distances need to be combined. By far the most popular aggregation method is the weighted linear sum that we encountered earlier in the context of multi-point queries:

$$D(x,q) = \sum_{i=1}^{k} w_i d_i (x,q),$$

(1.2)

in which we now sum over k features rather than over query points. The great majority of relevance feedback methods is concerned with adjusting weights of each individual feature component either in a flat (Ishikawa et al., 1998; Peng, Bhanu, & Qing, 1999) or in a hierarchical feature model (Sclaroff, Taycher, & La Cascia, 1997; Rui, Huang, Ortega, & Mehrotra, 1998; Schettini, Ciocca, & Gagliardi, 1999; Rui & Huang, 2000; Heesch & Rüger, 2003; Urban & Jose, 2004b). We shall refer to the two types of weights as *component weights* and *feature weights*, respectively.

Early Models

In the hierarchical model proposed by Rui et al. (1998), the weight of a component is taken to be inversely proportional to the standard deviation of that component among relevant images. This heuristic is based on the intuition that a feature component that shows great variation among the relevant images does not help to discriminate between relevant and nonrelevant images. Although any function that monotonically decreases with the variance would appear to be a good candidate, it turns out that dividing by the standard deviation agrees with the optimal solution that later was derived in the optimization framework of Ishikawa et al. (1998). The feature weights are adjusted by taking into account both negative and positive examples using simple heuristics. Although experiments suggest a substantial improvement in retrieval performance on a database containing more than 70,000 images, the figures should be treated with great caution, as the number of images on which relevance feedback is given lies in the somewhat unrealistic range of 850 to 1,100.

Optimizing Generalized Euclidean Distances

An elegant generalization of the relevance feedback method of Rui et al. (1998) was developed by Ishikawa et al. (1998). It is motivated by the observation that relevant images may not necessarily align with one of the feature dimensions, and so the weighted Euclidean distance used in Rui et al. (1998) cannot account fully for their distribution (its iso-distance lines form ellipsoids whose axes are parallel to the coordinate axes). This limitation can be addressed by considering a generalization of the Euclidean distance introduced by Chandra Mahalanobis under the name *D-statistic* in the study of biometrical data and now simply known as the Mahalanobis distance. It is more conveniently written in matrix notation as:

$$D(x, y) = (x - q)^T M (x - q)$$

in which M is any square matrix. If M is a diagonal matrix, then the expression reduces to equation (1.1) with the weights corresponding to the diagonal elements. If M is a full matrix, then the expression contains products between the differences of any two components. In two dimensions with:

$$M = \begin{bmatrix} a & b \\ c & d \end{bmatrix}$$

This writes as:

$$D(x,q) = a(x_1 - q_1)^2 + (b + c)(x_1 - q_1)(x_2 - q_2) + d(x_2 - q_2)^2$$

and similarly for higher dimensions. The iso-distance lines of the general Mahalanobis metric are ellipsoids that need not align with the coordinate axes. The components of M are found by minimizing the sum of the distances between the images marked relevant and the query. The interesting twist of the model is that the query itself is re-estimated at each step. The optimization thus determines not only M but also the query with respect to which the distances are minimized. The method integrates the two techniques of query point moving and metric update in one optimization framework. The objective function is:

$$\min_{M,q} \sum_{i=1}^{N} v_i (x^i - q) \ M (x^i - q),$$

(1.3)

in which N is the number of relevant images and v_i a relevance score given by the user. Note that x^i here denotes the vector of the ith image, not the ith component of some vector x. Under the additional constraints that $\det(M) = 1$ and that M is symmetric, the solutions for q and M are:

$$q = \frac{\sum_{i=1}^{N} v_i x_i}{\sum_{i=1}^{N} v_i}$$

and

$$M = [\det(C)]^{\frac{1}{n}} C^{-1},$$

in which C is the covariance matrix of the positive examples. In order for the inverse of C to exist, relevance feedback needs to be given on at least as many images as there are feature components. If this is not the case, a pseudo-inverse can be used instead.

Based on our preceding discussion, some of the limitations of the approach taken by Ishikawa et al. (1998) should be evident: First, the approach tackles the problem of query point moving but does not support multi-point queries; second, it exploits only positive feedback, which might be rather scarce at the beginning of the search; and third, it assumes a flat image representation model with all features for one image concatenated into one single vector. This inflates the number of parameters to be learned with the effect of rendering parameter estimation less robust.

To address the last shortcoming, Rui and Huang (2000) extend the optimization framework of Ishikawa et al. (1998) by adding feature weights. For each feature, distances are computed using the generalized Euclidean metric and the overall similarity are obtained according to equation (1.2). Like in Ishikawa et al. (1998), the aim is to minimize the summed distances between relevant images and the query. The objective function takes the form of equation (1.3) except for an additional inner sum:

$$\min_{M,q,w} \sum_{i=1}^{N} v_i \sum_{j=1}^{k} w_j (x_{ij} - q_j)^T M(x_{ij} - q_j),$$

in which, as before, v are relevance scores, w are feature weights, and x_{ij} is the jth feature vector of the ith relevant image. The optimal solutions for q and M are the same as in Ishikawa et al. (1998), while the feature weights are given by

$$w_j \propto \frac{1}{\sqrt{\sum_{i=1}^{N} v_i d(x_j, q_j)}},$$

in which the squared denominator is the sum of the weighted distances between the query and all relevant images under feature j.

Optimization with Negative Feedback

The previous methods only make use of positively labeled examples despite the fact that negative feedback repeatedly has been shown to prevent the retrieval results from converging too quickly toward local optima (Heesch & Rüger, 2002; Müller, Marchand-Maillet, & Pun, 2002; Müller, Müller, Squire, Marchand-Maillet, & Pun, 2000; Vasconcelos & Lippman, 2000). An innovative method that takes explicit account of negative examples is by Aggarwal et al. (2002). They adopt the general framework of Ishikawa et al. (1998) by minimizing equation (1.3), but the extra constraint is added that there are no nonrelevant images within some small neighborhood of q. This is achieved by automatically modifying the relevance scores v. In particular, given some solution q and M of equation (1.3) with an initial uniform set of relevance scores, the relevance score of the relevant image that is farthest from the current query point q is set to zero and the scores of any other positive image set to the sum of its quadratic distances from the negative examples. Minimizing the objective function again with the thus altered scores yields a new solution q and M, which is more likely to contain only relevant images. This scheme is iterated until the neighborhood contains only relevant images.

Another example of metric optimization involving negative feedback is by Lim, Wu, Singh, and Narasimhalu (2001). Here, users are asked to rerank retrieved images, and the system subsequently minimizes the sum of the differences between the user-given ranks and the computed ranks. Because of the integral nature of ranks, the error function is not analytic, and numerical optimization is required to find the feature weights.

A method that admits to an analytic solution is proposed in Heesch, Pickering, Yavlinsky, and Rüger (2003). Relevance feedback is given by positioning retrieved images closer to or further away from the query that originally is situated at the center (Figure 4 left and middle). The user implicitly provides a real-valued vector of new distances, and the objective

Figure 4. In search for blue doors: left: initial display with default distance metric; middle: display after user feedback; right: results with newly learned distance metric

function is the sum of the squared errors between the distances computed by the system and the new distances supplied by the user. The distance function that minimizes the objective function is used for the next retrieval step (Figure 4 right).

Multi-Dimensional Scaling

The methods discussed thus far retrieve a ranked list of images that often are organized on a two-dimensional grid or, as in Heesch et al. (2003), in the form of a spiral around the query. Crucially, however, the mutual distances within the set of retrieved images are not taken into account for the display (i.e., returned images that are visually similar may not necessarily be displayed close to each other). Rubner, Guibas, and Tomasi (1997) apply multi-dimensional scaling (Kruskal, 1964) to the search results in order to achieve a more structured view. Given a set of objects and their mutual distances, we can place each object in a high-dimensional metric space such that the distances are preserved exactly. For practical purposes, the preferred dimensionality of the space is two, for which distances only can be approximated. The result is an approximate two-dimensional embedding that preserves as far as possible the distances between objects. The technique can be applied to the set of retrieved images, but it also can be used as a means to display the entirety of small collections in a perceptually meaningful way. Navigation through the collection can be achieved by letting the user select one of the retrieved images as the new query.

Another attempt of a synthesis between automated search and browsing is described in Santini and Jain (2000) and Santini, Gupta, and Jain (2001). Similar to Rubner et al. (1997), the proposed system seeks a distance-preserving projection of the images onto two dimensions. As well as selecting an image from the display as the new query, users move images to new positions. In Santini and Jain (2000), the system finds feature and component weights that minimize the mismatch between the relations imposed by the user and the computed distances.

Similarity on Manifolds

Up to now, we only have considered global metrics. Distances for all images are computed using the same, possibly parameterized distance metric. On the assumption that relevant images fall on some manifold in the Euclidean feature space, a better approach would be to find the best local metric. He, MA, Zhang. (2004) propose to approximate the metric structure of the manifold at the location of the query. The approximation makes use of positive examples, which are assumed to be close to the query under the geodesic distance. The algorithm proceeds by computing the k-nearest neighbors of each of the positive examples. The union of these sets constitutes the set of candidates from which we eventually will retrieve. The geodesic distance is approximated by the topological distance on a graph whose vertices correspond to elements of the candidate set along with the query and the positive examples. Edges are constructed between any two images if their unweighted Euclidean distance does not exceed some threshold. The geodesic distance then is approximated by the topological distance on the graph; that is, the length of the shortest path between two images. Retrieval on the manifold returns the set of images with the smallest topological distance to the query.

Similarity Search as Classification

A third class of techniques treats the problem of similarity search as one of classification. The techniques are similar to the class of metric optimization discussed in the preceding section, and some can be interpreted as estimating parameters of some similarity function.

Probabilistic Approaches

Methods that approach the classification problem from a Bayesian perspective explicitly model probability densities. The aim of these methods is to assign class probabilities to an image based on the class-specific feature densities estimated from relevance feedback. Let p be an image, x its feature representation, and R and N the sets of relevant and nonrelevant images. By Bayes' rule we have:

$$P(p \in R \mid x) = \frac{P(x \mid p \in R) P(p \in R)}{P(x)}.$$

In Nastar, Mitschke, and Meilhac (1998), the feature density of relevant images $P(x|p \in R)$ is assumed to be Gaussian, and features are assumed to be independent so that $P(p \in R|x)$ is a product of Gaussians:

$$P(p \in R \mid x) \propto \prod_{i=1}^{k} P(x_i \mid p \in R) .$$

If we only were to consider relevant examples, the mean and standard deviation could readily be found using the principle of maximum likelihood. Nastar et al. (1998) suggest an iterative technique that takes into account negative examples. It does this by determining the proportion of negative examples falling into a 3σ confidence interval around the current mean and the proportion of positive examples falling outside of it. The error is simply the sum of the two terms. To account better for multi-modality, a mixture of Gaussians can be used, an extension that has the slight disadvantage of requiring numerical optimization for parameter estimation (Vasconcelos & Lippman, 2000; Yoon & Jayant, 2001).

Meilhac and Nastar (1999) drop the assumption of Gaussianity of feature densities and use a Parzen window for nonparametric density estimation. Feature densities are estimated for both relevant and nonrelevant images, and the decision rule is

$$I(x_i) = -\log[P(x_i \mid p \in R)] + \log[P(x_i \mid p \in N)]$$

for each feature. Assuming independence of features we obtain

$$I(x) = \sum_{i=1}^{k} (-\log[P(x_i \mid p \in R)] + \log[P(x_i \mid p \in N)])$$

The additiveness of this density estimation method allows incremental update of the decision function over a number of feedback rounds.

The Bayesian framework developed by Cox, Miller, Omohundro, and Yianilos (1998) and Cox, Miller, Minka, Papathomas, and Yianilos (2000) for target search is based on an explicit model of what users would do, given the target image they want. The system then uses Bayes' rule to predict the target, given their action.

Discriminant Classifiers

An alternative approach to classification that does not require an explicit modeling of feature densities involves finding a discriminant function that maps features to class labels using some labeled training data.

An increasingly popular classifier is the support vector machine or SVM (Vapnik, 1995). SVMs typically map the data to a higher-dimensional feature space using a possibly nonlinear transform associated to a reproducing kernel. Linear discrimination between classes then is attempted in this feature space. SVMs have a number of advantages over other classifiers that make them particularly suitable for relevance feedback methods (Chen, Zhou, & Huang, 2001; Crucianu, Ferecatu, & Boujemaa, 2004; He, Tong, Li, Zhang & Zhang., 2004; Hong & Huang, 2000; Jing, Zhang, Zhang, & Zhang, 2003; Tong & Chang, 2001). Most notably, SVMs avoid too restrictive distributional assumptions regarding the data and are flexible, as prior knowledge about the problem can be taken into account by guiding the choice of the kernel.

In the context of image retrieval, the training data consist of the relevant and nonrelevant images marked by the user. Learning classifiers reliably on such small samples is a particular challenge. One potential remedy is active learning (Cohn, Atlas, & Ladner, 1994). The central idea of active learning is that some training examples are more useful than others for training the classifier. It is guided by the more specific intuition that points close to the hyperplane; that is, regions of greater uncertainty regarding class membership are most informative and should be presented to the user for labeling instead of a random subset of unlabeled points. Applications of active learning to image retrieval are found in Tong & Chang (2001) and He, Tong, Li, Zhang & Zhang. (2004). In the former work, a support vector machine is trained over successive rounds of relevance feedback. In each round, the system displays the images closest to the current hyperplane. Once the classifier has converged, the system returns the top *k* relevant images farthest from the final hyperplane. Although the method involves the user in several rounds of potentially ungratifying feedback, the performance of the trained classifier improves over that of alternative techniques such as query point moving and query expansion.

Table 1. Overview of relevance feedback systems (Grouped according to objective and sorted chronologically within each group

Author	Objective	Type	Range
Rui et al., 1997	Query point moving	+/-	Binary
Rui et al., 1998	Query point moving	+	Real
Ishikawa et al., 1998	Query point moving	+	Real
Rui & Huang, 2000	Query point moving	+	Real
Urban et al., 2003	Query point moving	+	Binary
Porkaew et al., 1999	Query expansion	+	Binary
Kim & Chung, 2003	Query expansion	+	Binary
Urban & Jose, 2004a	Query expansion	+	Binary
Ishikawa et al., 1998	Metric optimization	+	Real
Rui et al., 1998	Metric optimization	+	Discrete
Lim et al., 2001	Metric optimization	+/-	Discrete
Aggarwal et al., 2002	Metric optimization	+/-	Real
Heesch & Rüger, 2003	Metric optimization	+/-	Real
He, Ma, & Zhang, 2004	Metric optimization	+	Binary
Nastar et al., 1998	Modeling distributions	+/-	Binary
Meilhac & Nastar, 1999	Modeling distributions	+/-	Binary
Tong & Chang, 2001	Discriminant classifier	+/-	Discrete

A summary can be found in Table 1. For each system, we note the kind of information communicated through feedback (positive/negative) and the part of the system that is modified in response.

Interlude

Let us now take a step back and assess the merit of the general methodology already described. The reported performance gains through relevance feedback are often considerable, even though any performance claims must be judged carefully against the experimental particulars, especially the database size, the performance measures, and the type of queries. Next, we suggest two major problems with the relevance feedback methodology.

Parameter Initialization

The utilization of relevance feedback for query expansion and multi-modal density estimation has attracted much attention and appears justified on the ground that the feature distributions of most relevance classes tend to be multi-modal and form natural groups in feature space. But unless the query itself consists of multiple images representing these various groups, we reasonably should not expect images from various groups to be retrieved in response to the query. If anything, the retrieved images will contain images from the cluster to which the query image is closest under the current metric.

But not only do relevance classes often form visually distinct clusters, images also often belong to a number of relevance classes. This is an expression of the semantic ambiguity that pertains, in particular, to images and that relevance feedback seeks to resolve. With queries consisting of single images, the question to resolve is which natural group does the query image belong to, not so much the various natural groups belonging to the relevance class of the query. But while some systems cater to multi-modality, none explicitly deals with polysemy. By initializing parameter values, systems effectively impose a particular semantic interpretation of the query.

The problem of parameter initialization has received insufficient attention so far. One notable exception is the work by Aggarwal et al. (2002), which we mentioned earlier in a different context. The system segments the query image, modifies each segment in various ways, and displays a set of modified queries to the users, who mark segments that continue to be relevant. The feature weights then are computed similarly to Rui et al. (1997) by considering the variance among the relevant segments.

Another method of parameter initialization that is very similar in spirit to that of Aggarwal et al. (2002) is developed in Heesch (2005). The method seeks to expose the various semantic facets of the query image by finding all images that are most similar to it under equation (1.2) for some weight set w. As we vary w, various images will become the nearest neighbor of the query. For each nearest neighbor, we record its associated w, which we may regard as a representation of one of the semantic facets in which users may be interested. Users select a subset of these nearest neighbors and, thereby, implicitly select a set of weights. These weights then are used to carry out a standard similarity search. The method outperforms relevance feedback methods that retrieve with an initially uniform weight set w but are not inexpensive computationally. NN^k Networks, which we will discuss in the next section, provide another attempt to tackle the initialization problem.

Exploratory Search

With very few exceptions, the methods described rely on the assumption that users know what they are looking for. The methods are designed to home in on a set of relevant items within a few iterations and do not support efficient exploration of the image collection. We will see in the second half of this chapter that more flexible interaction models may address this issue more successfully.

Search through Browsing

Browsing offers an alternative to the conventional method of query-by-example but has received surprisingly little attention. Some of the advantages of browsing follow.

Image browsing requires only a mental representation of the query. Although automated image annotation (Lavrenko, Manmatha, & Jeon, 2003; Feng, Manmatha, & Lavrenko, 2004; Zhang, Zhang, Li, Ma, & Zhang, 2005; Yavlinsky, Schofield, & Rüger, 2005) offers the possibility to reduce visual search methodologically to traditional text retrieval, there often may be something about an image that cannot be expressed in words, leaving visually guided browsing a viable alternative.

Retrieval by example image presupposes that users already have an information need. If this is not the case, then enabling users to navigate quickly between various regions of the image space becomes of much greater importance.

For large collections, time complexity becomes an issue. Even when hierarchical indexing structures are used, performance of nearest neighbor searches has been shown to degrade rapidly in high-dimensional feature spaces. For particular relevance feedback techniques, approximate methods may be developed that exploit correlations between successive nearest neighbor searches (Wu & Manjunath, 2001), but there is not a universal cure. Meanwhile, browsing structures can be precomputed, allowing interaction to be very fast.

The ability of the human visual system to recognize patterns reliably and quickly is a marvel yet to be fully comprehended. Endowing systems with similar capabilities has proved to be an exceedingly difficult task. Given our limitations in understanding and emulating human cognition, the most promising way to leverage the potential of computers is to combine their strengths with those of users and achieve a synergy through interaction. During browsing, users continuously are asked to make decisions based on the relevance of items to their current information needs. A substantial amount of time is spent, therefore, engaging users in what they are best at, while exploiting computational resources to render interaction quickly.

Hierarchies

Hierarchies have a ubiquitous presence in our daily lives; for example, the organization of files on a computer, the arrangement of books in a physical library, the presentation of information on the Web, employment structures, postal addresses, and many more.

To be at all useful for browsing, hierarchical structures need to be sufficiently intuitive and allow users to predict in which part of the tree the desired images may reside. When objects are described in terms of only a few semantically rich features, building such hierarchies is relatively easy. The low-level, multi-feature representation of images renders the task substantially more difficult.

Agglomerative Clustering

The most common methods for building hierarchies is by way of clustering either by iteratively merging clusters (agglomerative clustering) or by recursively partitioning clusters (divisive clustering). See Duda, Hart, and Stork (2001) for an overview.

Early applications of agglomerative clustering to image browsing are described in Yeung and Liu (1995), Yeung and Yeo (1997), Zhang and Zhong (1995), and Krishnamachari (1999). The first two papers are concerned with video browsing and clustering involving automated detection of topics and, for each topic, the constituent stories. Stories are represented as video posters, a set of images from the sequences that associate with repeated or long shots and act as pictorial summaries. In Zhang and Zhong (1995) and Yang (2004), the self-organizing map algorithm (Kohonen, 2001) is applied to map images on a two-dimensional grid. The resulting grid subsequently is clustered hierarchically. One of the major drawbacks of the self-organizing map algorithm (and neural network architectures, in general) is its computational complexity. Training instances often need to be presented multiple times, and convergence has to be slow in order to achieve good performance, particularly so for dense features. Chen et al. (1998, 2000) propose the concept of a similarity pyramid to represent image collections. Each level is organized such that similar images are in close proximity on a two-dimensional grid. Images first are organized into a binary tree through agglomerative clustering based on pairwise similarities. The binary tree subsequently is transformed into a quadtree that provides users a choice of four instead of two child nodes. The arrangement of cluster representatives is chosen so that some measure of overall visual coherence is maximized. Since the structure is precomputed, the computational cost incurred at browsing time is slight.

Divisive Clustering

Agglomerative clustering is quadratic in the number of images. Although this can be alleviated by sparsifying the distance matrix, this method becomes inaccurate for dense feature representations and is more amenable to keyword-based document representations.

A computationally more attractive alternative is divisive clustering in which clusters are split recursively into smaller clusters. One popular clustering algorithm for this purpose is *k*-means. In Pecenovic, Do, Vetterli, and Pu (2000), it is applied to 6,000 images with cluster centroids displayed according to their position on a global Sammon map. However, compared to agglomerative clustering, the divisive approach has been found to generate less intuitive groupings (Yeung & Yeo, 1997; Chen et al., 2000), and the former has remained the method of choice in spite of its computational complexity.

Networks

Nearest-Neighbor Networks

A significant work on interlinked information structures dates back to the mid-1980s (Croft & Parenty, 1985). It proposes to structure a collection of documents as a network of documents and terms with accordingly three types of weighted edges. The authors suggest keeping only links between a document and the document most similar to it, and similarly, for terms. Term-term and document-document links thus connect nearest neighbors, and each document gives rise to a star cluster that comprises the document itself and all adjacent nodes. Although the structure is intended for automated search, the authors are aware that "as well as the probabilistic and cluster-based searches, the network organization could allow the user to follow any links in the network while searching for relevant documents. A special retrieval strategy, called browsing, could be based on this ability" (Croft & Parenty, 1985, p. 380). However, the number of document-document edges does not exceed by much the number of documents, and star clusters are disconnected, rendering browsing along document-document nodes alone impractical.

Importantly, the work has inspired subsequent work by Cox (1992, 1995). Cox motivates associative structures for browsing by observing that "people remember objects by associating them with many other objects and events. A browsing system on a static database structure requires a rich vocabulary of interaction and associations." His idea is to establish a nearest-neighbor network for each set of the various object descriptors. Being aware that different features may be important to different users, Cox realizes the importance of interconnecting nearest-neighbor networks to allow multi-modal browsing.

Unfortunately, Cox's work has not become as widely known as perhaps it should have. What may account in part for this is that content-based image retrieval was then in its very early beginning, and the first research program that grew out of the initial phase of exploration happened to be that of query-by-example, which pushed browsing somewhat to the periphery.

NN^k Networks

The problem with many of the aforementioned structures is that the metric underlying their construction is fixed. The advantage of fast navigation, therefore, comes at a price: users no longer are in a position to alter the criterion under which similarity is judged. The structures thus deride the principal tenet that motivates relevance feedback techniques. Zhou and Huang (2003) arrive at a similar conclusion when they observe that "the rationale of

Figure 5. The set of NN^k in a network of 32,000 Corel images for two positions

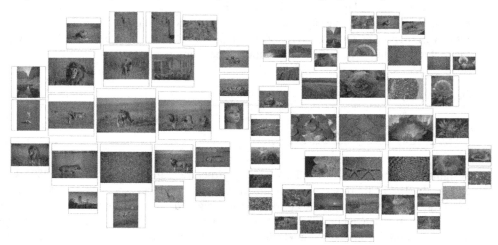

relevance feedback contradicts that of pre-clustering."

A browsing structure that has been designed with this in mind is NN^k Networks (Heesch, 2005; Heesch & Rüger, 2004a; Heesch & Rüger, 2004; Heesch, 2005). The structure is a directed graph in which an arc is established from p to q, if q is the nearest neighbor of p under at least one combination of features which are represented in terms of index i in equation (1.2). Instead of imposing a particular instance of the similarity metric, NN^k Networks expose the various semantic facets of an image by gathering all top-ranked images under different metrics. During browsing, users select those neighbors in the graph that match their target best. NN^k Networks exhibits small-world properties (Watts & Strogatz, 1998) that make them particularly well-suited for interactive search. Relevant images tend to form connected subgraphs so that a user who has found one relevant image is likely to find many more by following relevance trails through the network. The screenshots in Figure 5 illustrate the diversity among the set of neighbors for two positions in a network of 32,000 Corel images. The size of the image is a measure of the number of different metrics under which that image is more similar than any other to the currently selected image.

Pathfinder Networks

For browsing, at least parts of the network need to be visualized. The large number of links in a network may prevent users from recognizing structural patterns that could aid navigation. A practical strategy is to reduce the number of links. The pathfinder algorithm is one example of a link-reduction algorithm (Dearholt & Schvaneveldt, 1990). It is not concerned with constructing the original network but rather with converting a network of any kind to a sparser network. The pathfinder algorithm removes an edge between vertices if there is another path of shorter length. An application of pathfinder networks to the problem of organizing image collections is found in Chen et al. (2000), but the scope for interaction is limited. Indeed, it seems that the principal application domain of pathfinder networks has been visual data mining so far, not interactive browsing. The reason is quite likely to be found in the computational complexity that is prohibitive for collection sizes of practical significance. Moreover, visualization and navigation places somewhat different structural demands on the networks. While visualization requires the extraction of only the most salient structure, retaining some degree of redundancy renders the networks more robust for navigation purposes.

Dynamic Trees: Ostensive Relevance Feedback

The ostensive model of Campbell (2000) is an iterated query-by-example in disguise, but the query only emerges through the interaction of the user with the collection. The impression for the user is that of navigating along a dynamically unfolding tree structure. While originally developed for textual retrieval of annotated images, the ostensive model is equally applicable to visual features (Urban et al., 2003). It consists of two components: the relevance feedback model (the core component) and the display model.

Relevance feedback takes the form of selecting an image from those displayed. A new query is formed as the weighted sum of the features of this and previously selected images. In Urban et al. (2003), images are described by color histogram. Given a sequence of selected images, the color representation of the new query is given as the weighted sum of individual histograms with weights taking the form of $w_i = 2^{-i}$ ($i = 0$ indexing the most

Figure 6. An interface for ostensive relevance feedback

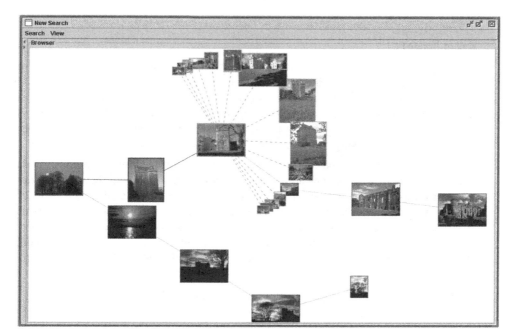

Table 2. Overview of browsing models

Author	Structure	RF	Flexible Metric	Off-line	Online	# img
Zhang & Zhong, 1995	Hierarchies	No	No	$O(n^2)$	$O(1)$	n/a
Krishnamachari et al., 1999	Hierarchies	No	No	$O(n^2)$	$O(1)$	3,856
Peceonivc et al., 2000	Hierarchies	No	No	$O(n)$	$O(1)$	6,100
Chen, Bouman, & Dalton, 2000	Hierarchies	No	No	$O(n^2)$	$O(1)$	10,000
Cox, 1995	Networks	No	Yes	$O(n^2)$	$O(1)$	< 100
Chen, Gagaudakis, & Rosin, 2000	Networks	No	No	$O(n^4)$	$O(1)$	279
Heesch & Rüger, 2004	Networks	No	Yes	$O(n^2)$	$O(1)$	32,000
Urban et al., 2003	Trees	Yes	Yes	$O(1)$	$O(n)$	800

recent image).

The display model is that of an unfolding tree structure: images closest under the current query are displayed in a fanlike pattern to one side of the currently selected image. Users can select an image from the retrieved set, which is placed in the center, and a new set of images are retrieved in response. Since previous images are kept on the display, the visual impression of the user is that of establishing a browsing path through the collection. In Urban et al. (2003), the browsing path is displayed in a fisheye view, as shown in Figure 6.

The ostensive model attempts to track changing information needs by continuously updating the query. Which set of images are retrieved depends on which path the user has traveled in order to arrive at the current point. Because the number of such different paths grows quickly with the size of the image collection, it is impractical to compute a global structure beforehand. Nonetheless, for the user, the impression is one of navigating in a relatively unconstrained manner through the image space. Unlike many other relevance feedback systems, users do not have to rank or label images or change their relative location. The interaction is, thus, light and effective.

Again, a summary of the browsing models is given in Table 2.

Conclusion

It has become clear over the past decade that content-based image retrieval can benefit tremendously from letting the user take on a greater role in the retrieval process. In this chapter, we have examined the different forms of user involvement in two contexts: query-by-example and interactive browsing. In the former setting, users initiate a search by submitting a query image and wait for images to be retrieved for them. A standard method of involving users in the subsequent stages of the process is to ask for relevance feedback on the retrieved images. The relevance information can be used to automatically modify the representation of the original query (query update), to adjust the function that is used to compute similarities between images and the query, or to learn a classifier between nonrelevant and relevant images.

The query-by-example setting has a number of limitations. Most importantly, it assumes that users already have an information need and a query image at their disposal. Systems of this category generally do not support free exploration of a collection. The second part of this chapter examined a number of browsing models in which the user becomes the chief protagonist. In addition to requiring only a mental representation of the query, browsing structures have the advantage that they may be precomputed so user interaction is fast. Browsing structures often take the form of hierarchies or networks, and browsing takes place by moving between vertices of the graph. Hierarchies can be constructed readily through hierarchical clustering and support search from the more general to the more specific, thus affording an impression of progressive refinement. However, it equally may create a sense of lost opportunities if navigation is restricted to the vertical dimension.

Networks have the advantage over hierarchies of less constrained navigation. At the same time, it is more difficult to provide a global overview of the content so that it becomes increasingly important to organize objects in the network such that the local neighborhood of the currently selected object contains enough information for users to decide where to go next.

There is a more general problem with precomputed structures that affects most of the models discussed. By being precomputed, users generally are not in a position to remold the structure according to their own preferences. This seems necessary, however, as the structures are almost always constructed by fixing the distance metric and applying that same metric across the entire collection. The advantage of fast navigation comes at the price that users no longer can impose their own perceptions of similarity.

There remains a number of exciting and important problems, a solution to which should lead to a new generation of smarter, more versatile systems for visual search. For example, while searching interactively for images, users continuously provide implicit relevance feedback. In addition to exploiting this information for the current search session, one should clearly wish to endow systems with some form of long-term memory. Also, large collections will take an appreciable amount of time to be cast into a browsable structure. This seems acceptable, provided the effort needs to be expended only once, but many collections are dynamic with new images regularly being added and others removed. An update should not involve a complete recomputation of the structure, but the extent to which the above models lend themselves to an efficient update seldom is investigated. Finally, most of the systems we have discussed either involve a precomputed structure or initiate a new query at every step.

Systems of the first kind are often too rigid, and systems of the second are too slow for large collections. What may hold promise are hybrid structures that are partially precomputed but flexible enough to remain responsive to relevance feedback.

References

Aggarwal, G., Ashwin, T., & Ghosal, S. (2002). An image retrieval system with automatic query modification. *IEEE Transactions on Multimedia, 4*(2), 201–213.

Campbell, I. (2000). *The ostensive model of developing information-needs* [doctoral thesis]. Glasgow, UK: University of Glasgow.

Chen, C., Gagaudakis, G., & Rosin, P. (2000). Similarity-based image browsing. In *Proceedings of the IFIP World Computer Congress*, Beijing, China.

Chen, J.-Y., Bouman, C., & Dalton, J. (2000). Hierarchical browsing and search of large image databases. *IEEE Transactions on Image Processing, 9*(3).

Chen, Y., Zhou, X., & Huang, T. (2001). One-class SVM for learning in image retrieval. In *Proceedings of the International Conference on Image Processing*, Thessaloniki, Greece.

Cohn, D., Atlas, L., & Ladner, R. (1994). Improving generalization with active learning. *Machine Learning, 15*(2), 201–221.

Cox, I., Miller, M., Minka, T., Papathomas, T., & Yianilos, P. (2000). The Bayesian image retrieval system, PicHunter: Theory, implementation, and psychophysical experiments. *IEEE Transactions on Image Processing, 9*(1), 20–38.

Cox, I., Miller, M., Omohundro, S., & Yianilos, P. (1998). An optimized interaction strategy for Bayesian relevance feedback. In *Proceedings of the IEEE Conference on Computer Vision and Pattern Recognition*, Santa Barbara, CA.

Cox, K. (1992). Information retrieval by browsing. In *Proceedings of the International Conference on New Information Technology*, Hong Kong.

Cox, K. (1995). *Searching through browsing* [doctoral thesis]. Canberra, Australia: University of Canberra.

Croft, B., & Parenty, T. (1985). Comparison of a network structure and a database system used for document retrieval. *Information Systems, 10*, 377–390.

Crucianu, M., Ferecatu, M., & Boujemaa, N. (2004). Relevance feedback for image retrieval: A short survey. In *State of the art in audiovisual content-based retrieval, information universal access and interaction including datamodels and languages* (DELOS2 Report [FP6 NoE]).

Dearholt, D., & Schvaneveldt, R. (1990). Properties of pathfinder networks. In R. Schvaneveldt (Ed.), *Pathfinder associative networks: Studies in knowledge organization*. Norwood, NJ.

Duda, R., Hart, P., & Stork, D. (2001). *Pattern recognition*. New York: Wiley.

Fagin, R., Kumar, R., & Sivakumar, D. (2003). Efficient similarity search and classification via rank aggregation. In *Proceedings of the ACM International Conference on Management of Data*, San Diego, CA.

Feng, S., Manmatha, R., & Lavrenko, V. (2004). Multiple Bernoulli relevance models for image and video annotation. In *Proceedings of the International Conference on Computer Vision and Pattern Recognition*, Cambridge, UK.

He, J., Tong, H., Li, M., Zhang, H.-J., & Zhang, C. (2004). Mean version space: A new active learning method for content-based image retrieval. In *Proceedings of the International Workshop on Multimedia Information Retrieval in Conjunction with ACM Multimedia*, New York.

He, X., Ma, W.-Y., & Zhang, H.-J. (2004). Learning an image manifold for retrieval. In *Proceedings of the ACM Multimedia*, New York.

Heesch, D. (2005). *The NN^k technique for image searching and browsing* [doctoral thesis]. London: Imperial College London.

Heesch, D., Pickering, M., Yavlinsky, A., & Rüger, S. (2003). Video retrieval within a browsing framework using keyframes. In *Proceedings of the TREC Video Retrieval Evaluation*, Gaithersburg.

Heesch, D., & Rüger, S. (2002). Combining features for content-based sketch retrieval—A comparative evaluation of retrieval performance. In *Proceedings of the European Conference on Information Retrieval*, Glasgow, UK.

Heesch, D., & Rüger, S. (2003). Performance boosting with three mouse clicks—relevance feedback for CBIR. In *Proceedings of the European Conference on Information Retrieval*, Pisa, Italy.

Heesch, D., & Rüger, S. (2004a). NN^k Networks for content-based image retrieval. In *Proceedings of the European Conference on Information Retrieval*, Sunderland, UK.

Heesch, D., & Rüger, S. (2004b). Image browsing: A semantic analysis of NN^k Networks. In *Proceedings of the International Conference on Image and Video Retrieval*, Dublin, Ireland.

Hong, P., Tian, Q., & Huang, T. (2000). Incorporate support vector machines to content-based image retrieval with relevant feedback. In *Proceedings of the IEEE International Conference on Image Processing*, Vancouver, Canada.

Ishikawa, Y., Subramanya, R., & Faloutsos, C. (1998). Mindreader: Querying databases through multiple examples. In *Proceedings of the International Conference on Very Large Data Bases*, New York.

Jing, F., Li, M., Zhang, L., Zhang, H.-J., & Zhang, B. (2003). Learning in region-based image retrieval. In *Proceedings of the IEEE International Symposium on Circuits and Systems*, Bangkok, Thailand.

Kim, D.-H., & Chung, C.-W. (2003). *Qcluster: Relevance feedback using adaptive clustering for content-based image retrieval*. In *Proceedings of the ACM SIGMOD International Conference on Management of Data*, Madison, WA.

Kohonen, T. (2001*). Self-organizing maps* (Vol. 30). Berlin, Germany: Springer.

Kruskal, J. (1964). Multi-dimensional scaling by optimizing goodness-of-fit to a nonmetric hypothesis. *Psychometrika, 29*, 1–27.

Lavrenko, V., Manmatha, R., & Jeon, J. (2003). A model for learning the semantics of pictures. In *Proceedings of the International Conference on Neural Information Processing Systems*, Whistler, Canada.

Lim, J., Wu, J., Singh, S., & Narasimhalu, D. (2001). Learning similarity matching in multimedia content-based retrieval. *IEEE Transactions Knowledge and Data Engineering, 13*(5), 846–850.

Meilhac, C., & Nastar, C. (1999). Relevance feedback and category search in image databases. In *Proceedings of the International Conference on Multimedia Communications Systems*, Florence, Italy.

Müller, H., Marchand-Maillet, S., & Pun, T. (2002). The truth about Corel—evaluation in image retrieval. In *Proceedings of the International Conference on Image and Video Retrieval*, London.

Müller, H., Müller, W., Squire, D., Marchand-Maillet, S., & Pun, T. (2000). Strategies for positive and negative relevance feedback in image retrieval. In *Proceedings of the International Conference on Pattern Recognition*, Barcelona, Spain.

Nastar, C., Mitschke, M., & Meilhac, C. (1998). Efficient query refinement for image retrieval. In *Proceedings of the IEEE Conference on Computer Vision and Pattern Recognition*, Santa Barbara, CA.

Pecenovic, Z., Do, M., Vetterli, M., & Pu, P. (2000). Integrated browsing and searching of large image collections. In *Proceedings of the International Conference on Advances in Visual Information Systems*, Lyon, France.

Peng, J., Bhanu, B., & Qing, S. (1999). Probabilistic feature relevance learning for content-based image retrieval. *Computer Vision and Image Understanding, 75*(12), 150–164.

Porkaew, K., Chakrabarti, K., & Mehrotra, S. (1999). Query refinement for multimedia similarity retrieval in Mars. In *Proceedings of the ACM International Conference on Multimedia*, Orlando, FL.

Rocchio, J. (1971). *The SMART retrieval system. Experiments in automatic document processing*. Englewood Cliffs, NJ: Prentice Hall.

Rubner, Y., Guibas, L., & Tomasi, C. (1997). The earth mover's distance, multi-dimensional scaling, and color-based image retrieval. In *Proceedings of the DARPA Image Understanding Workshop*, New Orleans, LA.

Rui, Y., & Huang, T. (2000). Optimizing learning in image retrieval. In *Proceedings of the IEEE Conference on Computer Vision and Pattern Recognition*, Hilton Head Island, SC.

Rui, Y., Huang, T., & Mehrotra, S. (1997). Content-based image retrieval with relevance feedback in Mars. In *Proceedings of the IEEE International Conference on Image Processing*, Washington, DC.

Rui, Y., Huang, T., Ortega, M., & Mehrotra, S. (1998). Relevance feedback: A power tool for interactive content-based image retrieval. *IEEE Transactions Circuits and Video Technology, 8*(5), 644–655.

Salton, G., & McGill, M. (1982). *Introduction to modern information retrieval*. Columbus, OH: McGraw-Hill.

Santini, S., Gupta, A., & Jain, R. (2001). Emergent semantics through interaction in image databases. *IEEE Transactions on Knowledge and Data Engineering, 13*(3), 337–351.

Santini, S., & Jain, R. (2000). Integrated browsing and querying for image databases. *IEEE MultiMedia, 7*(3), 26–39.

Schettini, R., Ciocca, G., & Gagliardi, I. (1999). Content-based color image retrieval with relevance feedback. In *Proceedings of the International Conference on Image Processing*, Kobe, Japan.

Sclaroff, S., Taycher, L., & La Cascia, M. (1997). ImageRover: A content-based image browser for the World Wide Web. In *Proceedings of the IEEE International Workshop on Content-Based Access of Image and Video Libraries*, San Juan, Puerto Rico.

Tong, S., & Chang, E. (2001). Support vector machine active learning for image retrieval. In *Proceedings of the ACM International Conference on Multimedia*, New York.

Urban, J., & Jose, J. (2004a). Ego: A personalised multimedia management system. In *Proceedings of the International Workshop on Adaptive Multimedia Retrieval*, Valencia, Spain.

Urban, J., & Jose, J. (2004b). Evidence combination for multi-point query learning in content-based image retrieval. In *Proceedings of the IEEE International Symposium on Multimedia Software Engineering*, Miami, FL.

Urban, J., Jose, J., & van Rijsbergen, K. (2003). An adaptive approach towards content-based image retrieval. In *Proceedings of the International Workshop Content-Based Multimedia Indexing*, Rennes, France.

Vapnik, V. (1995). *The nature of statistical learning theory*. Springer.

Vasconcelos, N., & Lippman, A. (2000). Bayesian relevance feedback for content-based image retrieval. In *Proceedings of the IEEE Workshop Content-Based Access of Image and Video Libraries*, Hilton Head Island, SC.

Watts, D., & Strogatz, S. (1998). Collective dynamics of small-world networks. *Nature, 393*, 440–442.

Wolfram, S. (2004). *A new kind of science*. Wolfram Ltd.

Wu, L., Faloutsos, C., Sycara, K., & Payne, T. (2000). Falcon: Feedback adaptive loop for content-based retrieval. In *Proceedings of the International Conference on Very Large Data Bases*, Cairo, Egypt.

Wu, P., & Manjunath, B. (2001). *Adaptive nearest neighbour search for relevance feedback in large image databases*. Ottawa, Canada: ACM Multimedia.

Yang, C. (2004). Content-based image retrieval: A comparison between query by example and image browsing map approaches. *Journal of Information Science, 30*(3), 254–267.

Yavlinsky, A., Schofield, E., & Rüger, S. (2005): Automated image annotation using global features and robust nonparametric density estimation. In *Proceedings of the International Conference on Image and Video Retrieval*, Singapore.

Yeung, M., & Liu, B. (1995). Efficient matching and clustering of video shots. In *Proceedings of the IEEE International Conference on Image Processing*, Washington, DC.

Yeung, M., & Yeo, B. (1997). Video visualization for compact presentation and fast browsing of pictorial content. *IEEE Transactions on Circuits and Systems for Video Technology, 7*, 771–785.

Yoon, J., & Jayant, M. (2001). Relevance feedback for semantics based image retrieval. In *Proceedings of the International Conference on Image Processing*, Thessaloniki, Greece.

Zhang, H., & Zhong, D. (1995). A scheme for visual feature based image indexing. In *Proceedings of the SPIE/IS&T Conference on Storage and Retrieval for Image and Video Databases III*, La Jolla, CA.

Zhang, H.-J., & Su, Z. (2001). Improving CBIR by semantic propagation and cross modality query expansion. In *Proceedings of the Conference on Multimedia Content-Based Indexing and Retrieval*, Rocquencourt, France.

Zhang, R., Zhang, Z., Li, M., Ma, W.-Y., & Zhang, H.-J. (2005). A probabilistic semantic model for image annotation and multi-modal image retrieval. In *Proceedings of the International Conference on Computer Vision*, Beijing, China.

Zhou, X., & Huang, T. (2003). Relevance feedback in image retrieval: A comprehensive review. *ACM Multimedia Systems, 8*(6), 536–544.

Chapter IX

Semi-Automatic Ground Truth Annotation for Benchmarking of Face Detection in Video

Dzmitry Tsishkou, Ecole Centrale de Lyon, France

Liming Chen, Ecole Centrale de Lyon, France

Eugeny Bovbel, Belarusian State University, Belarus

Abstract

This work presents a method of semi-automatic ground truth annotation for benchmarking of face detection in video. We aim to illustrate the solution to the issue in which an image processing and pattern recognition expert is able to label and annotate facial patterns in video sequences at the rate of 7,500 frames per hour. We extend these ideas to the semi-automatic face annotation methodology in which all object patterns are categorized into four classes in order to increase flexibility of evaluation results analysis. We present a strict guide on how to speed up manual annotation process by 30 times and illustrate it with the sample test video sequences that consists of more than 100,000 frames, 950 individuals, and 75,000 facial images. Experimental evaluation of face detection using the ground truth data that was labeled semi-automatically demonstrates the effectiveness of the current approach both for learning and for test stages.

Introduction

Rapid growth of telecommunication data is caused by tremendous progress in information technology development and its applications. Modern figures of the volume of video data ranged from a hundred hours of home video for personal use to a TV company's million-hour archives. The condition of a successful practical use of these data is the stringent enforcement in the following areas: storage, transmission, preprocessing, analysis, and indexing. Despite the fact that information storage and transmission systems are capable of supplying video data to users, there is still a developmental gap between the storage and the indexing. A problem of manual video data preprocessing, analysis, and indexing has no practical solution in terms of human perception and physiological abilities. Therefore, the construction of the fully automatic video indexing system is the current research subject (Snoek & Worring, 2002). High-level semantic analysis and indexing systems generally exploit some known specific properties of the video (e.g., that the video is a hierarchical multi-layer combination of background/foreground objects). One can decompose the video into a set of such objects and the structure. That requires video structure understanding and object recognition solutions (Wang, Liu, & Huang, 2000). Examples of such object recognition solutions include face detection and recognition, among others (Jones & Viola, 2003; Zhao, Chellappa, Resonfeld, & Phillips, 2000). There are several reasons why one is interested in a use of facial information for video analysis and indexing. First, visual facial information serves for clue-up on a personal identity (Tsishkou, 2002). That is one of the most informative tags for a particular video invent (Eickeler, Wallhoff, Iurgel, & Rigoll, 2000). The second reason that face detection and recognition is important is that it has been given great attention from the scientific society until present. It's a strong advantage, taking into account the fact that the number of potential objects to be recognized by the video analysis and indexing system is nearly infinite. Finally, the most important reason that facial information is important is that it enables us to dramatically increase the number of applications of the video analysis and indexing (Tsekeridou & Pitas, 2001). In order to optimize a learning process and evaluate these methods in a broad range of diversity video genres, one needs to select representative enough data samples and manually index them (Karpova, Tsishkou, & Chen, 2003). These would be used as ground truth data. Since manual face annotation rate is around 200 to 600 frames per hour, depending on index structure, there is an issue with preprocessing time. In this chapter, we expend the scope of face annotation in video for optimization and benchmarking of face detection and recognition methods to include the semi-automatic face annotation methodology, which allows reaching a 7,500 frames per hour annotation rate. In particular, we present the theory of semi-automatic face annotation in video analysis and indexing applications from the simplest concepts of a video indexing and analysis system to the formal description of the annotation methodology. The organization of the chapter follows. In Section 2, we review related work on video annotation. A brief sketch of ideas presented in this chapter is discussed in Section 3. Manual classification of individual objects is described in Section 4. We extend these ideas to the semi-automatic object annotation methodology in Section 5. In Section 6, we review the current version of the semi-automatic annotation system and experimental setup that uses a comprehensive dataset. We summarize the ideas discussed throughout the chapter and discuss further research directions in Section 7.

Related Work on Video Annotation

Many research works on face detection have been developed since the early 1990s. However, most of them focused on general purpose face detection in still images (Rowley, Baluja, & Kanade, 1998). The datasets were composed of 500 to 1,000 test images, which were labeled manually for further benchmarking (Sung, 1996). A growing trend in the field of video indexing is driving a need for more effective and comprehensive evaluation of face detection systems on large amounts of digital video. Therefore, manual annotation has to be added by the new functionality that can make the annotation task more efficient. While recent advances in content analysis, feature extraction, and classification are improving capabilities for effectively searching and filtering digital video content, the process to reliably and efficiently index multimedia content is still a challenging issue (Snoek & Worring, 2002). IBM's VideoAnnEx system, also known as the IBM MPEG-7 Annotation Tool, assists users in the task of annotating video sequences with MPEG-7 metadata (Naphade, Lin, Smith, Belle, & Basu, 2002). Each shot in the video can be annotated with static scene descriptors, key object descriptions, event descriptions, and other lexicon sets. The annotated descriptions are associated with each video shot or regions in the key frames and are stored as MPEG-7 XML file (Tseng, Lin, & Smith, 2002). It is one of the first MPEG-7 annotation tools being made publicly available. Since the facial information is incorporated in the MPEG-7 standard, this tool can be considered a flexible solution to the face segmentation in video problems (Manjunath, Salembier, & Sikora, 2002). By now, the tool explores a number of interesting capabilities, including automatic shot detection; key-frame selection; automatic label propagation; template annotation propagation to similar shots; and importing, editing, and customizing ontology and controlled term lists (Lin, Tseng, & Smith, 2003). Given the lexicon and video shot boundaries, visual annotations can be assigned to each shot by a combination of labels predication and human interaction (Lin et al., 2003). By the time a shot is being annotated, the system predicts its labels by propagating the labels from the last shot in time within the same cluster. An annotator can accept these predicted labels or select new labels from the hierarchical controlled-term lists (Naphade et al., 2002). In spite of the fact that the VideoAnnEx tool has a template matching mechanism to help users to detect regions of interest in the shots, it has no particular method for dealing with human faces. Therefore, it cannot be used to automatically propagate facial labels and further reduce manual annotation complexity. The lexicon used in this annotation task was drafted by the IBM Research TREC Video team and includes the following human-related terms: human, male face, female face, person, people, and crowd. These descriptors are meant to be clear to annotators; however, they aren't meant to indicate how automatic detection should be achieved. There are at least three major components in the VideoAnnEx tool that can make the annotating task of facial descriptors more efficient.

- **User interface:** The VideoAnnEx is divided into four graphical sections. On the upper right-hand corner of the tool is the Video Playback window with shot information (Lin et al., 2003). On the upper left-hand corner of the tool is the Shot Annotation with a key frame image display (Lin et al., 2003). On the bottom portion of the tool are two View Panels of the annotation preview. The fourth component is the Region Annotation pop-up for specifying annotated regions (Lin et al., 2003).

- **Video segmentation:** Video segmentation is performed to cut up video sequences into smaller video units (Lin et al., 2003).

- **Annotations learning:** Annotations learning is a characteristic that helps speed up annotation speed. Right before the user annotates a video shot, predicted labels would have been shown on the keyword field of the VideoAnnEx (Lin et al., 2003). The prediction functionality propagates labels from the visually most similar annotated shot. This propagation mechanism has been shown quite effective and helpful in speeding up the annotation task (Lin et al., 2003).

Collaborative work of 111 researchers from 23 institutes on association of 197,822 ground truth labels at 62.2 hours of videos in the framework of the VideoAnnEx project has raised several important questions. Here, we will post some of them with answers to the problems related to the facial annotation in video.

- **After you were familiar with the VideoAnnEx Annotation Tool, on average, how long did you need to annotate a 30-minute news video?** On average, annotators use 3.39 hour per 30-minute video. This corresponds to 6.8x of the real time speed (Lin et al., 2003).

- **Did you use the template matching and annotating learning?** Nearly 55% of the annotators considered the template detection useful, and 33% had the same opinion on the label propagation feature.

Other MPEG-7 (Zhao, Bhat, Nandhakumar, & Chellappa, 2000) annotation tools are available publicly. The Ricoh's MovieTool is developed for creating video content descriptors conforming MPEG-7 syntax interactively (Ricoh, 2005). Automatically generated MPEG-7 information is limited to overall structure of the video; automatic extractions of content-based features, such as still image or audio features, are not provided. Facial annotation can be done only manually by adding a textual description such as a name, with no relevant indicators of facial position. The Know-Center is used as a MPEG-7 based annotation and retrieval tool for digital photos. Hierarchical video content description and summarization using unified semantic and visual similarity is presented in Zhu, Fan, Elmagarmid, & Wu, 2003). Authors adopt video description ontology in order to describe video content efficiently and accurately. A semi-automatic video content annotation framework was introduced. This framework allows integration of acquired content description data with group merging and clustering. The integration makes possible fast semi-automatic propagation of obtained results to unfold the video content in a progressive way. We must mention that the propagation makes possible semi-automatic association of labels with frame number only, while missing spatial interframe information. Hoogs, Rittscher, Stein, and Schmiederer (2003) addresses the problem of extracting visual content from broadcast video news by automatically annotating objects, events, and scene characteristics. Three types of visual information that encompass the large majority of all visual content were used, including people, vehicles and manmade and natural objects. Although the use of automatic face detector (Rowley et al., 1998) empowers processing capability of the content annotating system, there is no way to control quality of the annotation while its reported False Rejection Rate is higher then

10% for real-world applications. Video summarization and translation that uses techniques of multimedia annotation is given in Nagao, Ohira, and Yoneoka (2002). Their system allows users to create annotation including voice transcripts, video scene descriptions, and visual/auditory object descriptions. The last one is done by tracking and interactive naming of people and objects in video scenes. More specifically, the tool provides capability to the user to select a particular object on a frame, and an automatic object tracking technique is executed to index spatial coordinates of the region of interest for every single frame.

Therefore, we might conclude that most of the current video annotation systems have no special tools for facial annotation, while all faces are treated in the same manner as any other objects. This results in the fact that collaborative work of 111 researchers on Video-AnnEx project was able to proceed 197,882 ground truth labels in total, which cost about 421 hours of work (we must note that the total number of frames was around 5,300,000, so only 0.88 frames were annotated for each second of video), compared to the specific solution proposed in this chapter that allows to proceed 100,000 labels in total for every video frame within 13 hours. For others, which have face-specific tracking procedures, one must consider missing ground truth information provided or verified by a human on every frame of indexed video.

Overview of the Semi-Automatic Face Annotation Approach

Our work is concerned with the development of efficient mechanisms of facial annotation of video with a primary target that a human person verifies every single label of ground truth information on every single frame. Otherwise, even if we are using the most sophisticated automatic face detection algorithm (Tsishkou et al., 2002, there is no way to guarantee 100% of correct classification of facial images. So, further verification and correction steps will be required to double-check results produced by the automatic method. This procedure, as we have seen previously, requires enormously higher annotation time. Another problem resolved in this work refers to an extremely short amount of annotation information that currently is used to store along with facial labels. Most of them are restricted to coordinates of eyes and a head-bounding rectangle (Viola & Jones, 2001). Three closely related basic questions are immediately identifiable.

What can be done to preserve human verification of annotated labels on every frame in order to ensure the highest quality of ground truth data and match automatic face annotation systems in terms of labeling time?

How can the testing procedure evaluate ability of the face detector to correctly classify facial image of the same person during N next frames in a raw? This question is important for face tracking in video applications, which progressively update information on facial appearance of multiple persons at the same time.

How can we correctly compare difference in performance of two face detectors on the same dataset, which is composed of clear anchor person and complex football game shots? For example, if the number of the anchor person shots, which are much easier to detect, is 90%

of the entire facial population and the rest is composed of difficult long-range facial images with multiple rotations and changes of lighting conditions, it would be hard to correctly compare performance of two face detectors if both classification results on the anchor person and long-range variable facial images would be taken into account in an arbitrary manner.

The approach presented here shows that an image processing expert with a strict manual on how to categorize facial images on several complexities, semi-automatic facial labeling routine, and a user-friendly facial annotation framework is able to provide ground truth information for 7,500 frames of video (which may include several labels) within one hour of work. The approach for this purpose brings together several critical aspects.

- **Strict guide on manual classification of facial images into four categories according to their complexity of classification by the state-of-the-art facial detectors:** Our approach collects information on basic parameters that affect performance of the state-of-the-art face detectors, such as facial size, image quality, lighting conditions, resolution, orientation, and occlusions. The user is required to classify each parameter into several categories and further make fusion of the results according to the combination rule. Those operations have to be kept in mind in order to make facial annotation fast. Experimental session with 10 image processing experts proved that two hours is enough time to memorize all information necessary to complete this task. Selection of categories and combination rule was done in an experimental way and has proved its efficiency on a real-world test.

- **Semi-automatic facial labeling routine, which automatically propagates facial tags between user-selected key frames:** The nature of automatic propagation comes from linear facial trajectory computing within the interval that is manually selected and verified by the annotation expert. Binary searchlike usage of forward-rewind operation allows the expert to efficiently select key frames with approximately linear translation of facial position and stable facial complexity in between. Confirmed on practice, this concept makes it possible to speed up the annotating process by the factor of 30 or even more, depending on a genre of video.

- **User-friendly facial annotating framework:** It is designed in a way to decrease the number of manual operations that have to be done in order to label facial image by effectively grouping facial images of multiple persons currently under consideration together with specifically designed one-mouse move + button click sequences of operations, which significantly minimizes processing time.

Keeping the above issues in mind, our approach was applied successfully to annotate more then 100,000 frames in various genres including video with news, commercial, sci-fi, sport, music, Oscar show, and adventure. An average semi-automatic annotating rate for the image processing expert was 7,500 frames per hour. The obtained facial tags were verified for every video frame. Each of the tags is composed of personal ID (to distinguish between different persons), facial center coordinates, and facial complexity based on one of three categories. The ground truth data generated by the system described in this chapter were used successfully in series of testing and learning experiments on a database of 400 hours of video of various genres.

Manual Classification of
Individual Objects in Video

The purpose of semi-automatic face annotation in video is to extract the information that concerns facial appearance. Since not all the faces share the same complexity in terms of quality, lighting, resolution, orientation, and others, one needs to classify them manually. Therefore, we will not suffer from incomplete analysis of benchmarking results working with all faces in the same way. Our main goal here is to formulate the classification rules and to define the necessary criteria. Since the current issue concerns human activity that has to be completed in a short period of time (less than $1/5^{th}$ of second), there is no possibility of automatically measuring important parameters using the current level of image processing methods. This is true because those methods suffer to deal with all kinds of variations described here in an arbitrary way. So, in some cases, they may produce adequate results; for others, results with errors, which would make incorrect classification. As an example, we may consider the ability of an automatic system to correctly determine an angle of out-plane facial rotations in the presence of lighting variations and video artifacts caused by low bit-rate. Therefore, for some of these categorization issues, we can prescribe exact analytical solutions; for other issues, we only can provide seat-of-the-pants experience gained from working with face detection over the last several years. There are essentially five criteria in the manual face classification that were used in this work for the frontal face detector (Rowley et al., 1998):

- **Image quality:** It can be seen that image quality depends on two main factors such as compression and filming conditions. These factors have a major influence on the following effects: geometric artifacts (distortions, blur, out of focus, occlusions), quan-

Figure 1. Video compression artifacts (blur on the left face)

Figure 2. Multiply scales face detection

tization artifacts (macro block boundaries, sharp gradients), and photometric artifacts (extreme gamma correction, unnatural colors, over-lighting, over-darkness). Figure 1 illustrates video compression artifacts that are vital to face detection. Depending on the facial gallery that has been used for training of the face detector, one can decide whenever the combination of those effects (if any) is familiar to the face detector or not. If it's true, we would estimate the video quality of a particular face pattern as high, otherwise as low.

• **Size:** The size of the facial pattern is a crucial parameter for most face detection and recognition systems. It is characterized by the resolution of video and the original facial size. For those systems that are using coarse to fine (pyramidal) face detection, the higher size of the facial pattern increases the probability to correctly localize it. Figure 2 shows how the size of the facial pattern affects the false rejection rate.

The red rectangle indicates the result of the face detection. One can see that the central face was detected three times on the various scales, and the other two faces were detected only once; therefore, the FRR for this type of face pattern is lower compared to the patterns with minimum allowed size. There are three forms of the size complexity used in our manual face

Table 1. Facial size complexity classification

Facial Size Complexity	Range in Pixels
Low	> 64×64
Medium	32×32 to 64×64
High	0×0 to 32×32

Figure 3. Lighting conditions of high complexity

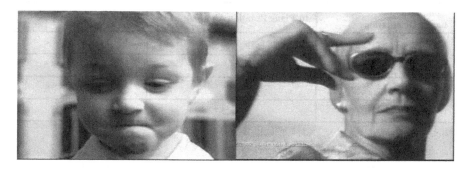

classification philosophy; namely, low size complexity, medium size complexity, and high size complexity. Table 1 shows how all three forms of size complexity are classified.

- **Lighting conditions:** The way in which lighting conditions affect performance of the face detector is similar to photometric artifacts. However, we decided to consider this issue an independent criterion because of its importance in theory of image processing and pattern recognition. Origin of lighting condition changes is the form of the light source. Generally, it's hard to implement the face detector with the gallery that includes large lighting variations, especially spotlight multi-source variations. Therefore, we classify lighting conditions complexity of a face pattern on a basis of lighting maps. By the lighting map, we understand the form of a product of interaction between a lighting source and face. For example, in Figure 4, each face has a uniform lighting map because of the uniform lighting source, however, Figure 5 illustrates the case with nonuniform lighting sources, so that the two lighting maps can be considered equal, if they share the same lighting sources and vice versa. Table 2 present the classification of lighting conditions complexity. Figure 3 illustrates some examples of lighting conditions of high complexity.

- **Orientation:** For the case in which we wish to use face detection with a frontal orientation of faces rather than profile, we classify facial orientation complexity into three

Table 2. Lighting conditions classification

Complexity of Lighting Conditions	Part of the Facial Pattern that Shares the Same Lighting Source
Low	>4/5
Medium	4/5 to 2/3
High	2/3<

Table 3. Facial orientations classification

Complexity of Facial Orientations	In-Plane Rotation About any Axis	Out-Plane Rotation about any Axis
Low	$\pm 25\,°\leq$	$\pm 25\,°\leq$
Medium	$\pm 25\,°\leq$	$>\pm 25\,°$
Medium	$>\pm 25\,°$	$\pm 25\,°\leq$
High	$>\pm 25\,°$	$>\pm 25\,°$

Table 4. Facial occlusions classification

Complexity of Facial Occlusion	Visible Area of the Informative Facial Part (IFP)/Total Area of IFP
Low	>4/5
Medium	4/5 to 2/3
High	2/3<

classes. This procedure basically refers to the evaluation of in-plane and out-plane facial rotations. The orientations can be classified coarsely by an expert in image processing according to their complexity using the rule shown in Table 3.

- **Occlusions:** The issue of occlusions is twofold. First, it can be caused by video arti-facts or filming conditions. Next, it depends on personal appearance. Nevertheless, we classify the occlusion complexity according to the visible area of the facial informative part. Table 4 illustrates the rule that is used to classify the occlusion complexity in our system.

Although this is something that has to be measured (facial size, lighting conditions, orienta-tion, visible area of the informative part), we suppose that an expert who is working in the image processing and pattern recognition domain is able to coarsely classify those issues to one of the categories that has been presented in real time.

We now come back to the issue that we have raised several time in this work; namely, how we label the facial pattern and how many categories we need. Our idea is to choose four clusters of facial patterns that can be localized by a face detector. This comes from the experi-ments so it gives a good balance between manual annotation time by reducing the number of choices that have to be analyzed by an expert, and informative complexity of benchmarking results that increases because of independent evaluation on each facial complexity category. Hereby, Table 5 illustrates the classification rules that have to be used to label each facial pattern at manual annotation process. We must mention that type 3 is excluded from our experimental setup, since none of the state-of-the-art face detectors (Rowley et al., 1998) currently is capable of accurately (FRR < 0.1, FAR < 0.0000001) detecting that kind of faces in automatic mode.

Table 5. Facial classification rules

Facial Categories	Points
Type 0 (low complexity)	<50
Type 1 (medium complexity)	>=50&<170
Type 2 (high complexity)	>=170&<350
Type 3 (ultra high complexity)	>350
Classification criteria	**Points**
Video quality: Low	170
Video quality: High	0
Size complexity: Low	0
Size complexity: Medium	30
Size complexity: High	60
Lighting conditions complexity: Low	0
Lighting condition complexity: Medium	50
Lighting conditions complexity: High	100
Orientation complexity: Low	0
Orientation complexity: Medium	40
Orientation complexity: High	80
Occlusions complexity: Low	0
Occlusions complexity: Medium	45
Occlusions complexity: High	90

The categorization formula was tuned to the possible use of a general state-of-the-art frontal face detector and an expert with basic skills in image processing and pattern recognition to minimize annotation time and errors.

Semi-Automatic Object Annotation Methodology

In this section, we deal with practical implementation issues of the semi-automatic face annotation software. Manual face segmentation is a frame-by-frame process that requires the following steps to be completed by an expert:

Figure 4. Frame-by-frame manual face annotation process

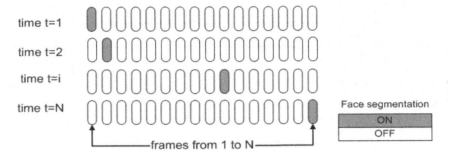

- **Reconnaissance:** An expert has to recognize all facial patterns within the particular video frame.
- **Identification:** Each facial pattern has to be identified using previous results of identification. If there is no corresponding record, then a new one has to be created.
- **Categorization:** All facial patterns have to be classified into predefined categories.
- **Labeling:** Coordinates of a facial tag have to be stored for each individual. Many facial tags can be used, including the informative part center and other facial features positions.
- **Indexing:** Results of the identification, categorization, and labeling have to be indexed and stored.

Basically, manual face segmentation can be replaced by semi-automatic face annotation in order to reduce processing time. In fact, most of those steps potentially could be completed by image processing and pattern recognition methods; however, in this chapter, we are limited to very simple and effective solutions. A key question, therefore, is how do we improve annotation effectiveness without use of sophisticated procedures. The possible solution is to group reconnaissance, identification, and categorization together. From this point, the manual face annotation is a three-steps process: visual analysis (reconnaissance, identification, and categorization), labeling, and indexing. Figure 4 illustrates a typical frame-by-frame manual face annotation process that is the same for all three steps described previously.

It can be seen that all frames are annotated one after another for the manual mode. So the total number of annotation iterations is proportional to the total number of frames and facial images inside of them.

There are three key factors that were used to create a semi-automatic face annotation routine. Each of them is connected with the optimization of user interaction with standard manual annotation software. The stages in the optimization are as follows:

Figure 5. Individual's informative part motion trajectory

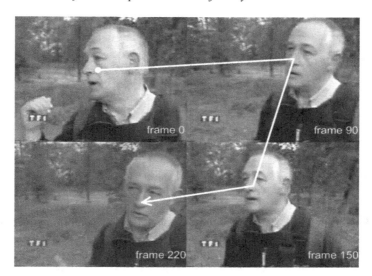

- **Fast visual analysis:** For the type of video in which a new individual appears, not every frame (general conditions) reconnaissance, identification, and categorization can be optimized. Once a new face pattern is recognized and identified, it became easier for an expert to repeat those actions for the same individual over a period of time. Slow facial complexity changes reduce categorization time as well by omitting the action over intermediate frames. Therefore, one can use fast search principles for visual analysis. Consider an expert who is working with an anchorperson's shot and can predict that the same individual will remain alone over the next 100 frames. Therefore he or she can reliably forward wind a video sequence by 50 frames, and if the situation doesn't change, he or she can guess that there were no changes at all in the intermediate frames. Otherwise, it's possible to rewind by 25 frames similar to the manner in which the binary search does and vice versa. Unfortunately, there is no simple, theoretically correct way to make such choices. However, all our past experience has proved the effectiveness of the fast visual analysis. This approach requires fast forward wind and fast rewind possibilities for the software that is used for semi-automatic face annotation.

- **Fast labeling:** Sharing the same philosophy as in the previous block, one can track changes in the individual appearance. The issue is to manually find the two remotest (according to the timeline) facial patterns of the same individual whose motion trajectory of the informative part center has a form of a straight line and facial complexity that remains stable. This process can be considered an approximation of the facial motion trajectory with straight segments. Figure 5 shows an individual's movement at a fixed shot over the timeline. The recorded details on individuals can be shown on a screen to enforce annotation rate while processing new frames. For example, the information on the informative part center helps in analyzing motion trajectory format in order to determine if it is straight or not.

Figure 6. Semi-automatic face annotation chart

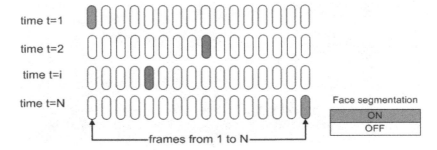

- **Fast indexing:** Using the information on the key motion trajectory points, one can compute all intermediate points automatically. This calculation procedure dramatically reduces annotation time and is a basic of fast indexing. Since the trajectory is a straight line, computational equation is as follows: $y_i = a * x_i + b, i = 1..N$, $a = (y_N - y_0)/(x_N - x_0), b = y_N - a * x_N$, in which $y_0, x_0,$ and y_N, x_N are coordinates of the informative part center for frames 0 and N, respectively.

Experimental use of all three optimization stages together on "news" test video sequence has shown that more than 20 times speedup was achieved by an experienced expert compared to the manual face annotation. Figure 6 illustrates the semi-automatic face annotation chart.

According to this figure, the difference between manual and semi-automatic facial annotations is in the data-management principle. The second solution manages the data in the more sophisticated way that ignores analysis of interframes in which information changes are not important for the final result; therefore, it greatly reduces total processing time. As we can seen on Figure 6, key frames 1 and i are selected within three iterations of visual analysis, compared to i iterations for the manual mode.

Experimental Setup of the
Semi-Automatic Face Annotation System

The semi-automatic face annotation tool is divided into three graphical sections, as illustrated in Figure 7. On top of the figure is the main window, and there are two pop-up windows on the bottom. These sections provide interactivity to assist authors of the annotation tool.

Typical face annotation process with the use of the tool that is described previously is the following. The *Open Video* button is used to load a new video sequence that has to be annotated. The annotator uses video navigation controls (5-15) in order to browse the video sequence doing manual classification of facial objects in it. Once the facial complexity of

Figure 7. Graphical user interface of the semi-automatic face annotation tool

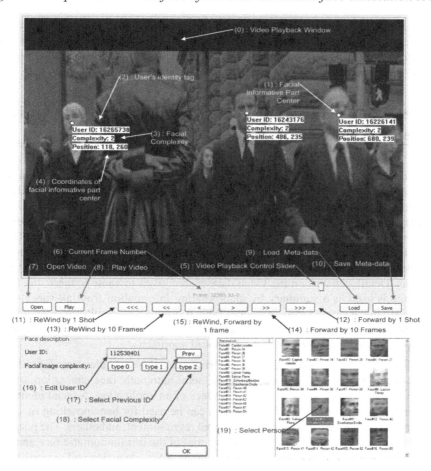

any of the faces within the video is lower than 3, the labeling procedure starts. The mouse cursor is used to mark the center of the informative part (1), and the right button is used to display the pop-up window with the facial annotation controls (16-18) in it. As the annotator types a name or leaves a default name that was generated randomly and selects relevant facial complexity, the labels (2-4) are displayed below the corresponding informative part center (1). Furthermore, the navigation continues until either the facial complexity is changed or the facial motion trajectory becomes nonlinear. In this case, the left mouse button is used to mark the informative part center on the last valid frame. The prediction functionality of the current tool propagates labeling for all intermediate frames. The keyboard alias allows the annotator to execute the personal image list population mode in which the mouse is used to manually select a facial region of the person, which would be used frequently for further annotation process. This allows replacing the typing of user's ID by the search of the relevant person in the image list (19). The *Select Previous ID* control (17) offers fast changing between two persons. Binary search strategy for the identification of the key frames in the video sequence with the concern of the time intervals of different duration (shot, 10 frames, one frame) must be used. For enhanced comprehension of the particular video segment, the *Play Video* option can be used. The *Load/Save Meta-Data* button can be used as backup/restore functionalities. After the annotations are identified for the whole video, each frame is associated with corresponding labels, if any.

Video Datasets Used in Experimental Semi-Automatic Face Annotation

This subsection describes the data sets used for semi-automatic face annotation, which we have used in order to illustrate effectiveness of the proposed approach in real scenarios. One common aspect of all videos used in the chapter is that they all are stored in MPEG-4 format with low-bit rate (100-300 Kbs). The dataset is composed of seven subsets: News, Commercial, Sci-Fi, Sport, Music, Oscar, and Adventure video sequences. The subsets were designed to encourage broad versatility of video genres for further evaluation. Facial complexity was classified according to the methodology that was presented in the previous sections. Detail statistics on the video datasets are given in Table 6; there, the Meta video sequence is used as a summary by incorporating information from all subsets discussed previously into a single video sequence. Table 6 summarizes results on semi-automatic annotating of the video sequences listed previously. It includes a broad range of experimental data, proving the effectiveness of the proposed approach.

Each level of computational complexity corresponds to a particular class for video analysis and indexing systems. Therefore, using the semi-automatic face annotation, one can find an optimal balance according to the tech specs of an application. A facial identifier that is used in the semi-automatic face annotation can be used for benchmarking of verification, recognition, open-world recognition, personal regroupment, and others. The purpose of this brief discussion is to point out the vast potential of the semi-automatic face annotation for optimizing and benchmarking the basic processes of face detection and recognition in video analyzing and indexing applications. Although use of semi-automatic face annotation has contributed greatly to face detection and recognition in video analysis, indexing optimization, and benchmarking, there are some inherited limitations of the current approach for facial analysis. A major limitation is the subjective facial categorization. Despite the fact that we have introduced the methodology to do this, it is highly unlikely that two different persons would identically segment the same video. Standard facial categorization algorithm can be used to completely solve this problem. Another limitation is the absence of a face detector that can be used for preliminary face annotation and nonlinear facial motion trajectories modeling. Finally, with the use of a facial features detector, one can store more complicated indexes; therefore, it allows improving flexibility of the benchmarking. However, in spite of these limitations, this type of semi-automatic face annotation has proved to perform reasonably well in practice for most face detection and recognition problems.

Table 6. Summary of annotating results on selected video sequences

Title of Video	News	Com-mercial	Sci-Fi	Sport	Music	Oscar	Adven-ture	Meta
Max duration of tracking a person	shot	shot	mixed	mixed	mixed	mixed	mixed	mixed
Total number of individuals	443	51	129	72	114	86	68	963
of type 0	111	7	4	1	5	0	3	131
of type 1	193	14	36	20	33	35	28	359
of type 2	139	30	89	51	76	51	37	473
Total number of facial images	35325	1987	10351	9821	4861	8160	4933	75438
of type 0	12698	314	361	62	501	-	174	14110
of type 1	17665	646	5063	6915	1645	3969	2320	38223
of type 2	4962	1027	4927	2844	2715	4191	2439	23105
Average number of facial images per frame	1.02	0.31	0.66	0.49	0.35	0.90	1.01	0.725
Average number of facial images per individual	79.74	38.96	80.24	136.40	42.64	94.88	72.54	78.33
of type 0	114.39	44.85	90.25	62	100.2	-	58	107.70
of type 1	91.52	46.14	140.63	345.75	49.84	113.4	82.85	106.47
of type 2	35.69	34.23	55.35	55.76	35.72	82.17	65.91	48.84
Number of frames	34307	6387	15600	19913	13796	9062	4878	103943
Semi-automatic annotation time (min)	270	50	146	72	85	68	45	736
Semi-automatic annotation FPH	7623	7664.4	6410.95	16594.17	9738.35	7995.88	6504	8473
Resolution	352x288	352x288	512x272	352x288	352x288	352x288	544x304	-
Frames per second	25	25	25	25	25	25	25	25
Size (Mb)	137	18	65	64	46	29	16	375

Future Trends and Conclusion

By now, it is agreed that among the necessary ingredients for the widespread deployment of semantic video retrieval applications services are standards for layout reconstruction, people recognition, objects recognition, settings detection, and multimodal integration (Haskell et al., 1998). Without standards, real-time services suffer because of different multimedia content providers that may not be able to communicate with each other (Haskell et al., 1998). Nonreal-time services using stored semantic indexes also may be disadvantaged because of either service providers' unwillingnesses to encode their content in a variety of formats to match customer capabilities or the reluctance of customers themselves to choose among a large number of digital media content provider devices to be able to handle a plethora of content formats (Haskell et al., 1998). As new technologies offer ever-greater functionality and performance, the need for standards to reduce the enormous number of possible permutations and combinations becomes increasingly important. Unlike video information coding, compression, representations, and transporting, which can take advantage of a well-understood and highly-developed (comparatively to the issue of interest) needs and solutions, semantic video retrieval has no such complete solution to rely on yet (Snoek & Worring, 2005). As such, in order to be able to exploit a wide range of semantic information contained within the video, it is essential to take advantage of any observable features of interest that can be recognized reliably and accurately (Colombo, Del Bimbo, & Pala, 1999). The four most important criteria of performance evaluations of any semantic video content understanding solution are semantic content description reliability, semantic content description accuracy (also known as false rejections and acceptance capacity of a system), computational complexity, and diversity of media content comprehension and interpreting capabilities.

- **Semantic video content description reliability:** It goes in front of the accuracy because of enormous data complexity for the unconstrained video (Satoh, Nakamura, & Kanade, 1999). Arguably, the most important decision to be made when developing the semantic video content understanding system is choosing priorities among accuracy and reliability. This selection leads directly to the technical specifications of potential users' demands. According to our years of experience in this field, we conclude that it is much more important to have a system that produces a minimum number of false acceptance errors (i.e., is highly reliable) rather than have a more accurate but less stable system in a general multimedia content understanding issue; therefore, the usage of semi-automatic video annotation techniques greatly contributes in the setup of comprehensive experiments with tuning of system performance balance.

- **Semantic video content description accuracy:** This guarantees an average number of false rejections in the case in which a semantic object wasn't recognized correctly. A greater number of false rejections leads to lower accuracy and, thus, makes either a selected part of the layout reconstruction, people recognition, objects recognition, settings detection, or multimodal integration unacceptable to be included in the semantic video understanding system. We must admit that the semi-automatic annotation presented in this chapter allows significant extension of experimental data involved;

thus, it makes it possible to work on real-world applications without spending a huge amount of human resources on data-preprocessing.

- **Computational complexity:** The prediction of semantic video retrieval methods' parameters guaranties the accuracy of accumulation of various descriptors' results and continuous video processing. Managing computational complexity helps to stabilize control over recognition routines and to make it less sensitive to changes in unconstrained video data. The multimodal integration units also are used to handle control over complexity for particular semantic descriptors according to the updated priorities. The strategy differs for real-time applications and off-line multimedia indexing and analysis systems. Usage of semi-automatic object annotation that includes profiles for various object detection/recognition complexities makes it possible to track and pay significant attention to every subproblem of the general issue, leading to a more balanced, accurate, and fast solution.

- **Diversity of media content comprehension and interpreting capabilities:** This takes advantage of the number of semantic descriptors involved in the video understanding problem and multimodal integration and reconstruction capabilities. Through various testing experiments, we have learned several ways to exploit a human's ability to integrate mid-level information from different sources into a high-level semantic knowledge. At the same time, a general concept of object annotation in video presented in this chapter supports all those descriptors in their diversities and results in a huge increase in the annotating performance for most of them.

In this chapter, we have attempted to present the theory of semi-automatic face annotation in semantic-based visual information retrieval applications from the simplest concepts of video indexing and analysis system to the formal description of the annotation methodology. It has been our purpose to focus on practical implementation issues; hence, we have avoided long, drawn-out proofs and derivations of the key results and have concentrated primarily on trying to interpret the meaning of the math and how it could be used in real-world systems. We have also attempted to illustrate some applications of the semi-automatic face annotation to the real-world experimental conditions with the sample test video sequences that consist of more than 100,000 frames, 950 individuals, and 75,000 facial images. We illustrated the solution to the issue in which an image processing and pattern recognition expert is able to segment and annotate facial patterns in video sequences at the rate of 7,500 frames per hour, which means a 30-time speedup compared to a conventional annotation mode. This chapter concludes with a discussion of future trends in semantic-based visual information retrieval from the perspective of this topic. Those researchers who are interested in the use of the test sequences with ground truth data should contact the first author of this chapter.

Acknowledgments

The authors are sincerely grateful to the reviewers for the numerous suggestions and critiques that enabled us to improve the chapter.

References

Colombo, C., Del Bimbo, A., & Pala, P. (1999). Semantics in visual information retrieval. *IEEE Multimedia, 6*(3), 38–53.

Eickeler, S., Wallhoff, F., Iurgel, U., & Rigoll, G. (2000). Content-based indexing of images and video using face detection and recognition methods. In *Proceedings of the IEEE International Conference on Acoustics, Speech, and Signal Processing ICASSP*. Retrieved September 26, 2006, from http://citeseer.ist.psu.edu/eickeler01contentbased.html

Haskell, B.G., et al. (1998). Image and video coding-emerging standards and beyond. *IEEE Transactions on Circuits and Systems for Video Technology, 8*(7), 814–837.

Hoogs, A., Rittscher, J., Stein, G., & Schmiederer, J. (2003). Video content annotation using visual analysis and large semantic knowledgebase. In *Proceedings of the Conference on Computer Vision and Pattern Recognition*. IEEE.

Jones, M. J., & Viola, P. (2003). Fast multi-view face detection. In *Proceedings of the IEEE Conference on Computer Vision and Pattern Recognition*. Retrieved September 26, 2006, from http://www.merl.com/publications/TR2003-096/

Karpova, E., Tsishkou, D., & Chen, L. (2003). The ECL skin-color images from video database. In *Proceedings of the IAPR International Conference on Image and Signal Processing* (pp. 47–52).

Lin, C.-Y., Tseng, B., & Smith, J. R. (2003). *Video collaborative annotation forum: Establishing ground-truth labels on large multimedia datasets*. Retrieved from http://www-nlpir.nist.gov/projects/tvpubs/tvpapers03

Manjunath, B. S., Salembier, P., & Sikora, T. (2002). *Introduction to MPEG 7: Multimedia content description language*. New York: Wiley.

Nagao, K., Ohira, S., & Yoneoka, M. (2002). Annotation-based multimedia summarization and translation. In *Proceedings of the International Conference on Computational Linguistics*.

Naphade, M., Lin, C.-Y., Smith, J. R., Belle, L.T., & Basu, S. (2002). Learning to annotate video databases. In *Proceedings of the SPIE Electronic Imaging 2002—Storage and Retrieval for Media Databases*. Retrieved from http://www.research.ibm.com/VideoAnnEx

Ricoh Image Communication. (2005). *Movie Tool*. Retrieved from http://www.ricoh.co.jp/src/multimedia/MovieTool/about/index.html

Rowley, H. A., Baluja, S., & Kanade, T. (1998). Neural networkbased face detection. *IEEE Trans. on Pattern Analysis and Machine Intelligence, 20*(1), 23–38.

Satoh, S., Nakamura, Y., & Kanade, T. (1999). Nameit: Naming and detecting faces in news videos. *IEEE Multimedia, 6*(1), 22–35.

Snoek, C. G. M., & Worring, M. (2002). A state-of-the-art review on multimodal video indexing. In *Proceedings of the 8th Annual Conference of the Advanced School for Computing and Imaging* (pp. 194–202).

Snoek, C. G. M., & Worring, M. (2005). Multimodal video indexing: A review of the state-of-the-art. In *Multimedia tools and applications*. Retrieved from http://carol.science.uva.nl/~cgmsnoek/publications.html

Sung, K. K. (1996). *Learning and example selection for object and pattern detection* [doctoral thesis]. Massachusetts Institute of Technology, Electrical Engineering and Computer Science. Retrieved September 26, 2006, from http://historical.ncstrl.org/tr/pdf/mitai/AITR-1572.pdf

Tsekeridou, S., & Pitas, I. (2001). Content-based video parsing and indexing based on audio-visual interaction. *IEEE Transactions on Circuits and Systems for Video Technology, 11*(4), 522–535.

Tseng, B. L., Lin, C.-Y., & Smith, J.R. (2002). Video personalization and summarization system. In *Proceedings of SPIE Photonics East 2002—Internet Multimedia Management Systems*. Retrieved from http://www.research.ibm.com/VideoAnnEx

Tsishkou, D. (2002). Face detection, tracking and recognition in video using genetic algorithms. In *Proceedings of the GECCO 2002 Graduate Student Workshop* (pp. 304–306).

Tsishkou, D., Chen, L., & Peng, K. (2002). Face detection, tracking and recognition in video. *Journées Francophones sur l'Accès Intelligent aux Documents Multimédias*, 83–90.

Viola, P., & Jones, M. (2001). Rapid object detection using a boosted cascade of simple features. In *Proceedings of the Second International Workshop on Statistical and Computational Theories of Vision-Modeling, Learning, Computing and Sampling*. Retrieved September 26, 2006, from http://research.microsoft.com/~viola/Pubs/Detect/violaJones_CVPR2001.pdf

Wang, Y., Liu, Z., & Huang, J. (2000). Multimedia content analysis using both audio and visual clues. *Signal Processing Magazine, 17*, 12–36.

Zhao, W., Bhat, D., Nandhakumar, N., & Chellappa, R. (2000). A reliable descriptor for face objects in visual content. *Journal of Signal Processing: Image Communications, Special Issue on MPEG-7 Proposals, 16*, 123-136.

Zhao, W., Chellappa, Resonfeld, A., & Phillips, P. J. (2000). *Face recognition: A literature survey* (UMD CfAR Tech. Rep. No. CAR-TR-948). Retrieved September 26, 2006, from http://citeseer.ist.psu.edu/zhao00face.html

Zhu, X., Fan, J., Elmagarmid, A. K., & Wu, X. (2003). Hierarchical video summarization and content description joint semantic and visual similarity. *ACM/Springer Multimedia Systems Journal, 9*(1), 31–53.

Chapter X

An Ontology-Based Framework for Semantic Image Analysis and Retrieval

Stamatia Dasiopoulou, Aristotle University of Thessaloniki, Greece

Charalampos Doulaverakis, Centre for Research & Technology Hellas, Greece

Vasileios Mezaris, Centre for Research & Technology Hellas, Greece

Ioannis Kompatsiaris, Centre for Research & Technology Hellas, Greece

Michael G. Strintzis, Aristotle University of Thessaloniki, Greece

Abstract

To overcome the limitations of keyword- and content-based visual information access, an ontology-driven framework is developed. Under the proposed framework, an appropriately defined ontology infrastructure is used to drive the generation of manual and automatic image annotations and to enable semantic retrieval by exploiting the formal semantics of ontologies. In this way, the descriptions considered in the tedious task of manual annotation are constrained to named entities (e.g., location names, person names, etc.), since the ontology-driven analysis module automatically generates annotations concerning common domain objects of interest (e.g., sunset, trees, sea, etc.). Experiments in the domain of outdoor images show that such an ontology-based scheme realizes efficient visual information access with respect to its semantics.

Introduction

As a result of the recent advances in numerous key technologies that enable content creation, processing, and delivery, huge amounts of digital visual content are currently available in unstructured and nonindexed forms on the Web in proprietary commercial repositories and in personal collections. This ever-increasing growth of available content gave rise to a plethora of visual information-related applications, which, in turn, established the urgent need for tools to intelligently analyze, index, and manage the available visual content in a manner that is capable of satisfying end users' real information needs.

A key functionality, therefore, is the ability to access the content in terms of the semantics it represents rather than relying on mere denotative aspects, and to allow users to interact with it in a meaningful way. However, there exists a significant gap between the desired semantic level of access and the one currently provided by the existing tools for automatic visual content analysis and retrieval. Manual annotation approaches, despite ensuring conceptual descriptions of a high level of abstraction, suffer from subjectivity of descriptions and are extremely expensive in terms of labor and time resources. Hence, given the volume of information to deal with, it is imperative that the process of extracting (i.e., analyzing) such semantic descriptions (i.e., annotations) takes place in an automatic manner, or with minimum human intervention.

Acknowledging machine perception of the rich semantics that characterize visual content as the only viable solution, the so-called *semantic gap* between visual descriptions, user queries, and actual meaning conveyed (Smeulders, Worring, Santini, Gupta, & Jain, 2000) has received a huge amount of research attention from a variety of research disciplines. Seen from the analysis perspective, the challenge lies in how to associate low-level measurable signal-based properties of visual content with the connotative meaning in a (semi-) automated manner. Early analysis approaches focused on automatic extraction of numerical descriptions based on low-level features (e.g., color, motion, shape, etc.), and retrieval was performed based on similarity metrics attempting to imitate the way humans evaluate image similarity. However, such data-driven analysis and subsequent retrieval fail to capture the underlying conceptual associations, while performance is bound to the application-specific context of use (Brunelli, 1999). The use of domain knowledge appears to be a promising way by which higher-level semantics can be incorporated into techniques that capture the semantics through automatic analysis, and therefore, numerous approaches to knowledge-driven analysis have emerged.

Clearly, incorporating prior knowledge is a prerequisite for the retrieval as well in order to allow understanding of the semantics of user search and, subsequently, aligning and matching with the content annotation semantics. Proprietary knowledge representation solutions, although effective within the predefined usage context, impose serious limitations considering reusability and interoperability. Underpinned by the emergence of the Semantic Web, the need for sharable, sufficiently rich, knowledge representation formalisms has revived interest and has given rise to advanced knowledge modeling solutions. Ontologies (Staab & Studer, 2004), promising a common and shared understanding of a domain, have turned up into key enabling technologies providing machine understandable semantics. Furthermore, due to the well-defined semantics, ontologies provide automated inference support that can further enhance retrieval by exploiting hidden conceptual associations discovery and derivation of new ones for realizing semantic-based recommendation services.

In this chapter, a knowledge-driven framework using ontologies as the means for knowledge representation is investigated for still image semantic analysis and retrieval. An appropriately defined domain ontology provides the vocabulary for the annotation, drives the extraction of automatically produced annotations, and provides the conceptualization utilized in retrieval. In the sequel, a comprehensive review of the relevant background in semantic image annotation and retrieval is presented, followed by a detailed description of the proposed ontology infrastructure and the ontology-based semantic analysis and retrieval components. Experimental results in the domain of outdoor images and comparative evaluation against both plain content-based and keyword-based retrieval illustrate the contribution of the proposed approach.

Background and Related Work

Semantic Image Retrieval

Semantic-based visual information retrieval approaches have witnessed different trends while evolving to meet user needs and application requirements. The first generation considered manual descriptions in the form of textual or keyword annotations, and retrieval took place in the textual domain. Although the employed annotations entailed a high level of abstraction, they lacked formal semantics, leading to low precision and recall. Additional weaknesses, namely, the different terminologies employed among users and annotators and insufficient user familiarity with the subject area, in combination with the high cost of manual annotation, rendered such approaches inefficient.

As a result, the next generation of retrieval approaches turned to annotations that could be extracted automatically from the visual content. Low-level visual features, such as color, texture and shape, as well as higher-abstraction structural information formed the content annotation metadata. Indexing was performed using the automatically extracted numerical descriptors, and retrieval was based on similarity measures that try to match the way humans perceive visual similarity. Four broad categories can be identified, depending on the chosen content and indexing paradigm: query by example, iconic, textual, and hybrid. The relevant literature considers a huge number of diverse approaches for a variety of application domains (Smeulders et al., 2000). Although such approaches performed satisfactorily within their context of usage, retrieval was defined solely on syntactic similarity, and no insight into the perceptual and conceptual reasons underlying visual similarity assessment could be induced and exploited. Techniques such as relevance feedback and incremental machine learning enabled user intervention in the process of knowledge acquisition, and promising semi-automated and semi-supervised retrieval systems emerged (Zhou & Huang, 2003).

Naturally, the consequences of the semantic gap between the high-level conceptual user queries and the low-level annotations automatically extracted from visual content remotivated research in analysis, aiming this time at automatic extraction of high-level semantic-based annotations, as discussed in the following section. However, coupling the semantic annotations produced by such an automatic analysis module with query-by-example or keyword-based retrieval approaches is still far from semantic-based access. Consequently, apart from the necessity of having content automatically annotated according to the meaning conveyed, the need for semantic-enabled annotations representation and manipulation during retrieval was established.

The emerging Semantic Web vision and the efforts toward standardized knowledge modeling solutions that enable semantics sharing and reuse naturally affected the multimedia retrieval community as well. Ontologies, being the leading-edge technology for providing explicit semantics in a machine-processable rather than just a readable form, were espoused for semantically annotating, browsing, and retrieving visual content. Ontology-based frameworks for manual image annotation and semantic retrieval include the ones presented in Schreiber, Dubbeldam, Wielemaker, and Wielinga (2001) and (Hollink, Schreiber, Wielemaker, & Wielinga (2003), considering photographs of animals and art images, respectively. In Reidsma, Kuper, Declerck, Saggion, and Cunningham (2003), ontology-based information extraction is applied to improve the results of information retrieval in multimedia archives through a domain-specific ontology, multilingual lexicons, and reasoning algorithms to integrate cross-modal content annotations. Ontologies also have been applied successfully for handling museum collections, as in Sinclair, Goodall, Lewis, Martinez, and Addis (2005), in which they are used to attach semantics to the multimedia museum objects and support interoperable, concept-based browsing and display.

Semantic Image Analysis

The severe limitations in accessing visual content effectively when only low-level numerical descriptions are considered resulted in shifting focus from the selection of suitable descriptors and matching metrics to the exploitation of knowledge in the automated analysis and annotation process. A variety of knowledge-driven approaches have been proposed within the last years, a rough categorization of which yields two classes depending on the adopted knowledge acquisition and representation process (i.e., implicit, as realized by machine-learning techniques, and explicit, as in the case of model-based approaches). The former is advantageous when addressing ill-posed problems for which knowledge is incomplete or too complex to be represented explicitly. However, usually they require large amounts of training data, retraining is required to adapt to different applications or to further extend the current framework, and more importantly, their internal representation does not allow for transparency in the learned associations. On the other hand, model-based approaches consider the available knowledge in the form of formally defined facts and rules, and enable further inference.

Among the most commonly applied machine learning techniques are hidden Markov models (HMMs), support vector machines (SVMs), and neural networks (NNs). In the COBRA model described in Petkovic and Jonker (2001), HMMs are combined with rules and appropriate grammars to formalize object and event descriptions, while in Naphade, Kozintsev, and Huang (2002), a factor graph network of probabilistic multimedia objects (multijects) is defined using HMMs and Gaussian mixture models. Visual and audio information using Controlled Markov Chains also is employed in Leonardi, Migliorati, and Prandini (2004) for the purpose of structural and semantic classification in soccer videos, while in Xie, Xu, Charn, and Divakaran (2004), HMMs are used for soccer structure analysis.

In Bose and Grimson (2003), an object SVM classifier is presented, which adapts automatically to arbitrary scenes by exploiting context features. An SVM-MRF framework also is introduced in Wang and Manjunath (2003) to model features and their spatial distributions, while in Li, Wang, and Sung (2004), an SVM active learning approach is presented to address multilabel image classification problems. SVMs also have been employed for the detection of concepts such as goal, yellow card, and substitution in the soccer domain, as in Snoek and Worring (2003), and for the semantic classification of news videos (Lin & Hauptmann, 2002). NNs have been proposed among others for locating clothed people in photographic images (Sprague & Luo, 2002) and for contextual edge-based car detection in real-time video streams (Nescovic & Cooper, 2003). Neural learning and fuzzy inference have been investigated for event semantics recognition in Tzouvaras, Tsechpenakis, Stanou, and Kollias (2003) and Kouzani (2003) for locating human faces within images.

Considering model-based semantic analysis, the approach taken for knowledge representation, varying from proprietary formats to logic-based formalisms, is the key differentiating factor. In Yoshitaka, Kishida, Hirakawa, and Ichikawa (1994), an ad-hoc, object-oriented data model is adopted. In Lingnau and Centeno (2003), image classification is performed, based on fuzzy rules provided by domain experts through a graphical interface. In the approach presented in Dorado, Calic, and Izquierdo (2004), fuzzy logic and rule mining techniques are utilized to approximate humanlike reasoning, while in Bertini, Del Bimbom and Nunziati (2005), finite state machines are used for encoding the temporal evolution of sports highlights.

In accordance with recent Semantic Web advances, several analysis approaches emerged utilizing the formal semantics and inference capabilities of ontologies. Description logic-based approaches have been proposed in Meghini, Sebastiani, and Straccia (1997) for describing both the form and the content of multimedia documents, and in Schober, Hermes, and Herzog (2004) for inferring semantic objects in images. In Mezaris, Kompatsiaris, and Strintzis (2004), a priori ontological knowledge is used to assist semantic-based image retrieval, while in Hudelot and Thonnat (2003), an ontology-based cognitive vision platform for the automatic recognition of natural complex objects is presented. RuleML rules are introduced in Hunter, Drennan, and Little (2004) to associate MPEG-7 low-level descriptors with high-level concepts, and in Dasiopoulou, Mezaris, Kompatsiaris, Papastathis, and Strintzis (2005), ontological knowledge augmented with F-Logic rules determine both the analysis steps and the identification of semantic objects in video. The user-assisted approach for automatic image annotation reported in Little and Hunter (2004) also is enhanced by rules on top of a domain ontology.

The presented background overview clearly shows that both semantic-based retrieval and automatic analysis and annotation of visual content turn toward knowledge-driven solutions. Furthermore, ontology appears to be the prevalent choice for knowledge representation and management among the computer vision community.

Ontology-Based Visual Content
Semantic Analysis and Retrieval

In this section, the proposed ontology-based semantic image analysis and retrieval framework is detailed and demonstrated in the domain of outdoor photographs. It must be noted that the domain of outdoor photographs is only an example in which the proposed framework is applicable. The presented ontology-based framework easily can be adapted to support different domains, provided the appropriate domain knowledge definitions are available. In the following, the motivation behind coupling semantic analysis and retrieval using an ontology-based framework is discussed first, and then the developed ontology infrastructure (i.e., the domain- and analysis-related definitions) is presented. Subsequently, the implementation details of the proposed ontology-driven analysis and retrieval components are given, and the section concludes with experimental results and comparative evaluation against plain content- and keyword-based retrieval, demonstrating the contribution and potential of the approach proposed.

Motivation

As implied by the state-of-the-art overview, making visual content semantics explicit is a prerequisite for enabling true semantic-based image retrieval and translates into two main challenges: the extraction of the conveyed semantics and their semantic-enabled processing to determine relevance with user queries. Acknowledging the disadvantages of manual annotation and taking into account the need for incorporating prior knowledge into the analysis itself, we propose an ontology-based framework, as illustrated in Figure 1, for enabling and coupling semantic image analysis and retrieval.

Under the proposed approach, a domain ontology provides the conceptualization and vocabulary for structuring content annotations. Thus, the annotation terminology is made transparent, and semantic browsing based on the ontology model is enabled, facilitating query formulation, while matching the user query semantics with the annotation metadata semantics becomes straightforward. This domain ontology drives both the restricted to named entities manual annotation and the automatic generation of lower-level descriptions by the semantic analysis module. To accomplish the latter, an analysis ontology is defined appropriately to model the analysis process and the knowledge required for proceeding with the domain concepts detection. The added value lies in the improvement entailed in both retrieval and analysis as individual components, and in the potential brought through their coupling under a unified ontology infrastructure.

Considering the retrieval component, the approach to ontology-based annotation and retrieval follows the relevant paradigm of the contemporary literature, taking advantage of the advances ontologies bring in knowledge sharing and reuse. Hence, from the retrieval perspective, the proposed framework is a practical example in this area. However, it must be noted that contrary to the typical ontology-based retrieval approaches in which the actual annotations upon which retrieval takes place either are assumed to be already available or need to be produced manually, under the proposed framework part of the needed annotations is generated automatically.

Figure 1. Ontology-based semantic image analysis and retrieval framework

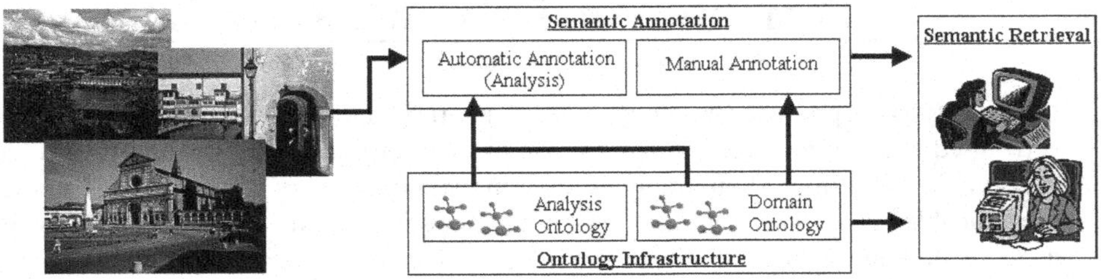

Ontology, being a formal specification of a shared conceptualization (Gruber, 1993), provides by definition the formal framework required for exchanging interoperable knowledge components. By making semantics explicit to machines, ontologies enable automatic inference support, thus allowing users, agents, and applications to communicate and negotiate over the meaning of information. Typically, an ontology identifies classes of objects that are important for the examined subject area (domain) under a specific viewpoint and organizes these classes in a taxonomic (i.e., subclass/super-class) hierarchy. Each such class is characterized by properties that all elements (instances) in that class share. Important relations between classes or instances of the classes are also part of the ontology. Advocated by the emergent Semantic Web vision, a number of initiatives, such as the RDFS, DAML+OIL, and lately OWL, have been defined by the World Wide Web Consortium (W3C) in order to attach meaning to information on the Web and to support knowledge sharing and reuse.

Consequently, following the ontology-based paradigm for the representation of the annotation metadata entails significant advantages for the purpose of semantic search and retrieval. More specifically, unlike keyword-based annotations, ontology-based annotations do not rely on plain word matching and, thus, manage to overcome the syntactic limitations from which keyword-based search and retrieval systems suffer (Kiryakov & Simov, 1999). Synonyms and homonyms, two common reasons accounting for low recall and precision, respectively, in traditional keyword-based approaches, do not have any implication when content descriptions are structured based on ontologies.

However, despite the potential brought by ontologies in semantic visual content annotation and retrieval, as briefly discussed previously and in the relevant overview section, there are still severe limitations due to the assumption of manual annotation on which these approaches build. Additionally, since the domain ontologies with which annotation and retrieval is performed are developed independently of the knowledge-modeling solutions employed by knowledge-assisted analysis approaches, their integration becomes rather difficult to handle. Under the presented framework, the practical issues related to the high cost of manual annotation are overcome partially by restricting manual annotations only to higher-level domain semantics and proper names. Furthermore, as described in the future trends section, following such an approach allows for further reducing the effort required for manual annotation by extending the current framework with appropriate rules to enable automatic inference of higher-level semantics based on the already existing, manually and automatically generated ones.

More specifically and likewise to retrieval, ontologies enhance traditional knowledge-assisted multimedia content analysis methodologies through the formal, inference-enabled, knowledge-modeling framework that they provide. The obtained advantages are twofold. Analysis and extraction of high-level content descriptions benefit from the formally encoded semantic relations and the supported inference capabilities, while the employed knowledge can be shared easily among different applications due to the explicit well-defined semantics of the modeled domain. Thereby, complex concepts or concepts with similar visual characteristics can be inferred, based on contextual knowledge, partonomic relations, and so forth. Additionally, the generated annotation metadata can be used directly in semantic Web-enabled retrieval applications by making available the corresponding ontology schema. Under the proposed framework, we exploit the formal semantics for achieving reliable detection of domain concepts that exhibit similar visual characteristics and additionally for determining the most appropriate detection steps from a pool of low-level processing algorithms in accordance with the domain concepts definitions.

In order to exemplify better the motivation and contribution of the proposed unified ontology-based analysis and retrieval framework, let us consider as an example of the current domain of experimentation a photograph depicting St. Peter's Basilica. Information such as that photograph is of a religious building and, more specifically, a catholic church located in Rome, which is a European country, is expected to be included in the annotations produced manually for this image. Additional annotations could refer to the architecture, the date the photograph was taken, and so forth. Using such annotations and relying solely on the ontology-based retrieval paradigm, as presented in the relevant literature, would support queries on images semantically related to St. Peter's Basilica, religion, Italy, and so forth. Hence, improved precision and recall rates compared to traditional keyword- and content-based retrieval approaches would be attained. However, if a query referred images of religious buildings located in Italy taken on a cloudy day, at sunset, or surrounded by trees, then the retrieved results would not differ from the ones retrieved in the case of a query about religious buildings located in Italy.

Such performance limitations are to be expected with regard to the existing ontology-based retrieval approaches, since the effort required for manual annotation does not allow exploiting the complete ontology conceptualization. Unavoidably, less attention will be given to information about semantics of ambivalent importance, such as if the church surroundings are cityscape, mountainous, and so forth. Consequently, the added value of the proposed framework lies in the introduction of an automatic semantic analysis to account for semantic domain concepts that are common but tend to be left out during the manual annotation. Thereby, considering the examined example case, queries about photographs "depicting a church with vegetation around" or "depicting a beach scene at sunset" are supported, enhancing the effectiveness and usefulness of the provided semantic-based analysis and retrieval framework.

Furthermore, the analysis performance is improved since the domain ontology definitions allow differentiating among domain objects when visual features alone are not sufficient. In the current implementation this functionality considers only spatial contextual knowledge at a local level (i.e., neighboring regions) and not global image context. Considering the examined domain, this means that objects such as Sea and River are successfully discriminated from Sky; Trees are distinguished from Grass, and so forth.

Ontology Infrastructure

In order to implement the proposed ontology-based framework for performing analysis and retrieval of visual content at a semantic level, two ontologies—analysis and domain—had to be defined and integrated appropriately. The domain ontology formalizes the domain semantics, providing the conceptualization and vocabulary for the visual content annotations and the subsequent semantic retrieval. The analysis ontology is used to guide the analysis process and support the detection of certain concepts defined in the domain ontology. Both ontologies are expressed in RDF(S), and their integration takes place using the conceptually common classes between the two ontologies as attachment points, thus resulting in a unified ontology-based framework for handling visual content at a semantic level. It is interesting to note that similarly to the visual content analysis approach used to automatically extract part of the underlying content semantics, additional annotations could be extracted automatically using an ontology-based natural language processing approach, as in Popov, Kiryakov, Ognyanoff, Manov, and Kirilov (2004), Domingue, Dzbor, & Motaa (2004), to analyze the textual descriptions accompanying the visual content and, thus, to further reduce the effort and cost of manual annotation.

Domain Ontology

As illustrated in the domain ontology snapshot of Figure 2, the main classes of the domain ontology account for the different types of objects and events depicted in an outdoor photograph. Relations are defined to model additional information regarding the person who took

Figure 2. Domain ontology

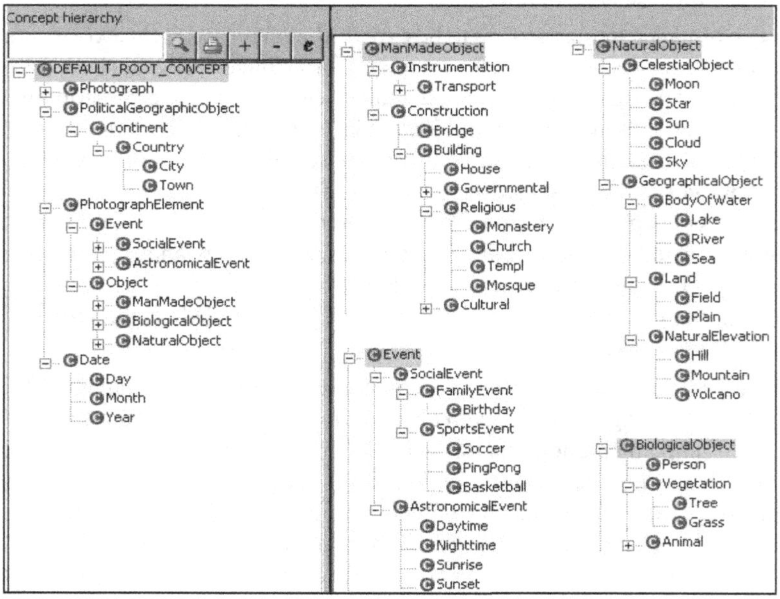

the photograph, the date the photograph was taken, and the corresponding location, as well as the way the various objects and events relate to each other and to the different classes of outdoor photographs. However, since in the current stage of development the semantic recommendation functionalities are not implemented, these relations are not utilized.

The top-level object and event classes are quite generic and follow the semantic organization espoused in DOLCE (Gangemi, Guarino, Masolo, Oltramari, & Schneider, 2002), which was designed explicitly as a core ontology. As such, DOLCE is minimal, in the sense that it includes only the most reusable and widely applicable upper-level categories, and rigorous in terms of axiomatization. Following the modeling specifications of DOLCE and structuring the developed domain ontology accordingly was based on the premise that due to the nature of core ontologies, such a choice would result in well-structured and consistent representations of the domain semantics. Hence, using a different outdoor photograph ontology, retrieval still would be valid as long as the associations to the corresponding DOLCE classes were available, ensuring extensibility and interoperability across applications.

The main classes of the ontology are briefly discussed next:

- **Class photograph:** Several properties are associated with this class in order to describe the features of every photograph instance. All instances of class photograph hold at least the type of photograph (indoors/outdoors) and the depicted objects and events. Furthermore, appropriate relations are defined to represent information about the person who took the photograph and the date and place that the photograph was taken.

- **Class event:** This class models the variety of events that could be depicted in a photograph taken outdoors, thus consisting of social events subclassed into family-related, sport- related, and so forth, as well as events associated with natural phenomena such as rain, snow, sunset, sunrise, and so forth.

- **Class object:** This class represents the main categories of objects that commonly are found in outdoor photographs. Thus, it is divided into three subcategories: (1) manmade objects that include, among others, buildings, constructions such as bridges and roads, and transportation-related artifacts; (2) natural objects that include various geological formations (e.g., mountain, volcano); and (3) biological objects that include the various living organisms encountered in such a context (i.e., persons, animals, and vegetation). The corresponding subclass hierarchies are shown in Figure 2.

- **Class political geographic object:** This class is used to model the various political geographic categorizations (i.e., countries, cities, etc.). In combination with class place, it provides the necessary definitions to represent the location of a depicted place, object, or event.

Given the domain ontology definitions, the annotation of the provided visual content can be accomplished in quite a straightforward way using any of the publicly available ontology-based image annotation tools, such as PhotoStuff (http://www.mindswap.org/2003/PhotoStuff/) and AKTive Media (http://www.dcs. shef.ac.uk/~ajay/html/cresearch.html).

Analysis Ontology

The goal of knowledge-assisted semantic visual content analysis is to extract semantic descriptions from low-level image descriptions by exploiting prior knowledge about the domain under consideration. Therefore, such domain knowledge needs to include proto-typical descriptions of the important domain concepts (i.e., objects and events) in terms of their visual properties and spatial context of appearance in order to allow for their identification. Detection then can be described roughly as a process based on appropriately defined matching criteria between the low-level features that are automatically extracted from the content and the predefined models that comprise the domain knowledge, plus possibly some additional decision-making support. Consequently, under this perspective, the developed analysis ontology can be seen as extending the initial domain ontology with qualitative and quantitative descriptions for those domain concepts that are to be detected automatically. Additionally, the developed analysis ontology models the analysis process itself (i.e., the low-level processing algorithms employed for the defined concepts detection), how they relate to the qualitative objects features, as well as issues regarding possible interdependences in object detection (i.e., certain objects detection maybe be facilitated significantly if other objects are detected first).

As illustrated in the analysis ontology snapshot of Figure 3, the main classes of the developed analysis ontology are the following:

- **Class object:** It is subclassed to the various types of domain concepts that need to be supported by the analysis. Consequently, in the current implementation, class object reduces to the subclasses of the domain ontology PhotographElement class that can be detected automatically, (i.e., the concepts sky, vegetation, sunset, and body of water, as detailed in the experimental results section). All object instances comprise models (prototypes) for the corresponding semantic concepts. Through the definition of ap-

Figure 3. Analysis ontology

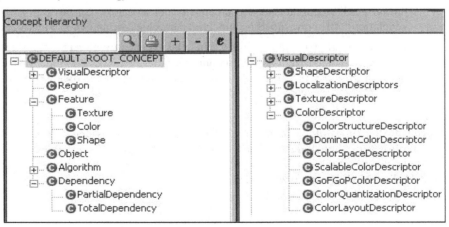

propriate properties, each object instance is linked to its visual description in terms of low-level features and spatial characteristics.

- **Class feature:** It is subclassed into the various visual content information modalities (i.e., color, texture, shape) and allows associating each object instance with the actual low-level descriptors that form its corresponding prototypical description.

- **Class visual descriptor:** It is subclassed into the various low-level descriptors that can be used to represent each visual feature, as illustrated in the right part of Figure 3. In order to further support interoperability of the proposed framework, the definition of the employed low-level descriptors followed the specifications of the MPEG-7 standard MPEG-7 Visual (2001). However, any other proprietary descriptor could be used instead or in combination with the standardized ones, showing another benefit of the formal knowledge representation framework provided by ontologies.

- **Class algorithm:** It includes the low-level processing algorithms used for spatial relation extraction and for low-level feature extraction and matching. Through the definition of appropriate rules, the selection of the most appropriate processing steps is determined on the basis of the detected objects visual features.

- **Class dependency.** It is used to model the possible dependencies between the concepts to be detected so that analysis can benefit from exploiting them. More specifically, this concept addresses the possibility that the detection of one object may depend on the detection of other objects due to spatial correlation.

The domain knowledge modeled by the analysis ontology, apart from qualitative and quantitative descriptions, is further augmented with F-Logic (2004) rules in order to determine the analysis steps required for the detection of the defined domain concepts. Three types of rules are currently used: rules to associate the available low-level processing algorithms to their corresponding low-level features and respective descriptors, while additionally determining the algorithms relevant execution order; rules to determine each algorithm input parameter values; and rules to deal with object dependencies as previously explained.

By building this unifying model of all aspects of the semantic visual content analysis process, all its parts are treated as ontological concepts. Thus, under the proposed framework, various visual content low-level descriptors and processing algorithms can be incorporated easily just by defining corresponding instances in the ontology and by providing appropriate interfaces. Furthermore, additional domain concepts can be detected and automatically extracted by describing their attributes and corresponding low-level features in the ontology and populating the knowledge base with appropriate instances. Using additional rules about object partonomic relations and rules to determine complex concepts that cannot be detected directly from visual features (e.g., the concept of Island or SubMarineScene), higher-level semantics extraction can be achieved, leading to more complete automatically produced annotations. As already mentioned, in the current implementation, the automated inference support has been used only for control strategy tasks (i.e., for determining the most appropriate detection steps from a pool of low-level processing algorithms). It is within future intensions to extend the current framework with domain-specific rule definitions to automatically infer higher-level semantics.

Semantic Analysis

The architecture of the implemented ontology-driven semantic image analysis system is presented in Figure 4. As illustrated, analysis starts by segmenting the input image and by extracting low-level visual descriptors and spatial relations in accordance with the domain ontology definitions. Then, an initial set of hypotheses is generated by matching the extracted low-level descriptors against the ones of the objects prototypes included in the domain knowledge definitions. In order to evaluate the plausibility of the produced hypothetical annotations for each region and to reach the final semantic annotation, the spatial context domain knowledge is used, whereby the image semantics are extracted and the respective annotation metadata are generated.

The segmentation used is an extension of the well-known recursive shortest spanning tree (RSST) algorithm (Adamek, O'Connor, & Murphy, 2005). The extraction of the supported visual descriptors, currently the Dominant Color and the Region Shape, is performed following the MPEG-7 (2001) standard guidelines. As mentioned previously, in the domain knowledge definitions, apart from low-level descriptions, it is necessary to include information about objects spatial context as well, since this is the only way to discriminate different objects with similar visual characteristics. Apart from adjacency and inclusion information, the presently employed spatial context knowledge includes four directional relations: above-of, below-of, left-of, and right-of. Additionally, two absolute relations—above-all and below-all—were introduced.

Having extracted the low-level visual descriptors and the spatial context for the examined image regions, the next step is to calculate the degree of matching between the objects prototype descriptors that are included in the domain knowledge and the ones extracted from the segmented image regions in order to determine which domain concepts are plausible annotations for each region. To accomplish this, it is essential to form for each region a single measure that accounts for all employed low-level descriptors. This combined distance then is used to estimate the degree of matching with the domain concepts prototype instances. Since MPEG-7 does not provide a standardized method for estimating a single distance based on more than one visual descriptor, the simple approach of weighted summation was followed. The weight of the Dominant Color descriptor was selected to be greater than the one used for the Region Shape descriptor, since the former has exhibited better discriminative performance for the current experimentation domain.

Figure 4. Ontology-based semantic image analysis

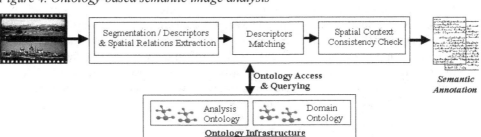

Since during the descriptor matching step, only low-level visual features are taken into account, a set of possible semantic annotations with different degrees of confidence is generated for each region of the analyzed image. To evaluate the plausibility of each of these hypotheses in order to reach the final semantic interpretation, spatial context is used. More specifically, for each pair of adjacent regions, the system checks whether the extracted spatial relation matches the spatial context that is associated with the domain concepts included in each region hypotheses set.

The determination of the exact sequence of analysis steps (i.e., low-level processing algorithms) required for each of the supported concepts detection is based on appropriately defined rules. As mentioned in the previous section, three types of rules have been employed in order to associate qualitative features with processing algorithms, to determine their corresponding input arguments, and to define execution order taking into account the possible dependencies among the examined concepts. Taking as an example the Body of Water concept of the current implementation, the resulting detection steps include color-based segmentation, extraction of the produced segments Dominant Color descriptors, matching against this concept prototypes to generate initial hypotheses, detection of concept Sky, and extraction and checking of consistency of the below-of spatial relation between segments hypothesized as Body of Water and those annotated as Sky. This sequence of steps was a direct result of the domain definitions associated with the Body of Water concept (i.e., a color homogeneous object characterized using the Dominant Color descriptor and respective prototype values and associated to the concept Sky through the below-of spatial relation). In Figure 5, exemplary results of the developed semantic analysis system are illustrated for the outdoor photographs domain.

The retrieval of the domain concepts prototype instances required during analysis for generating the initial hypotheses set as well as the process of inferring which low-level processing algorithms are the most appropriate for each object detection are performed using the OntoBroker (http://www.ontoprise.de/products/ontobroker_en) engine. OntoBroker supports RDF(S) as input ontology language; hence, all appropriate ontology files can be loaded easily. Furthermore, its internal representation is based on F-Logic, thus offering a convenient way to add rules on top of an ontology and to reason over them.

Semantic Retrieval

In order to realize the ontology-based retrieval component, the semantic-enabled functionalities of an existing RDF knowledge base are exploited. To this end, Sesame was selected for storing and querying the RDFS ontology and metadata. Sesame (http://www.openrdf.org/) is an open source RDF database that allows persistent storage and at the same time provides support for RDF schema inference and querying. A Web-based interface has been developed to allow communicating user queries to the knowledge base, retrieving the query results, and accessing the content repository.

As shown in Figure 6, in the developed interface, the subclass hierarchies of the four main classes of the domain ontology (i.e., Object, Event, Political Geographic Object and Date classes) are depicted. For the Political Geographic Object domain class, the Location naming convention was followed to enhance user-friendliness. The selection of these four classes is a direct result of the specifications followed during ontology building; namely, the assump-

Figure 5. Semantic analysis results

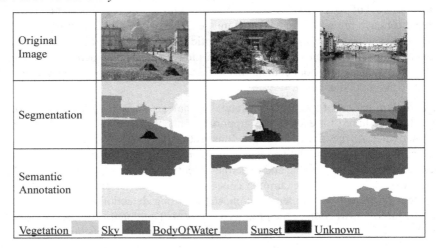

tion that the image semantics address the objects and events depicted as well as the location and date the photograph was taken. Thereby, the user can navigate through the respective concept hierarchies, explore the underlying content semantics organization and familiarize with the vocabulary and the associations among the modeled domain concepts.

Under the developed semantic-based retrieval, concept-based queries that can be further augmented with named entity queries supported. Concept-based queries are formulated by expanding the domain concepts hierarchy and by selecting the class or the classes that correspond to the content in which the user is interested. The resulting query is the conjunction of the selections made in the expanded ontology concept hierarchies. In order to support semantic queries enhanced by additional named-entities information, a textbox appears when, during query formulation, the user selects a concept whose instances are proper names.

Experimental Results

For the purpose of experimental evaluation, a collection of heterogeneous outdoor photographs was assembled, and ground truth annotations were generated manually using the developed domain ontology. In order to populate the domain knowledge with low-level features and spatial context information for the domain concepts targeted by semantic analysis, prototype instances were created for each concept using part of the assembled collection as a training set. After experimentation, it proved that having 10 prototype instances for each concept is sufficient to achieve its detection with satisfactory performance. In the current implementation, four domain concepts are supported by the automatic semantic analysis component: Vegetation, Body of Water, Sunset, and Sky.

In Figure 6, results for different queries are presented, demonstrating the added value that results from coupling the domain ontology taxonomy with the complementary annotations produced by the proposed automatic semantic analysis component. As shown in the first example query (Figure 6(a)), the class Religious in the class Object taxonomy has been checked (i.e., the user is searching for photographs depicting religious buildings), and the retrieved images are photographs of churches and temples. Practically, this means that the metadata associated with the retrieved images are instances of either the class Religious or any of each supported subclass.

Figure 6. Ontology-based semantic image analysis and retrieval results

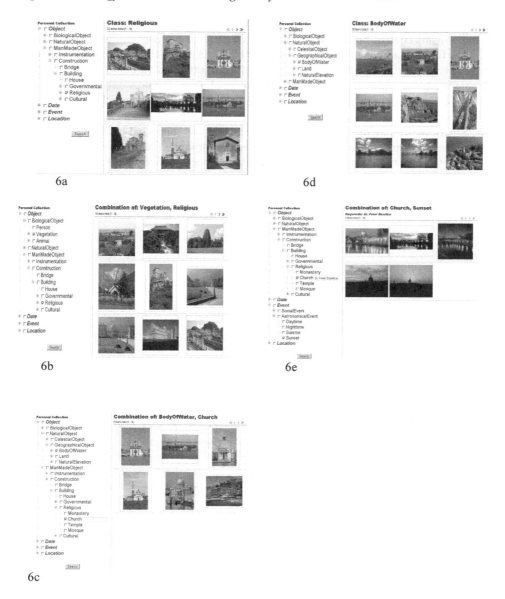

The next examples (Figure 6b and Figure 6c) illustrate two refinements of the initial query in which domain concepts supported by the semantic analysis component also have been employed. The corresponding queries consider photographs that depict religious buildings with vegetation around them and photographs that depict a church near water, respectively. In Figure 6d, a retrieval example using solely automatically produced annotations is presented in which the user queries about photographs depicting water. Finally, an example of a concept-based query extended by named entities information is illustrated in Figure 6e, in which the user is searching for photographs that depict St. Peter's Basilica at sunset, which again is a concept supported due to the semantic analysis functionality.

In order to evaluate the performance of the proposed ontology-based framework, additional experiments were conducted to assess the attained performance against the one obtained when using content-based and keyword-based retrieval approaches. The recall and precision rates for each approach are illustrated in Table 1. As expected, the content-based approach results in poor performance, since the significant intraclass variability does not allow determining semantic relevance based on low-level visual features only. On the other hand, the performance of the keyword-based approach fluctuates substantially, depending on the keyword(s) used each time, which shows its limitations. Consequently, even at this early stage of development in which only four concepts are supported by the semantic analysis and the automated inference support provided by ontologies is employed only for controlling the detection steps, the proposed framework introduces noticeable improvement.

Table 1. Comparative evaluation results

Query	Content-Based Approach		Keyword-Based Approach			Proposed Approach	
	Precision	Recall	Query Keywords	Precision	Recall	Precision	Recall
Religious building	18.5	26.8	religious	81.2	100.0	100.0	100.0
Religious building' and Vegetation	8.0	21.1	vegetation, religious	47.3	100.0	77.0	91.0
			tree, grass, religious	53.1			
Church and Body of Water	4.2	23.3	river, church	51.0	100.0	75.0	94.2
			water, river, sea, church	67.6			
Body of Water	40.8	56.6	water, sea, river, lake	80.5	100.0	75.0	94.2
St. Peter's Basilica and Sunset	4.4	44.0	church, sunset	0	100.0	72.0	53.6

Future Trends and Conclusion

In this chapter, an approach to semantic-based image retrieval was presented, addressing semantic image annotations generation, representation, and manipulation. A unified ontology-based framework was proposed for coupling semantic analysis and retrieval in order to partially alleviate the limitations entailed by the high cost of manual annotation through automatic generation of additional semantic annotations, thus enabling a realistic approach to effective image access at the semantic level. Under the proposed framework, a domain ontology was developed in order to capture the examined domain semantics and to provide the conceptualization and vocabulary for the annotation. Furthermore, an analysis ontology was designed to model the semantics extraction process and the required qualitative and quantitative low-level information definitions.

The benefits of following such a unified ontology-based approach are twofold. From the annotation and retrieval perspective, the formal machine-understandable ontology semantics improve recall and precision performance, while, through concept-based browsing, query formulation is facilitated significantly. Seen from the analysis perspective, domain concepts detection is assisted by the knowledge provided in the domain ontology definitions, while extensibility and interoperability are advanced as all analysis aspects are treated as ontological concepts. Thus, simply by enriching the employed analysis ontology with appropriate domain knowledge, flexible adaptation to different or extended domains is achieved.

However, there are still several challenges to be addressed. Clearly, a vital issue is the further reduction of the effort required for manual annotation. Consequently, apart from including prototypes and respective ontology definitions to support the automatic detection of more domain concepts, it is among the authors' intentions to investigate the potential of exploiting ontologies' automated inference support for the generation of higher-level semantics. Furthermore, since in certain cases the available visual content comes with some sort of textual metadata, an ontology-based text analysis component could be used to allow semi-automatic derivation of ontology-based annotations. Addressing these challenges would result in improved retrieval over a richer set of content annotations.

Acknowledgments

This research was supported partially by the European Commission under contracts FP6-001765 aceMedia and FP6-507482 KnowledgeWeb, and by the GSRT Greek project REACH.

References

Adamek, T., O'Connor, N., & Murphy, N. (2005). Region-based segmentation of images using syntactic visual features. In *Proceedings of the Workshop on Image Analysis for Multimedia Interactive Services (WIAMIS 2005)*, Montreux, Switzerland.

Bertini, M., Del Bimbo, A., & Nunziati, W. (2004). Highlights modeling and detection in sports videos *Pattern Analysis and Applications, 7*(4), 411–421.

Bose, B., & Grimson, E. (2003). Learning to use scene context for object classification in surveillance. In *Proceedings of the Joint IEEE International Workshop on Visual Surveillance and Performance Evaluation of Tracking and Surveillance (in conjunction with ICCV '03)*, Nice, France.

Brunelli, R., Mich, O., & Modena, C.M. (1999). A survey on video indexing. *Journal of Visual Communications and Image Representation, 10*, 78–112.

Dasiopoulou, S., Mezaris, V., Kompatsiaris, I., Papastathis, V. K., & Strintzis, M. G.. (2004). Knowledge-assisted semantic video object detection. *IEEE Trans.CSVT, 15*(10), 1210–1224.

Domingue, J., Dzbor, M., & Motaa, E. (2004). Collaborative semantic Web browsing with Magpie. In *Proceedings of the 1ˢᵗ European Semantic Web Symposium (ESWS 2004)*, Heraklion, Crete (pp. 388–401).

Dorado, A., Calic, J., & Izquierdo, E. (2004). A rule-based video annotation system. *IEEE Transactions on Circuits and Systems for Video Technology, 14*(5), 622–633.

Gangemi, A., Guarino, N., Masolo, C., Oltramari, A., & Schneider, L. (2002). Sweetening ontologies with DOLCE. In *Proceedings of the 13ᵗʰ International Conference on Knowledge Engineering and Knowledge Management (EKAW)*, Siguenza, Spain.

Gruber, T. R. (1993). Towards principals for the designing of ontologies used for knowledge sharing. *International Journal of Human-Computer Studies* [Special Issue on Formal Ontology in Conceptual Analysis and Knowledge Representation, Guest Eds. N. Guarino & R. Poli].

Hollink, L., Schreiber, A., Wielemaker, J., & Wielinga, B. (2003). Semantic annotations of image collections. In *Proceedings of the Workshop on Knowledge Capture and Semantic Annotation (KCAP)*, Sanibel, FL.

Hunter, J., Drennan, J., & Little, S. (2004). Realizing the hydrogen economy through Semantic Web technologies. *IEEE Intelligent Systems, 19*(1), 40-47.

Hudelot, C., & Thonnat, M. (2003). A cognitive vision platform for automatic recognition of natural complex objects. In *Proceedings of the International Conference on Tools with Artificial Intelligence (ICTAI)*, Sacramento, CA (pp. 398–405).

Kiryakov, A., & Simov, K. I. (1999). Ontologically supported semantic matching. In *Proceedings of the Nordic Conference on Computational Linguistics (NODALIDA'99)*, Trodheim, Norway.

Kouzani, A. Z. (2003). Locating human faces within images. *Computer Vision and Image Understanding, Elsevier, 91*, 247–279.

Leonardi, R., Migliorati, P., & Prandini, M. (2004). Semantic indexing of soccer audiovisual sequences: A multimodal approach based on controlled Markov chains. *IEEE Transactions on Circuits and Systems for Video Technology, 14*(5), 634–643.

Li, X., Wang, L., & Sung, E. (2004). Multi-label SVM active learning for image classification. In *Proceedings of the IEEE International Conference on Image Processing (ICIP),* Singapore (pp. 2207–2210).

Lin, W. H., & Hauptmann, A. (2002). News video classification using SVM-based multimodal classifiers and combination strategies. In *Proceedings of the ACM Multimedia,* Juan les Pins, France (pp. 323–326).

Lingnau, F. B. A. C., & Centeno, J. A. S. (2003). Object oriented analysis and semantic network for high resolution image classification. *Boletim de Ciencias Geodesicas, 9*, 233–242.

Little, S., & Hunter, J. (2004). Rules-by-example: A novel approach to semantic indexing and querying of images. In *Proceedings of the International Semantic Web Conference (ISWC),* Hiroshima, Japan (pp. 534–548).

Mezaris, V., Kompatsiaris, I., & Strintzis, M.G.. (2004). Region-based image retrieval using an object ontology and relevance feedback. *EURASIP Journal on Applied Signal Processing, 2004*(6), 886-901.

Naphade, M. R., Kozintsev, I. V., & Huang, T. S. (2002). A factor graph framework for semantic video indexing. *IEEE Transactions on Circuits and Systems for Video Technology, 12*(1), 40–52.

Neskovic, P., & Cooper, L. N. (2003). Providing context for edge extraction in application to detection of cars from video streams. In *Proceedings of the International Conference on Engineering Applications of Neural Networks* (pp. 222-229).

Meghini, C., Sebastiani, F., & Straccia, U. (1997). Reasoning about the form and content for multimedia objects. In *Proceedings of AAAI 1997 Spring Symposium on Intelligent Integration and Use of Text, Image, Video and Audio* (pp. 89-94). CA: Stanford University.

MPEG-7 Visual. (2001). *FCD information technology—multimedia content description interface—part 3* (ISO/IEC 15938-3).

Petkovic, M., & Jonker, W. (2001). Content-based video retrieval by integrating spatiotemporal and stochastic recognition of events. In *Proceedings of the IEEE Workshop on Detection and Recognition of Events in Video*, Vancouver, Canada (pp. 75–82).

Popov, B., Kiryakov, A., Ognyanoff, D., Manov, D., & Kirilov, A. (2004). KIM: A semantic platform for information extraction and retrieval. *Journal of Natural Language Engineering, 10*(3–4), 375–392.

Reidsma, D., Kuper, J., Declerck, T., Saggion, H., & Cunningham, H. (2003). Cross document annotation for multimedia retrieval. In *Proceedings of the 10th Conference of the European Chapter of the Association for Computational Linguistics (EACL)*, Budapest, Hungary.

Schober, J.-P., Hermes, T., & Herzog, O. (2005). Picturefinder: Description logics for semantic image retrieval. In *Proceedings of the IEEE International Conference on Multimedia and Expo (ICME),* Amsterdam, The Netherlands (pp. 1571–1574).

Schreiber, A., Dubbeldam, B., Wielemaker, J., & Wielinga, B. J. (2001). Ontology-based photo annotation. *IEEE Intelligent Systems, 16*(3), 66–74.

Sinclair, P. A. S., Goodall, S., Lewis, P. H., Martinez, K., & Addis, M. J. (2005). Concept browsing for multimedia retrieval in the SCULPTEUR project. In *Proceedings of the Multimedia and the Semantic Web Workshop in the 2nd Annual European Semantic Web Conference (ESWC),* Heraklion, Greece (pp. 28–36).

Smeulders, A. W. M., Worring, M., Santini, S., Gupta, A., & Jain, R. (2000). Content-based image retrieval at the end of the early years. *IEEE Transactions on Pattern Analysis and Machine Intelligence, 22*(12), 1349–1380.

Snoek, C. G. M., & Worring, M. (2003). Time interval based modelling and classification of events in soccer video. In *Proceedings of the 9th Annual Conference of the Advanced School for Computing and Imaging,* Heijen, The Netherlands.

Sprague, N., & Luo, J. (2002). Clothed people detection in still images. In *Proceedings of the IEEE International Conference on Pattern Recognition (ICPR),* Quebec, Canada (pp. 585–589).

Staab, S., & Studer, R. (2004). *Handbook on ontologies.* Heidelberg: Springer-Verlag.

Tzouvaras, V., Tsechpenakis, G., Stamou, G. B., & Kollias, S. D. (2003). *Adaptive rule-based recognition of events in video sequences.* In *Proceedings of the IEEE International Conference on Image Processing (ICIP),* Barcelona, Spain (pp. 607–610).

Wang, L., & Manjunath, B. S. (2003). A semantic representation for image retrieval. In *Proceedings of the IEEE International Conference on Image Processing (ICIP),* Barcelona, Spain (pp. 523–526).

Xie, L., Xu, P., Charn, S. F., Divakaran, A., & Sun, H. (2004). Structure analysis of soccer video with domain knowledge and HMM. *Pattern Recognition Letters, 25,* 767–775.

Yoshitaka, A., Kishida, S., Hirakawa, M., & Ichikawa, T. (1994). Knowledge-assisted content-based retrieval for multimedia databases. *IEEE Multimedia, 1*(4), 12–21.

Zhou, X. S., & Huang, T. S. (2003). Relevance feedback in image retrieval: A comprehensive review. *Multimedia Systems, 8*(6), 536–544.

Section V

Models and Tools for Semantic Retrieval

Chapter XI

A Machine Learning-Based Model for Content-Based Image Retrieval

Hakim Hacid, University of Lyon 2, France

Abdelkader Djamel Zighed, University of Lyon 2, France

Abstract

A multimedia index makes it possible to group data according to similarity criteria. Traditional index structures are based on trees and use the k-Nearest Neighbors (k-NN) approach to retrieve databases. Due to some disadvantages of such an approach, the use of neighborhood graphs was proposed. This approach is interesting, but it has some disadvantages, mainly in its complexity. This chapter presents a step in a long process of analyzing, structuring, and retrieving multimedia databases. Indeed, we propose an effective method for locally updating neighborhood graphs, which constitute our multimedia index. Then, we exploit this structure in order to make the retrieval process easy and effective, using queries in an image form in one hand. In another hand, we use the indexing structure to annotate images in order to describe their semantics. The proposed approach is based on an intelligent manner for locating points in a multidimensional space. Promising results are obtained after experimentations on various databases. Future issues of the proposed approach are very relevant in this domain.

Introduction

Data interrogation is a fundamental problem in various scientific communities. The database and statistics communities (with their various fields such as data mining) are certainly the most implied. Each community considers the interrogation from a different point of view. The database community, for example, deals with great volumes of data by organizing them in an adequate structure in order to be able to answer queries in the most effective way. This is done by using index structures. The statistics community deals only with data samples in order to produce predictive models that are able to draw conclusions on phenomena; these conclusions then are generalized to the whole items. This is achieved using various structures such as decision trees (Mitchell, 2003), Kohonen maps (Kohonen, 2001), neighborhood graphs (Toussaint, 1991), and so forth.

Dealing with multimedia databases means dealing with content-based retrieval. There are two fundamental problems associated with content-based retrieval systems: (a) how to specify a query and (b) how to access the intended data efficiently for a given query. The main objective is to capture the semantics of the considered data. For traditional database systems, the semantics of content-based access are finding data items that are match exactly the specified keywords in queries. For multimedia database systems, both query specification and data access become much harder (Chiueh, 1994).

To give the computer the ability to mimic the human being in scene analysis needs to explicit the process by which it moves up from the low level to the highest one. Multimedia processing tools give many ways to transform an image/video into a vector. For instance, MPEG-7 protocol associates a set of quantitative attributes to each image/video. The computation of these features is integrated and automated fully in many software platforms. In return, the labels basically are given by the user, because they are issued from the human language. The relevance of the image/video retrieval process depends on the vector of characteristics. Nevertheless, if we assume that the characteristics are relevant in the representation space, that it is supposed to be R^p, the images that are neighbors should have very similar meanings.

In order to perform an interrogation in a multimedia database, it must be structured in an adequate way. For that, an index is used. Indexing a multimedia database consists of finding a way to structure the data so that the neighbors of each multimedia document can be located easily according to one or more similarity criteria. The index structures used in databases are generally in a tree form and aim to create clusters, which are represented by the sheets of the tree and contain rather similar documents. However, in addition to the fact that a traditional index cannot support data with dimensions higher than 16, dealing with multimedia databases needs more operations such as classification and annotation. This is why the use of models issued from the automatic learning community can be very helpful.

The rest of this chapter is organized as follows. The next section introduces the point location and the database indexing problems. Section 3 presents the motivation and the contributions of our work. Section 4 describes the neighborhood graphs that are the foundation of this contribution. Our contributions are addressed in Section 5. The indexing method and the optimization of the neighborhood graphs are discussed in Section 5.1. Semi-automatic annotation is discussed in Section 5.2. Section 6 gives some experiments that were performed in order to evaluate and validate our approach. We conclude and give some future issues in Section 7.

Point Location and Databases Indexing Problems

Point Location Problem

Neighborhood search is a significant problem in several fields. It is handled in data mining (Fayyad, Piatetsky-Shapiro, & Smith, 1996), classification (Cover, & Hart, 1967), machine learning (Cost, & Salzberg, 1993), data compression (Gersho, & Gray, 1991), multimedia databases (Flickner et al., 1995), information retrieval (Deerwester, Dumais, Landauer, Furnas, & Harshman, 1990), and so forth. Several works in connection with the neighborhood search in databases exist, such as Berchtold, Böhm, Keim, and Kriegel (1997), Lin, Jagadish, and Faloutsos (1994), and White and Jain (1996). The point location problem is a key question in automatic multidimensional data processing. This problem can be defined as follows: Having a data set Ω of n items in a multidimensional space R^p, the problem is to find a way to preprocess the data so that if we have a new query item α, we will be able to find its neighbors n as little time as possible.

The point location problem in one-dimensional space can be solved by sorting the data and by applying a binary search that is rather fast and inexpensive in term of resources with a complexity of $O(n \log n)$. In a two-dimensional space, this problem can be solved by using a voronoi diagram (Preparata & Shamos, 1985), as illustrated in Figure 1.

Unfortunately, when the dimension increases, the problem becomes more complex and more difficult to manage. Several methods for point location in a multidimensional space were proposed. For example, we can quote the ones based on points projection on only one axis (Friedman, Baskett, & Shustek, 1975; Guan & Kamel, 1992; Lee & Chen, 1994) or the work based on partial distances calculation between items (Bei & Gray, 1985). In the database community, the point location problem is known as the indexing problem.

Figure 1. A Voronoi diagram in a bi-dimensional space

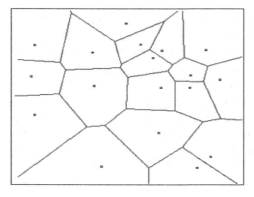

Database Indexing Problem

There are two main classes of databases indexing methods:

- **Point access methods (PAM):** In this class of methods, data are considered as points in a multidimensional space. The indexing structures generally are based on the *kd-Tree* principle (Bentley, 1975).
- **Space access methods (SAM):** In this class of methods, the data are in various geometrical forms (e.g., lines, rectangles, etc.). The structures used are alternatives of *R*-Tree (Guttman, 1984).

Several content-based information retrieval systems are based on these structures (trees) and use the *k*-nearest neighbors principle (Fix & Hudges, 1951; Velkamp & Tanase, 2000). This principle is based on sorting items by taking into account their distances to the query item; fixed *k* nearest items are turned over by the system as an answer to the query. For example, the QBIC system in its implementation for the Hermitage museum (Faloutsos et al., 1994) returns the 12 nearest images to a query submitted by a user.

Disadvantages of the *k-NN* Approach

There are two main disadvantages related to the use of the *k-NN* approach: (i) the symmetry (the most important one) and (ii) the subjectivity problems. The first one is due to the fact that the principle of the *k-NN* is based only on data sorting according to their distances to the query point.

Figure 2 illustrates the symmetry problem of the *k-NN* on one axis using five points (*a, b, c, d, e*).

Let us consider a *k-NN* with *k*=3. So, if we take point *a* in the first case as a query point, its three neighbors are {*b, c, d*}. In the other case, if we consider point *d* as a query point, then its three nearest neighbors are {*b, c, e*}. So, we can see clearly that the neighborhood is not coherent, because in the neighborhood of point *a*, we find point *d*, but in the neighborhood of point *d*, we do not find point *a*. The symmetry problem raises other problems in the decision-making process related to other important functions in image databases,

Figure 2. Illustration of the symmetry problem in the k-NN approach

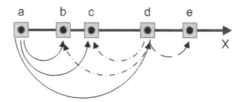

such as images classification and their automatic annotation. The navigation process also is affected by this problem.

The second problem, which is relatively a subjective one, concerns the manner of the determination of k. Indeed, some questions rise from that: Is the use of only a distance measure sufficient for the interrogation of an image database (neighborhood determination)? If this is the case, what is the maximum distance an item must respect to be a neighbor of a query item? What is the optimal number of neighbors of a query item? Dealing with image databases needs more operations (image classification, for example), so are the used techniques (trees and k-NN) sufficient and effective for that?

Motivation and Contribution

The structuration model of image databases is (or can be seen) as a graph that is based on similarity relations among items (e.g., k-NN [Mitchell, 2003] or a relative neighborhood graph [Scuturici, Clech, Scuturici, & Zighed, 2005]). The goal is to explore an image database through the similarity among images. Exploring the similarity can result in the search of the images' neighbors. The structuration model is very important, because the performances of a content-based information retrieval system depend strongly on the representation structure (i.e., index structure), which manages the data.

The main motivation of this work is the following: We believe that the problem of image database retrieval, navigation, automatic annotation, and other tasks in this field cannot be solved by using traditional approaches such as k-NN and trees. We believe that the problems related to the use of traditional approaches can be solved by using only more adequate structures. In addition, the combination among various methods seems important for us, which is why, in our case, we introduce neighborhood graphs (issued from the automatic learning community) for indexing, retrieving, and semi-automatically annotating image databases.

There are two main contributions in this chapter. The first one concerns the accommodation of neighborhood graphs for indexing image databases using low-level features. The second is the exploitation of the proposed structure in order to annotate an image database. So, the proposed approach presents multiple advantages. Indeed, we take into account the two possible retrieval functions; namely, retrieval by visual contents and retrieval by keywords. Hence, a user can submit his or her query using either an image or textual description in order to recover the images that relate to a same concept.

- **Retrieval by visual content:** In this case, the query is expressed in an image form and analyzed to extract its low-level characteristics. After that, it is inserted in the neighborhood graph, and its neighbors are turned over as an answer. In addition, with the graph representation of the database, the navigation becomes more interesting and more coherent for the user.

- **Retrieval by keywords:** In this case, the interrogation is ensured by the use of keywords that are expressed in a natural language and describing the wished semantics of the expected images.

Moreover, we show how we exploit the proposed approach for annotating image databases. The next section introduces the neighborhood graphs. These structures can help us to bring some answers to the previous questions and can serve as a solution for the disadvantages of the *k-NN* approach.

Neighborhood Graphs

Neighborhood graphs are used in various systems. Their popularity is due to the fact that the neighborhood is recovered by coherent functions that reflect, in some points of view, the human intuition mechanism. Their use is varied from information retrieval systems to geographical information systems.

Neighborhood graphs are geometrical structures that use the neighborhood concept in order to find the closest points to a given one in a multidimensional space R^p. For that, they are based on proximity measures (Toussaint, 1991). We will use the following notations throughout this chapter:

Let Ω be a set of points in a multidimensional space R^p. A graph $G(\Omega, \rho)$ is composed by a set of points Ω and a set of edges ρ. Then, for any graph we can associate a binary relation $\Re$ upon Ω in which two points $(\alpha, \beta) \in \Omega^2$ are in binary relation if and only if the pair $(\alpha, \beta) \in \rho$. In other words, (α, β) are in binary relation if and only if they are connected directly in graph G. From that, the neighborhood $N(\alpha)$ of a point α in the graph G can be considered as a subgraph that contains point α and all the points that are connected directly to it.

Several possibilities were proposed for building neighborhood graphs. Among them, we can quote the Delaunay triangulation (Preparata & Shamos, 1985), the relative neighborhood graph (Toussaint, 1980), the Gabriel graph (Gabriel & Sokal, 1969), and the minimum spanning tree (Preparata & Shamos, 1985). In this chapter, we consider only one of them: the relative neighborhood graph.

Relative Neighborhood Graph

In the relative neighborhood graph, $G_{rng}(\Omega, \rho)$, two points $(\alpha, \beta) \in \Omega^2$ are neighbors if they check the relative neighborhood property defined as follows: Let $H(\alpha, \beta)$ be the hypersphere of radius $\delta(\alpha, \beta)$ and centered on α, and let $H(\beta, \alpha)$ be the hypersphere of radius $\delta(\beta, \alpha)$ and centered on β ($\delta(\alpha, \beta)$, and $\delta(\beta, \alpha)$ are the distance measures between two points α and β. ($\delta(\alpha, \beta) = \delta(\beta, \alpha)$). So, α and β are neighbors if and only if the intersection of the two hyperspheres $H(\alpha, \beta)$ and $H(\beta, \alpha)$ is empty (Toussaint, 1980). Formally:

Consider $A(\alpha, \beta) = H(\alpha, \beta) \cap H(\beta, \alpha)$ So $(\alpha, \beta) \in \rho$ iff $A(\alpha, \beta) \cap \Omega = \phi$

Figure 3 illustrates the relative neighborhood graph.

Figure 3. Relative neighborhood graph in a bi-dimensional space

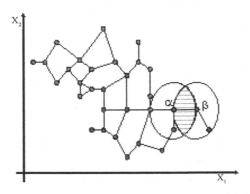

Advantages and Disadvantages

The advantages of the use of a model issued from the automatic learning community are primarily the possibility for a user to exploit the powerful decision-making functions of these structures. Also, their construction respects some geometrical properties that result from the use of both a similarity measure and the topology of the items in the multidimensional space for the neighborhood determination. The direct consequence is the number of neighbors associated with each item. Indeed, because of the geometry properties, the number of neighbors is not fixed by a user and is variable from one item to another. This avoids the repetitious calculation of the neighbors at each interrogation of the index structure. Another advantage of these structures is the neighborhood symmetry. This offers the effectiveness of the decision-making process related to different operations such as automatic annotation of images.

However, several problems concerning neighborhood graphs are under research and require more investigation to solve them. These problems primarily are related to their high construction cost and to their updating difficulties. For these reasons, optimizations are necessary for their construction and updating in order to make their application possible on image databases. In the following, we give a description of some neighborhood graph construction algorithms.

Neighborhood Graph Construction Algorithms

One of the common approaches to the various neighborhood graph construction algorithms is the use of a refinement technique. In this approach, the graph is built by steps. Each graph is built starting from the previous graph, containing all connections, by eliminating some edges that do not check the considered neighborhood property. Pruning (edge elimination) generally is done by taking into account the construction function of the graph or through geometrical properties.

The construction principle of neighborhood graphs consists of seeking for each point if the other points in the space are in its proximity. The cost of this operation has a complexity of $O(n^3)$ (n is the number of points in the space). Toussaint (1991) proposed an algorithm with a complexity of $O(n^2)$. He deduced the *RNG* starting from a Delaunay triangulation (Preparata & Shamos, 1985). Using the Octant neighbors, Katajainen (1988) also proposed an algorithm with the same complexity. Smith (1989) proposed an algorithm of complexity $O(n^{23/12})$ that is less significant than $O(n^3)$.

The major problem of these algorithms is the fact that they are not able to update the initial structure without rebuilding it. Indexing image databases using these structures needs to adapt them in order to support rapid interrogation and updating operations. The next section presents the optimization of these structures for indexing and annotating image databases.

Neighborhood Graphs for Image Indexing and Semi-Automatic Annotation

Neighborhood Graphs Indexing-Based Approach

Neighborhood graphs for navigation in image databases were used in Scuturici et al. (2005). The proposed approach focuses mainly on neighborhood graphs locally updating technique. This method really does not offer the anticipated results. Indeed, the graph actually is not updated, but the neighbors of each query point are considered those of the nearest neighbor. This approach is not effective, because actually, the neighbors of a point are not only/always those of its nearest neighbor. So, the use of this method deteriorates the graph.

We can consider two situations when dealing with the neighborhood graphs optimization problem. The first situation is when we have an existing graph. In this situation, if one uses an approximation method, he or she risks having another graph with other properties; we can obtain, for example, more or less neighbors for some items. In this case, we have to find a solution for effectively updating the graph when inserting or deleting a point without rebuilding it. The second situation is when the graph is not built yet. In this situation, we can apply an approximation method in order to obtain a graph that is as close as possible to the one that we can obtain using a standard algorithm. In this chapter, we are interested in the first case. We propose an effective method with a low complexity for locally updating neighborhood graphs. This algorithm can be extended for incrementally building neighborhood graphs.

A neighborhood graph locally updates task passes by the location of the inserted (or removed) point in the multidimensional space as well as the points that can be affected by the modifications. To achieve this, we proceed into two main stages: initially, we look for an optimal space area that can contain a maximum number of potentially close points to the query point. The second stage is done in the aim of filtering the items found beforehand in order to recover the real closest points to the query point, and this by applying an adequate neighborhood property. This last stage causes the effective updating of the neighborhood relations among the concerned points.

The main stage of this method is the *space area* determination. This can be considered a problem of determining a hypersphere of center α (the query point), maximizing the chance of containing the neighbors of α while minimizing the number of items that it contains.

We take advantage of the general neighborhood graph structure in order to establish the radius of the hypersphere. We focus especially on the nearest neighbor and the farthest neighbor concepts. So, two observations in connection with these two concepts seem to be interesting:

The neighbors of the nearest neighbor of α are potential candidates to the neighborhood of the query point α.

From that, by generalization, we can deduce that:

All the neighbors of a point are also candidates to the neighborhood of a query point for which it is a neighbor.

With regard to the first step, the radius of the hypersphere, which respects the above properties, is the one that includes all the neighbors of the nearest neighbor of the query point. So, considering that the hypersphere is centered at α, its radius will be the sum of the distances between α and its nearest neighbor and the one between this nearest neighbor and its furthest neighbor.

That is, let us consider α the query point and β its nearest neighbor with a distance δ_1; and let us consider λ the furthest neighbor of β with a distance δ_2. The radius *SR* of the hypersphere can be expressed as:

$$SR = \delta_1 + \delta_2 + \varepsilon$$

ε is a relaxation parameter that can be set according to the state of the data (e.g., their dispersion) or by a domain knowledge. We set experimentally this parameter to 1.

The content of the hypersphere is processed in order to check whether there are some neighbors (or all the neighbors). The second step constitutes a reinforcement step and aims to eliminate the risk of losing neighbors or including bad ones. So, we take all the neighbors of the query point, recovered beforehand (those returned in the first step), as well as their neighbors, and update the neighborhood relations among these points.

The computation complexity of this method is very low and meets perfectly our starting aims (i.e., locating the neighborhood of points in as short a time as possible). It is expressed by:

$$O(2n + n'^3)$$

in which *n* is the number of items in the database, and *n'* is the number of items in the hypersphere ($\ll n$).

Figure 4. Illustration of the principle of the proposed method

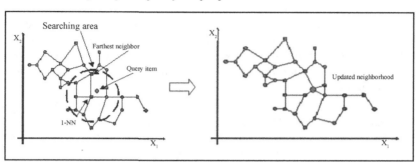

This complexity includes the two previously described stages; namely, the determination of the radius of the hypersphere and the determination of the points that are in it, corresponding to the term $O(2n)$. The second term corresponds to the necessary time for effectively updating the neighborhood relations.

This complexity constitutes the worst case complexity and can be optimized in several ways. The most obvious way is to use a fast nearest neighbor search algorithm. The example hereafter (Figure 4) illustrates graphically and summarizes the principle of the method.

With this method, neighborhood graphs are adequate to be used for indexing image databases. More details about this method are described in Hacid and Zighed (2005).

In the next section, we present the exploitation of this indexing structure for semi-automatic image annotation.

From Indexing to Semi-Automatic Annotating

Another important function when we deal with image database is the semantic association to its visual contents. This function is known as the annotation process. Image annotation is the process that consists of assigning for each image a keyword or a list of keywords that makes it possible to describe its semantic content. This function can be considered a function that allows a mapping between the visual aspects of the image and its low-level characteristics.

Image annotation is not an easy task. There are three types of image annotation: manual, semi-automatic, and automatic. The first one is carried out manually by a human who is charged with alloting a set of keywords for each image. The automatic annotation is carried out by a machine and aims to reduce the user's charge. The first annotation type increases precision and decreases productivity. The second type decreases precision and increases productivity. In order to make a compromise between these two tasks, their combination became necessary. This combination is named semi-automatic annotation.

Moreover, image annotation can be performed on two levels: the local level and the global level. In the local level, the image is regarded as a set of objects. The annotation aims to affect for each object a keyword or a list of keywords in order to describe it. The global level concerns the whole image and assigns a list of keywords to describe its general aspect. The two approaches, the first one more than the second one, depend considerably on the quality of the image segmentation. Unfortunately, the segmentation remains a challenge, and a lot of works concentrate on this topic (Shi & Malik, 2000). In this section, we are interested in the semi-automatic annotation of images at the global level. We consider the semi-automatic annotation because it requires user intervention in order to validate the system's decisions.

There is not a lot of work on image annotation. There are methods that apply a clustering of images and their associated keywords in order to make it possible to attach a text to images (Barnard, Duygulu, & Forsyth, 2001; Barnard & Forsyth, 2001). With these methods, it is possible to predict the label of a new image by calculating some probabilities. Picard and Minka (1995) proposed a semi-automatic image annotation system in which the user chooses the area to be annotated in the image. A propagation of the annotations is carried out by considering textures. Maron and Ratan (1998) studied the automatic annotation using only one keyword at a same time. Mori Takahashi, and Oka (1999) proposed a model based on co-occurrences between the image and keywords in order to find the most relevant keywords for an image. The disadvantage of this model is that it requires a large training sample in order for it to be effective. Duygulu, Barnard, De Freitas, and Forsyh (2002) proposed another model, the *translation model*, which is an improvement of the co-occurrence model suggested by Mori et al. (1999) and this by integrating a training algorithm. Probabilistic models such as the Cross Media Relevance Model (Jeon, Lavrenko & Manmatha, 2003) and the Latent Semantic Analysis (Monay & Gatica-Perez, 2003) also were proposed. Li and Wang (2003) use the two-dimensional hidden Markov chains to annotate images.

Our work as well as the work of Barnard et al. (2001) concerns the global level. We use a prediction model (neighborhood graphs) to annotate an image collection. The method that we propose can be adapted easily for a local level annotation. So, we will try to answer the following question: Having a set of annotated images, how can we proceed in order to annotate a new image introduced without annotations? For that, we exploit the indexing structure proposed beforehand in the previous section of this chapter.

Formally, let Ω be a set of n images $\Omega = \{I_1, I_2, \dots I_n\}$. Each image is described by a set of features $<f_1, f_2, \dots, f_m>$ that represents the low-level characteristics (color, texture, etc.) and a list of keywords $W = w_1, w_2, \dots, w_k$ ($m \neq k$). So, an image I_i can be described by a vector $I_i = <f_1, f_2, \dots, f_m, W_i>$ in which each image can have a different number of keywords. From that, having a new unlabeled image $I_x = <f_1, f_2, \dots, f_m>$, it is then a problem of finding a model that can assign to the image I_x the labels that can describe its semantics. In other words, the goal is to pass from a representation in the form of $I_x = <f_1, f_2, \dots, f_m>$ to a representation in the form of $I_x = <f_1, f_2, \dots, f_m, W_x>$.

In our approach, image annotation problem passes by two levels: data modeling (indexing) and decision making (effective annotation). The first level is described in detail in the previous sections. The result of this step is a graph representation of the image database based only on the low-level characteristics. This representation aims to keep the neighborhood relations among items.

After the data modeling process, the annotation phase can start. The main principle is the following: each point in the graph is considered a *judge*. The unlabeled image is then situated in the multidimensional space (using only its low-level features), and its neighbors are located. From there, the potential annotations of the image are deduced from the annotations of its neighbors. This is done by using vote techniques. So, we can decide, according to the decision of the judges (neighbors), which are the most suitable annotations to assign to an unlabeled image.

The effective annotation requires user intervention on two levels: setting up the decision parameters (decision threshold) and the validation of the machine's decisions. In the following, we detail the decision-making process.

Decision Making

At this stage, we consider that the database is indexed by using a neighborhood graph. The goal here is the effective association of annotations to a new unlabeled image. The main idea is the *heritage*. Indeed, we may inherit an image after its insertion in the neighborhood graph from the annotations of its neighbors by calculating scores for each potential annotation.

We can consider two simple ways to calculate the scores for the inherited annotations (the two scoring methods are given for illustration; other more sophisticated functions can be used for this purpose):

- **Naive method:** In this case, a score is calculated for each annotation by considering only the neighbors count of the unlabeled image. In other words, we give the same decision power for each judge. One then can

  se the following formula:

$$ S_t = \frac{\sum_{j=1}^{l} {}_{[t \in A(\beta_j) \ and \ \beta_j \in V(\alpha)]} 1}{\left| V(\alpha) \right|} $$

  in which $V(\alpha)$ is the set of the neighbors of the query image α; t is a specific annotation in the neighborhood; l is the number of neighbors of the query image α; and B_j is the neighbor j of the query image α.

The calculated score represents the number of close images that contain the annotation t compared to the total number of neighbors. The scores are calculated for each annotation that belongs to the neighborhood of the unlabeled image.

This approach is rather simple. However, it presents the disadvantage of allotting the same decision power for each judge that takes part in the decision-making process.

- **Weighting method:** The second possibility consists of considering the distances between the query point and its neighbors. In this case, we introduce a weighting function and give a more important decision power to the nearest neighbors. So, the more an image is near, the more its decision power is important. The weights affectation can be performed using the following formula:

$$W_i = 1 - \frac{\delta(\beta_i, \alpha)}{\sum_{j=1}^{l} \delta(\beta_j, \alpha)}$$

With W_i, the affected weight to the ith neighbor, $\delta(\beta_i, \alpha)$: the distance between the neighbor β_i and the query image α.

The scores assignation is performed in the same way as the previous one. However, in this case, we consider the weights instead of the items. The formula of the score calculation becomes:

$$S_i = \frac{\sum_{j=1}^{l} [_{i \in A(\beta_j)}\ and\ \beta_j \in V(\alpha)]} W_j}{\sum_{j=1}^{l} W_j}$$

At the end of the score calculation process, an annotation is assigned to an image if its score is equal to or higher than a threshold fixed by the user.

Example 2. In order to illustrate the different functions, let us consider the following example: consider a query image having four neighbors, I_1, I_2, I_3, I_4 (recovered using only the low-level characteristics from the graph). These images contain four concepts, c_1, c_2, c_3, c_4, as shown in Table 1.

The last column in Table 1 represents the weight of each neighbor according to its distance to the query point (needed in the second function). Using the naïve decision-making method, the scores one can obtain are illustrated in Table 2, which obtains the scores of Table 3 by using the second decision-making method.

If we consider a threshold of 75%, then the annotations of the query image using the first method will be c_1, c_2, and c_3 but will be only c_1 and c_3 using the weighting method.

Table 1. A set of image neighbors and their associated labels

Items	Concepts	Weights
I1	c1, c2, c3, c4	0.4
I2	c1, c3	0.3
I3	c1, c2	0.2
I4	c1, c2, c3	0.1

Table 2. Scores calculated using the naïve decision-making method

Concepts	Scores
c1	*100%*
c2	*75%*
c3	*75%*
c4	*25%*

Table 3. Scores calculated using the weighting decision-making method

Concepts	Scores
c1	100%
c2	70%
c3	80%
c4	10%

Propagation of the Annotations

The annotation of a new image depends on its neighborhood. This means that the annotation of a new image and its insertion in the graph can generate possible modifications in the labels of its neighbors (because a user can add new annotations to a query image). This is what we call the *annotations propagation*. In our case, we consider only the new inserted annotations.

We consider that the propagation scope follows the same schema as the neighborhood updating one (Hacid & Zighed, 2005). So, we use a hypersphere of ray $SR = \delta_1 + \delta_2 + \varepsilon$ to propagate the annotations and to recalculate the score for each new annotation.

That is, to update the annotations of the neighborhood, we calculate initially the ray of the hypersphere and recover the images in it. The next stage is the calculation of the scores for each image by considering each one as a query item. Annotation assignment follows the same principle of the previous one (i.e., the score must be equal to or higher than a given threshold). The new annotation, of course, is assigned to the inserted item without any calculation.

From that, we can notice that the annotation process that we propose is an incremental process. Indeed, the annotations of the images are updated incrementally as other images are inserted into the database. This enables us to have rather complete semantics at a given moment.

Experiments and Results

The interest of neighborhood graphs for content-based image retrieval is shown and discussed in Scuturici et al. (2005). The comparison tests done with several configurations of *k-NN* were conclusive and showed the utility of these structures in this field.

In what concern us, we are interested in this section in three different tests:

- Testing the validity of the obtained results using the suggested method.
- Testing the execution times of the proposed method.
- Testing the annotation performances.

Evaluation of the Validity of the Proposed Method

By the validity of the obtained results, we mean the ability of our method to find for each new inserted item in the graph the same neighbors as the ones we can find using the standard algorithm. For that, we carried out several tests on various data sets.

The principle of these experiments is as follows: We take m data sets S_1, S_2, ..., S_m with different items count n_1, n_2, ..., n_m, we build a relative neighborhood graph on each dataset, and we save the corresponding graph structure, which will serve as a reference graph. Once the reference graphs are built, we take each dataset and build new graphs using $n-1$ items. We then use our method in order to insert the remaining item, and we calculate the recall on the variously built graphs. This operation is repeated using several items on the m data sets.

We used various supervised datasets for these experiments; for example, Iris (Anderson, 1935; Fisher, 1936), UCI Irvine (Hettich, Blake, & Merz, 1998), and Breiman's waves (Breiman, Friedman, Olshen, & Stone, 1984). The graphic of Figure 5 illustrates the recall variation on one dataset with various insertions by considering three situations: the recall is calculated on the reference graph, then on a graph with $n-1$ items, and finally after the insertion of the remaining item.

The first experiment constitutes our reference. The experiments carrying numbers 2, 4, and 6 are carried out with an item in less ($n-1$ items) and experiments 3, 5, and 7 after the insertion of the remaining item in the corresponding previous experiment. That is, after the

Figure 5. Recall variations on one dataset using different items

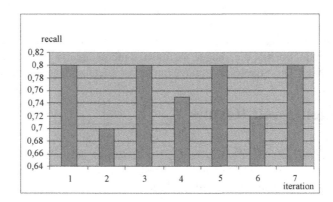

insertion of the remaining item, we always find the recall of the reference graph, which means that the method finds the good neighbors (i.e., exactly the same neighborhood as the reference graph) of the inserted item in each experiment, and the whole graph structure is well-updated.

Evaluation of the Execution Time

In this section, we are interested in the response times of the suggested method (i.e., the time that the method takes to insert a query item in an existing structure). The evaluation protocol is rather similar to the previous one, but instead of recovering the recall, we recover the execution time. One of the used datasets in these experiments is an artificial dataset containing 30,000 items represented in 24 dimensions. We also used a machine with an INTEL Pentium 4 processor (2.80 GHz) and 512 Mo of memory. The response times for 20 items, always arbitrarily taken from the same dataset, are shown in the graphic of Figure 6. Note that we do not give a comparison with the execution time using the standard algorithm for legibility facilities.

The response times (expressed in milliseconds) are interesting according to the volume of the used data set. They vary between 20 and 60 milliseconds per item (average of the execution time). This is very interesting, considering that by using a standard method, the same neighborhood can be obtained in four hours approximately in the same conditions. The variation of the execution times from one item to another is due especially to the fact that various amounts of candidate items for the neighborhood determination are used at each iteration.

Figure 6. Illustration of the execution time variations using a sample of 20 local insertions into a dataset containing 30,000 items

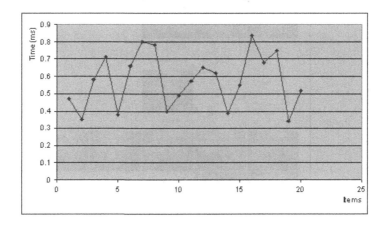

Semi-Automatic Annotation Performances

In order to show the interest and performances of the proposed method for images annotation, we use an image database (Nene, Nayar, & Murase, 1996). This image database contains 7,200 images that represent 100 objects taken from various views. After the transformation of the image database into a vector representation using some image analysis techniques and the low-level features extracted, we have annotated the entire image database with a list of keywords that describe the semantic content of each image. So, each image is represented by a vector containing a set of descriptive features (24 low-level features) and from two to six keywords (we used a simple image database because the interest here is not to show

Figure 7. Retrieving a database using only the low-level characteristics indexed with a neighborhood graph

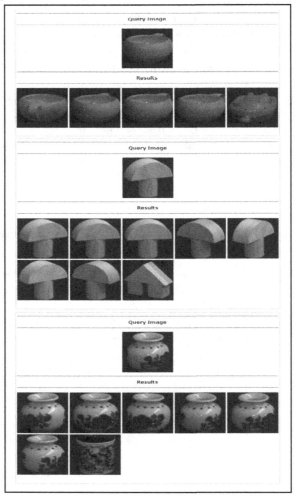

Table 4. The initial configuration of the neighborhood of the first query image

Images	Weights	Existing Annotations
1	0.35	Cup, Brown, Pottery, Bowl
2	0.25	Cup, Brown, Pottery, Bowl
3	0.15	Cup, Brown, Pottery, Bowl
4	0.1	Cup, Brown, Pottery, Bowl
5	0.15	Boat, Red

the performances of segmentation methods but only the utility of the proposed approach for annotating images).

We use approximately the same evaluation protocol as the previous tests. We first build a neighborhood graph using $n-1$ images considering only the low-level features (Figure 7 illustrates three retrieval examples using low-level characteristics). The second step consists of the insertion of the remaining item (the query images in Figure 7) in the graph previously built using the low-level characteristics. By applying the weighting annotation technique, we can predict the most suitable annotation that can be associated with the query image.

In order to illustrate the principle, let us consider the case of the first image of Figure 7. The associated weights and annotations for all images are summarized in Table 4.

According to this configuration, we calculate scores for each annotation. We obtain the scores illustrated in Table 5. Considering that the threshold is fixed to 75%, the image will be annotated as follows: Cup, Brown, Pottery, and Bowl with a score of 85%. These annotations are exactly the affected annotations to the query image. At the end, the user intervenes in order to accept the predicted annotations, discarding them or adding new free annotations. These last ones will be propagated in order to find out if other images can be annotated using them.

Generally speaking, the system has about 90% correct annotations of the images in the database. Of course, the images of the used database are rather simple; using a more complex image databases can affect the performances of the annotation approach but not considerably.

Table 5. Calculated scores for the candidate annotations

Annotations	Scores
Cup	85%
Brown	85%
Pottery	85%
Bowl	85%
Boat	15%
Red	15%

Conclusion

Content-based image retrieval is a complex task, primarily because of the nature of the images and the attached subjectivity to their interpretation. The use of an adequate index structure is primordial. In order to fix some problems related to the traditional indexing and retrieval approaches, we introduced neighborhood graphs as a substitution of the traditional tree structures and the *k-NN* approach. The introduction of neighborhood graphs in image databases indexing and retrieval is motivated by (a) the need for improving the neighborhood quality and (b) the necessity for automatic learning methods integration into the indexing process, which seems to us extremely important for mining image databases and for being able to discover possible hidden information in this type of data.

We also proposed a method for locally updating neighborhood graphs. Our method is based on the location of the potential items, which can be affected by the updating task, allowing an effective interaction with the index by supporting the most frequently applied operations on an index (insertion and deletion of an item). We also exploited the proposed indexing structure to help the user in the annotation of an image collection. The experiments performed on various datasets show the effectiveness and the utility of the proposed approach.

As future work, we plan to fix the problem of the relaxation parameter determination by setting up an automatic determination function. This can be achieved by taking into account some statistical parameters on the data, such as dispersion. On another side, we plan to use the proposed approach to annotate a more complex image collection. Finally, we plan to find a way to integrate user feedback in the system in order to make automatic the annotation process.

References

Anderson, E. (1935). The irises of the Gaspé Peninsula. *Bulletin of the American Iris Society, 59*, 2–5.

Barnard, K., Duygulu, P., & Forsyth, D. A. (2001). Clustering art. *Computer Vision and Pattern Recognition (CVPR), 2*, 434–441.

Barnard, K., & Forsyth, D. A. (2001). Learning the semantics of words and pictures. In *Proceedings of the International Conference on Computer Vision (ICCV)* (pp. 408–415).

Bei, C.-D., & Gray, R. M. (1985). An improvement of the minimum distortion encoding algorithm for vector quantization. *IEEE Transactions on Communications, 33*, 1132–1133.

Bentley, J. L. (1975). Multidimensional binary search trees used for associative searching. *Communication of the ACM, 18*(9), 509–517.

Berchtold, S., Böhm, C., Keim, D.A., & Kriegel, H.-P. (1997). A cost model for nearest neighbor search in high-dimensional data space. In *Proceedings of the 16th ACM SIGACT-SIGMOD-SIGART Symposium on Principles of Database Systems* (pp. 78–86).

Breiman, L., Friedman, J. H., Olshen, R. A., & Stone, C. J. (1984). *Classification and regression trees*. Belmont, CA: Wadsworth International Group.

Chiueh, T.-C. (1994). Content-based image indexing. In *Very large databases* (pp. 582–593).

Cost, R. S., & Salzberg, S. (1993). A weighted nearest neighbor algorithm for learning with symbolic features. *Machine Learning, 10*, 57–78.

Cover, T. M., & Hart, P. E. (1967). Nearest neighbor pattern classification. *IEEE Transaction in Information Theory, 13*, 57–67.

Deerwester, S. C., Dumais, S. T., Landauer, T. K., Furnas, G. W., & Harshman, R. A. (1990). Indexing by latent semantic analysis. *Journal of the American Society for Information Science (JASIS), 41*(6), 391–407.

Duygulu, P., Barnard, K., De Freitas, J. F. G., & Forsyth, D. A. (2002). Object recognition as machine translation: Learning a lexicon for a fixed image vocabulary. In *Proceedings of the European Conference on Computer Vision, (ECCV)* (pp. 97–112).

Faloutsos, C., et al. (1994). Efficient and effective querying by image content. *Journal of Intelligent Information Systems, 3*(3/4), 231–262.

Fayyad, U. M., Piatetsky-Shapiro, G., & Smyth, P. (1996). From data mining to knowledge discovery: An overview. In *Advances in Knowledge Discovery and Data Mining* (pp. 1–34). Menlo Park: AAAI Press.

Fisher, R. (1936). The use of multiple measurements in taxonomic problems. *Annals of Eugenics, 7*, 179–188.

Fix, E., & Hudges, J. L. (1951). *Discriminatory analysis: Non parametric discrimination: Consistency properties* (Tech. Rep. No. 21-49-004). Randolph Field, TX: USAF School of Aviation Medicine.

Flickner, M., et al. (1995). Query by image and video content: The QBIC system. *IEEE Computer, 28*(9), 23–32.

Friedman, J. H., Baskett, F., & Shustek, L. J. (1975). An algorithm for finding nearest neighbors. *IEEE Transactions on Computers, 24*(10), 1000–1006.

Gabriel, K. R., & Sokal, R. R. (1969). A new statistical approach to geographic variation analysis. *Systematic Zoology, 18*, 259–278.

Gersho, A., & Gray, R. M. (1991). *Vector quantization and signal compression*. Boston: Kluwer Academic.

Guan, L., & Kamel, M. (1992). Equal-average hyperplane partitioning method for vector quantization of image data. *Pattern Recognition Letters, 13*(10), 693–699.

Guttman, A. (1984). R-trees: A dynamic index structure for spatial searching. In *Proceedings of the SIGMOD Conference* (pp. 47–57).

Hacid, H., & Zighed, A.D. (2005). An effective method for locally neighborhood graphs updating. *Database and Expert Systems Applications* (LNCS 3588, 930-939).

Hettich, S., Blake, C., & Merz, C. (1998). UCI repository of machine learning databases.

Jeon, J., Lavrenko, V., & Manmatha, R. (2003). Automatic image annotation and retrieval using cross-media relevance models. In *Proceedings of the SIGIR* (pp. 119–126).

Katajainen, J. (1988). The region approach for computing relative neighborhood graphs in the LP metric. *Computing, 40,* 147–161.

Kohonen, T. (2001). *Self-organizing maps* (Vol. 30). New York: Springer.

Lee, C.-H., & Chen, L.H. (1994). Fast closest codeword search algorithm for vector quantisation. In *IEEE Proceedings: Vision, Image, and Signal Processing* (Vol. 141, pp. 143–148).

Li, J., & Wang, J. Z. (2003). Automatic linguistic indexing of pictures by a statistical modeling approach. *IEEE Transactions on Pattern Analysis and Machine Intelligence, 25*(9), 1075–1088.

Lin, K.-I., Jagadish, H. V., & Faloutsos, C. (1994). The TV-tree: An index structure for high-dimensional data. *Very Large Databases Journal, 3*(4), 517–542.

Maron, O., & Ratan, A. L. (1998). Multiple-instance learning for natural scene classification. In *Proceedings of the International Conference on Machine Learning* (pp. 341–349).

Mitchell, T. M. (2003). Machine learning meets natural language. In *Proceedings of the Progress in Artificial Intelligence 8th Portuguese Conference on Artificial Intelligence, EPIA '97* (p. 391).

Monay, F., & Gatica-Perez, D. (2003). On image auto-annotation with latent space models. In *Proceedings of the ACM International Conference on Multimedia (ACM MM),* Berkeley, CA (pp. 275–278).

Mori, Y., Takahashi, H., & Oka, R. (1999). Image-to-word transformation based on dividing and vector quantizing images with words. *Proceedings of the International Workshop on Multimedia Intelligent Storage and Retrieval Management* (pp. 341–349).

Nene, S. A., Nayar, S. K., & Murase, H. (1996). (Tech. Rep. No. CUCS-006-96). Columbia Object Image Library (coil-100).

Picard, R. W., & Minka, T. P. (1995). Vision texture for annotation. *Multimedia Systems, 3*(1), 3–14.

Preparata, F., & Shamos, M. I. (1985). *Computational Geometry: Introduction.* New York: Springer-Verlag.

Scuturici, M., Clech, J., Scuturici, V. M., & Zighed, A. D. (2005, January-June). Topological representation model for image databases query. *Journal of Experimental and Theoretical Artificial Intelligence (JETAI), 17*(1-2), 145–160.

Shi, J., & Malik, J. (2000). Normalized cuts and image segmentation. *IEEE Transactions on Pattern Analysis and Machine Intelligence, 22*(8), 888–905.

Smith, W. D. (1989). *Studies in computational geometry motivated by mesh generation* [doctoral thesis]. Princeton University.

Toussaint, G. T. (1980). The relative neighborhood graphs in a finite planar set. *Pattern Recognition, 12,* 261–268.

Toussaint, G. T. (1991). Some unsolved problems on proximity graphs. In *Proceedings of the First Workshop on Proximity Graphs*, Las Cruces, NM.

Veltkamp, R. C., & Tanase, M. (2000). *Content-based image retrieval systems: A survey* (Tech. Rep. No. UU-CS-2000-34). Utrecht University.

White, D. A., & Jain, R. (1996). Similarity indexing: Algorithms and performance. *Storage and Retrieval for Image and Video Databases (SPIE)*, 62–73.

Chapter XII

Neural Networks for Content-Based Image Retrieval

Brijesh Verma, Central Queensland University, Australia

Siddhivinayak Kulkarni, University of Ballarat, Australia

Abstract

This chapter introduces neural networks for content-based image retrieval (CBIR) systems. It presents a critical literature review of both the traditional and neural network-based techniques that are used to retrieve images based on their content. It shows how neural networks and fuzzy logic can be used in the interpretation of queries, feature extraction, and classification of features by describing a detailed research methodology. It investigates a neural network-based technique in conjunction with fuzzy logic to improve the overall performance of CBIR systems. The results of the investigation on a benchmark database with a comparative analysis are presented in this chapter. The methodologies and results presented in this chapter will allow researchers to improve and compare their methods, and it also will allow system developers to understand and implement the neural network and fuzzy logic-based techniques for content-based image retrieval.

Introduction

Recent technological advances in computer-based storage, transmission, and display of image data and the Internet have made it possible to process and store large amounts of image data. Perhaps the most impressive example is the fast accumulation of image data in scientific applications such as medical and satellite imagery. The Internet is another excellent example of a distributed database containing several million images. In recent years, there has been a rapid increase in the size of digital image collections. Retrieving images from large collections effectively and efficiently based on their content has become an important research issue for database, image-processing, and computer-vision communities. Creating and storing digital images nowadays is easy and getting cheaper all the time as the needed technologies are becoming available. There already is a vast number of digital visual data sources (e.g., various kinds of sensors, digital cameras, and scanners) in addition to the various image collections and databases for all kinds of purposes. Furthermore, the fast development of computing hardware has enabled us to switch from text-based computing to graphical user interfaces (GUIs) and multimedia applications. This transaction has fundamentally changed the use of computers and made visual information an inseparable part of everyday computing. As a result, the amount of information in visual form is increasing, and we need effective ways to process it. The existing and widely adopted methods for text-based data usually are inadequate for these purposes. Visual information generally is said to include both still images and moving videos. It is well-known that a successful CBIR system will have a large number of applications. Although many CBIR systems have been built, the performance of these systems is far from the demands of various applications. The aim of CBIR research is to retrieve desired images from such large collections automatically. Toward this goal, automatic CBIR systems have become an active research area in the last decade. One of the solutions is the development of automated methods that are able to recognize the composition and content of the images and to make use of the recognized image content in order to achieve content-based image retrieval. Such kinds of image retrieval systems involve a great deal of image understanding and machine intelligence. In addition, an ideal system should allow both a keyword and a concept search in conjunction with a content-based search. This already can be seen in many image retrieval systems. However, most of them do not provide flexibility in formulating these queries. At the start, the user should be able either to issue a direct query or to query by example. There is no good or proven way to search for and retrieve images from databases without annotation. Given the enormous amount of images stored on computers, annotation is becoming more and more impractical, and the need for an automatic feature identification based on image retrieval is growing.

Techniques that are based on neural networks also have been developed for CBIR. The main reason for using neural networks in CBIR is due to their adaptive learning capability. Neural networks offer unsupervised clustering as well as supervised learning mechanisms for the retrieval of images. Recent research in image retrieval suggests that significant improvement in retrieval performance requires techniques that, in some sense, "understand" the content of the image and queries. Recently, image retrieval researchers have used the application domain knowledge to determine relevant relationships between images and queries. The aim of the research presented in this chapter is to show the possibility of learning and use of domain knowledge in an image retrieval application by means of the learning and generalization capabilities of neural networks.

This chapter reviews the prominent CBIR systems along with neural network-based techniques and proposes a technique based on neural networks and fuzzy logic for retrieving images by using natural language query. This chapter is organized as follows: the first section details the review of existing CBIR techniques; the next section describes CBIR using neural networks and fuzzy logic; next we describe the experimental results conducted for various queries; then the results are compared and analyzed; and finally, the chapter is concluded.

Review of Content-Based
Image Retrieval (CBIR) Techniques

The problems of image retrieval are becoming widely recognized, and the search for solutions is becoming an increasingly active area for research in various areas of computer science discipline. Problems with traditional methods of image indexing have led to the rise of interest in techniques for retrieving images on the basis of automatically derived features. Several such systems have been developed (Flickner et al., 1995). These systems model image data using features such as color, texture, and shape (Fukushima & Ralescu, 1995). Feature extraction plays an important role in CBIR techniques. Many researchers extract features such as color, texture, shape, and so forth from images and store them in a database (Santini & Jain, 1996). After a decade of intensive research, CBIR technology now is beginning to move out of the laboratory and into the marketplace in the form of commercial products such as QBIC (Faloutsos et al., 1994; Lee et al., 1994) and Virage (Gupta, 1996).

IBM's Query By Image Content (QBIC) was the first commercial CBIR system (Faloutsos et al., 1994). It allows queries on large image databases based on example images, user-drawn sketches, selected color, and texture patterns. It uses common features of the image such as color, texture, and shape. QBIC supports various types of queries based on the example images, user-constructed sketches and drawings, and selected color and texture patterns. Visual Information Retrieval (VIR) Image Engine is a commercial product of Virage Inc.

Virage supports visual queries based on color, composition of color layouts, texture, and structure (object boundary information). The user can pose a query by selecting one of the features or a combination of them. The system returns the most similar images and displays them to the user (Pala & Santini, 1999). Blobworld system was developed at the University of California, Berkeley by Carson, Belongie, Greenspan, and Malik (1998). Blobworld is a technique of converting raw pixel data into a small set of localized coherent regions of color and texture. This is motivated by the observation that images can be characterized by localized regions of coherent color and texture, defined in terms of their relative size and position in an image plane. Netra is a prototype image retrieval system developed at the University of California, Santa Barbara Alexandria Digital Library (ADL) (Ma & Manjunath, 1999; Manjunath & Ma, 1996). It uses color, texture, shape, and spatial location information in the image regions to search and retrieve images with similar regions from a database. The main motivation for the development of this system is that region-based search improves the quality of image retrieval. Therefore, the system incorporates an automated region identification algorithm. There are other existing techniques in the literature for content-based retrieval, such as satellite image browsing using automatic semantic categorization

(Parulekar, Datta, Li, & Wang, 2005), partial relevance in interactive facial image retrieval (Yang & Laaksonen, 2005a), and region-based image clustering and retrieval using multiple instance learning (Zhang & Chen, 2005). A recent survey published in 2005 on content-based image retrieval can be found in Datta, Li, and Wang (2005).

Computational intelligence-based techniques such as neural networks also have been applied by some researchers to develop a prototype for CBIR. Neural networks also have been proposed for feature extraction (Oja, Laaksonen, Koskela, & Brandt, 1997), similarity measurement (Muneesawang and Guan, 2002a), relevance feedback technique (Lee & Yoo, 2001a, 2001b), and so forth. Various developed algorithms, such as error-back propagation, self-organizing map, radial-basis function, have been applied successfully to solve the various problems in image retrieval as well as to bridge the semantic gap used for processing a high-level query posed by a user and low-level features extracted by the system (Tsai, 2003a). Lee and Yoo proposed a human computer interaction approach to CBIR that is based on a radial basis function neural network (Lee & Yoo, 2000). This approach determines a nonlinear relationship among features so that a more accurate similarity comparison can be supported and will allow the user to submit a coarse initial query and continuously refine this information need via relevance feedback. Images are compared through a weighted dissimilarity function that can be replaced as a "network of dissimilarities." The weights are updated via an error-back propagation algorithm using the user's annotations of the successive set of result images (Fournier, Cord, & Philipp-Foliguet, 2001). It allows an iterative refinement of the search through a simple interactive process. In video retrieval by still image analysis technique (Kreyss, Roper, Alshuthe, Hermes, & Herzog, 1997), the average value of the texture features extracted using co-occurrence matrix over the four orientations are the inputs of a neural network trained with back propagation algorithm. An unsupervised learning neural network is explored to incorporate a self-learning capability (Koskela & Laaksonen, 2005; Koskela, Laaksonen, & Oja, 2002; Laaksonen, Koskela, & Oja, 1999a, 1999b; Muneesawang & Guan, 2002b; Yang & Laaksonen, 2005b) into an image retrieval system.

PicSOM is a content-based image retrieval system based on a neural network algorithm called Self-Organizing Map (SOM). The SOM is used to organize images into map units in a two-dimensional grid so that the similar images are located near each other. PicSOM uses a tree-structured version of the SOM algorithm (TS-SOM) to create a hierarchical representation of the image database (Laaksonen et al., 1999a, 1999b). In the PicSOM CBIR system, a Self Organizing Map (SOM) has been used as a relevance feedback technique (Koskela et al., 2002). The technique is based on the SOM's inherent property of topology-preserving mapping from a high-dimensional feature space to a two dimensional grid of artificial neurons. The self-organized tree map also is proposed to minimize user participation in an effort to automate interactive retrieval.

The automatic learning mode has been applied in order to optimize the relevance feedback (RF) method using a single radial basis function-based RF method (Muneesawang & Guan, 2002b). A two-stage mapping model to bridge the semantic gap between the low-level image features and high-level concepts in a divide-and-conquer manner is presented (Tsai, McGarry, & Tait, 2003b).

Neural networks have been used to calculate the similarity in the view of their learning capability (Naqa, 2000; Wood, Campbell, & Thomas, 1998) and their ability to simulate universal mapping. At the same time, these models usually require large volumes of training data for each individual query. The utilization of a nonlinear model based on Gaussian-shaped RBF has been introduced in Muneesawang and Guan (2001) to cope with the complex decision boundaries. One advantage of this model is that it requires only a small volume of training samples, and it converges very fast. Conventional predefined metrics such as Euclidean distance usually are adopted for the ranking of images based on the example image query in simple CBIR systems. They are not sufficient to deal with a complex evaluation of image contents. In a learning-based approach (Duan, Gao, & Ma, 2000; Zhuang, Liu, & Pan, 2000) users directly teach the system what they regard as significant image features and their notions of image similarities. A model proposed by Ikeda and Haiwara (2000) takes advantages of association ability of multilayer NNs as matching engines that calculate similarities between a user's drawn sketch and the stored images. The NNs memorize pixel information of every size-reduced image (thumbnail) in the learning phase. In the retrieval phase, pixel information of a user's drawn rough sketch is inputted to the learned NNs, which estimate the candidates. Thus, the system can retrieve candidates quickly and correctly by utilizing the parallelism and association ability of NNs. Yu and Aslandogan (2000) combine multiple evidences for searching facial images. The features used are faces in images and

Figure 1. Neural network and fuzzy logic based CBIR technique

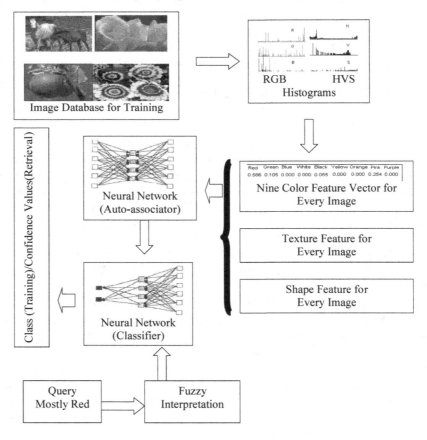

text in HTML pages. The neural network approach is used to detect faces in an image. Once detected, a face is isolated and submitted to face a recognition module based on Eigen-faces, which use a set of known facial images associated with a person's name.

CBIR Using Neural Networks and Fuzzy Logic

An overview of neural network and fuzzy logic-based technique for content-based image retrieval is presented in Figure 1.

From the previous section, it is seen that there are number of ways to improve techniques for retrieving images from a database. The research proposed in this chapter focuses on retrieving images based on their features from databases using a neural network technique. This new research area is becoming attractive and commercially based due to its application in various fields. Applications that have received particular attention include medicine, fashion and graphic design, publishing and advertising, architectural and engineering design, and so forth. This research area is very challenging due to the need for the development of feature extraction techniques and for specifying user-friendly queries. The following sections report the techniques that are integrated for use in a CBIR system.

Color Image Retrieval

This section describes the proposed technique for retrieving images based on the color feature of the image. The section is divided further into two stages in order to describe the technique in detail. The first stage concentrates on preprocessing, which includes building an image database, feature extraction technique, and preparing the feature database. The second stage describes database retrieval, which contains fuzzy interpretation of the queries, a neural network for learning those queries, and fusion of multiple queries of color feature to improve the confidence factor. The neural networks show the possibility of learning and use domain knowledge in an image retrieval application. In particular, the use of the error-back propagation learning algorithm was proposed.

Color Feature Extraction

Color is a very important cue in extracting information from images. Color feature extraction forms the basis for color image retrieval. Color distribution, which is represented best as a histogram of intensity values, is more appropriate as a global property that does not require knowledge of how an image is composed of various objects. A color histogram technique is proposed for extracting colors from images. The feature representation set is motivated by the research results of Carson and Ogle (1996), who identified nine colors that fall within the range of human perception. The feature representation set of colors is rep (color) = {red, green, blue, white, black, yellow, orange, pink, purple}.

Color Feature Database

The color feature is extracted from the image and stored in a database. N_p represents the total number of pixels in an image. For each color value (red, green, blue, etc.), the number of pixels that belong to the value are recorded and denoted by N_f where $f \in F_u$. In the methodology, the feature representation set F_u is the set of nine colors. The number of pixels is calculated for each color. The feature component F_R is calculated for each color in an image by the following formula:

$$F_R = \frac{N_R}{N_P}$$

N_R represents the number of red pixels in an image. Similarly, the feature component is calculated for all nine colors.

Fuzzy Interpretation for Queries

In most of the current CBIR systems, a query is submitted in the form of an example image, and subsequently, similar images are obtained from the database. This research proposes queries in terms of natural language using fuzzy logic. In some applications, fuzzy systems often perform better than traditional systems because of their capability to deal with non-linearity and uncertainty.

Referring to Figure 2, the query to retrieve the images from the database is prepared in terms of natural language, such as *mostly*, *many*, and *few* content of some specific color. This approach is used to make CBIR systems intelligent so that they can interpret human language.

The first step is to define a simple syntax for queries. In the real world, users do not want to enter many numbers or percentages for colors, shapes, and so forth. Usually, to retrieve images, users prefer to say that they would like to retrieve images that are "mostly red" or "mostly green and a little bit red." The technique describes these colors in three expressions such as mostly, many, and few. The user can pose a composite query in terms of colors and content types. For example, if the user is searching for the images that contain large amounts of red color and a little bit of green color, then the syntax for the query is "mostly red and few green." The general syntax for queries is as follows:

QUERY = {{<Content><Color>}}

where,

Content = {<mostly | many | few>}

Color = {<red | green | blue | white | black | yellow | orange | pink | purple>}

Figure 2. Query interpretation (mostly, many, and few)

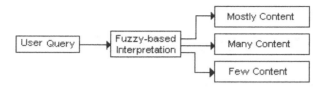

For example: A color query to retrieve images can be defined as follows:

QUERY = {mostly red AND few green}

The query to retrieve images from a database is defined in terms of natural language, such as mostly <content>, many <content>, and few <content> of some specific color.

$$\mu\left(content\right)=\begin{cases} mostly & if\ content \in <0.9\ \ 1> \\ many & if\ content \in <0.4\ \ 0.5> \\ few & if\ content \in <0.15\ \ 0.25> \end{cases}$$

Learning of Queries

In order to retrieve the required images from the database, it is necessary to learn the meaning of the query, which is in terms of nine colors and three content types such as mostly, many, and few. The error-back propagation neural network is proposed to learn the meaning of those queries. This approach is very novel and overcomes the problem of retraining of neural networks in real-world online applications in which databases get larger everyday. The range of minimum and maximum values for the three attributes mostly, many, and few are shown as follows:

Mostly: min = 0.9 +Δ; max = 1.0; $\Delta = \Delta + 001$;

Many: min = 0.4 +Δ; max = 0.5; $\Delta = \Delta + 0.001$;

Few: min = 0.15 +Δ; max = 0.25; $\Delta = \Delta + 0.001$;

The number of training pairs depends on the number of inputs and outputs. Considering the nine colors, the number of inputs are 11×9 (99). This process is repeated for the other content types such as many and few, giving the total number of 297 training pairs.

Fusion of Multiple Queries

Most existing CBIR systems simply treat the composite query as many separate queries and execute them one-by-one. This search can be improved by using neural network fusion.

The fusion of the two queries is described next, but the approach can be used for more than two queries.

Fusion of Queries Using Binary AND

For example, if the composite query is "mostly red AND few green," then the output of this query is shown using a binary AND. This approach will work if the possible output of the neural network is 1 or 0. But if the output of the neural network is in the range of 0 to 1 (e.g., 0.8 and 0.4) for the corresponding queries, then this approach will fail to give the final confidence factor. Another approach may be considered to fuse the outputs of the neural network (i.e., by the use of Fuzzy Logic AND).

Fusion of Queries Using Fuzzy AND

The possibility of getting the final confidence factor for Fuzzy AND on the example query "mostly red AND few green" is shown in Table 1.

This approach will work better than the binary AND if the outputs of the neural network for the two queries are similar or have very small differences. But if the difference between the neural network outputs is significant, then this approach will fail to give an adequate final confidence factor.

In Table 2, it is shown that for both cases, the final confidence factor is 0.3, although in the first case, the confidence factor for the first query is 0.9. This interpretation is not very accurate for different possible outputs of the neural network. The final approach is based on the fusion of the multiple queries using a neural network.

Table 1. Final confidence factor using fuzzy AND (first case)

Confidence factor for First query (mostly red)	Confidence factor for second query (few green)	Final confidence factor
0.6 0	.8	0.6 (60%)
0.85 0	.9 0	.85 (85%)

Table 2. Final confidence factor using fuzzy AND (second case)

Confidence factor for first query (mostly red)	Confidence factor for second query (few green)	Final confidence factor
0.9	0.3	0.3 (30%)
0.3	0.3	0.3 (30%)

Table 3. Final confidence factor using neuro-fuzzy AND

Confidence factor for first query (mostly red)	Confidence factor for second query (few green)	Final confidence factor
0.9	0.3 A	
0.3	0.3 B	(%A>%B)

Fusion of Queries Using Neuro-Fuzzy AND

The fusion algorithm is based on a two-layer, feed-forward neural network. The number of inputs to the neural network is equivalent to the number of queries, and the number of outputs is equivalent to one that is the actual output of the neural network. The actual output of the neural network is called the *confidence factor*. The network is trained for all possible outputs from the first neural network. The neural network decides the final confidence factor for the combination of the queries. This approach works better than the aforementioned binary AND and fuzzy AND.

In the first case, if the confidence factors are 0.9 and 0.3, then the final confidence factor will be decided by the neural network, say A.

In the second case, if the confidence factors are 0.3 for both queries, then the final confidence factor is B. But the value of B must be smaller than the value of A. This neural network fusion technique has been implemented, and the results are shown in the experimental results section.

Texture Image Retrieval

In order to retrieve images based on texture features, it is important to preprocess these images. The preprocessing includes the formation of a texture image database, extraction of texture features from these images, classifying these features in appropriate classes, and using these classes to retrieve images. The proposed technique is divided into two stages. Stage 1 deals with feature extraction from texture subimages. An auto-associator was designed to extract features. Stage 2 deals with the classification of features into texture classes. A Multi-Layer Perceptron (MLP) was designed to classify texture classes.

Preprocessing Texture Images

The texture database used to verify the proposed technique consists of 96 texture images. Images D1 through D96 are from the Brodatz (1966) album. These texture images can be categorized into different classes that look visually similar. Each image is divided into 16 nonoverlapping subimages that are each 128×128 pixels in size, thus creating a database of 1,536 (96×16) texture images. The first 12 subimages were used for the training of the auto-associator, and the last four images were used as a testing data set.

In order to reduce the size of the input vector provided to the neural network, the mean and standard deviation was calculated for each row of the subimage (128 pixels) as follows:

$$\mu = \frac{1}{n} \sum_{i=1}^{n} x_i$$

$$\sigma = \sqrt{\frac{1}{n} \sum_{i=1}^{n} (x_i - \mu)^2}$$

in which μ and σ are changed for mean and standard deviation, n is the number of pixels (in this case, 128).

Stage 1. Auto-Associator as Texture Feature Extractor

The goal of the feature extraction technique is to transform raw image data into a much lower dimensional space, thus providing a compact representation of the image. It is desirable that the computed feature vectors preserve the perceptual similarity; that is, if two image patterns are visually similar, then the corresponding feature vectors are also close to each other. The main idea of the auto-associator feature extractor is based on input:hidden:output mapping in which inputs and outputs are the same patterns (Kulkarni & Verma, 2001). The AAFE learns the same patterns and provides a characteristic through its hidden layer as a feature vector. An auto-associator feature extractor that uses a single hidden layer feed-forward neural network is designed. It has n inputs, n outputs, and p hidden units. The input and output of the AAFE are the same texture patterns (n in number), and the network is trained using a supervised learning algorithm. The p is the number of hidden units in the hidden layer, different from the number of inputs and outputs. After training is completed, the output of the hidden layer is extracted and used as a feature vector. The feature vector is fed to the MLP feature classifier in order to classify the features into separate classes.

Stage 2. MLP as a Texture Feature Classifier

This stage describes the capability of the Multi-Layer Perceptron (MLP) for the classification tasks and the implementation of a trainable MLP and presents a suitable kind of MLP structure for classification of texture data (Kulkarni & Verma, 2002). Before applying the output of the hidden layer to the input of the classifier, it is necessary to form the classes of these texture patterns. There were 96 texture patterns in the database, which were grouped into 32 similar classes, each of them containing one to five texture classes. All of the texture subimages belonging to the same similarity class were visually similar. The classifier has n inputs, which is the same as the number of hidden units in an auto-associator feature extractor. The output of the hidden layer that was obtained from the auto-associator was used as input to the classifier. There were 32 texture classes, so the number of outputs was 32.

Experimental Results

This section outlines all relevant experimental results that were conducted using the proposed neuro-fuzzy technique. The first section deals with the experiments based on the color feature of an image, and the next section gives the detail analysis for texture feature extraction, classification, and retrieval.

Image Retrieval Based on Color Feature

In order to test the effectiveness of the proposed system, the preliminary experiments were conducted on a single color feature and content type (Kulkarni, Verma, Sharma, & Selvaraj, 1999). Before retrieving the images based on the single color and content type, images were preprocessed. The real-world images were downloaded from the World Wide Web to form the image database. The color feature was extracted from all of the images and stored into a database. Color features are applied as input to the neural network and content type, such as mostly, many, and few, and form the output of the neural network.

Neural Network Architecture

The neural network was designed by considering nine inputs as colors and three outputs as content type. Table 4 indicates all the parameters that were used to train the neural network and the RMS error obtained after training. The experiments were conducted by varying the number of hidden units and iterations.

The number of inputs indicate all nine colors: red, green, blue, white, black, yellow, orange, pink, and purple. The number of outputs was three, which indicates the content type such as mostly, many, and few. The learning rate (η) and momentum (α) were varied for the experiments.

Fusion of Multiple Queries

The user could pose a query in terms of a single color and content type such as mostly red, few green, and so forth (Verma & Kulkarni, 2004). This technique was extended to pose a query in terms of multiple colors and content types. In order to obtain the final confidence

Table 4. Neural network parameters for experiments

Inputs	Outputs	Hidden Units	Training Pairs	η	α	Iterations	RMS Error
9	3	5	297	0.7	0.2	100	0.00393
9	3	5	297	0.7	0.2	1000	0.00124
9	3	6	297	0.8	0.3	1000	0.00107

Figure 3. Results of the query "mostly green" and "few red" using fuzzy AND technique

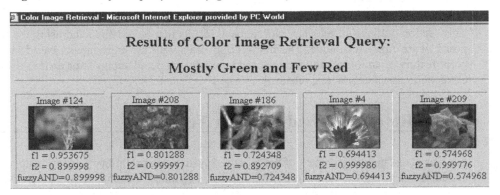

factor for binary AND, the input should be in terms of 0 and/or 1. This technique fails if the confidence factors have values other than 1 and 0 for the individual query, and it was found that it is not adequate to use this technique in real-world applications.

Fuzzy AND

The confidence factors are fused using a fuzzy AND technique. The fuzzy AND selects the smaller value of the two. Figure 3 shows the images retrieved using the Fuzzy AND technique. For the query "mostly green and few red," five images were retrieved. The confidence factors are indicated by *f1* and *f2* for the query mostly green and few red, respectively. The final confidence factor was indicated by Fuzzy AND. This technique was tested for different combinations of the queries.

Neuro-Fuzzy AND

A number of experiments were conducted in order to test the effectiveness of the neuro-fuzzy AND technique. The neural network was designed for the queries to have two colors and two content types. This technique is also suitable for more than two queries. Table 5 shows the different parameters used to train neural network.

Table 5. Neural network parameters for fusion

Number of inputs	Number of outputs	Hidden units	Learning rate	Momentum	RMS error
2	1	4	0.7	0.6	0.003244

Figure 4. Results of the query mostly red and few orange

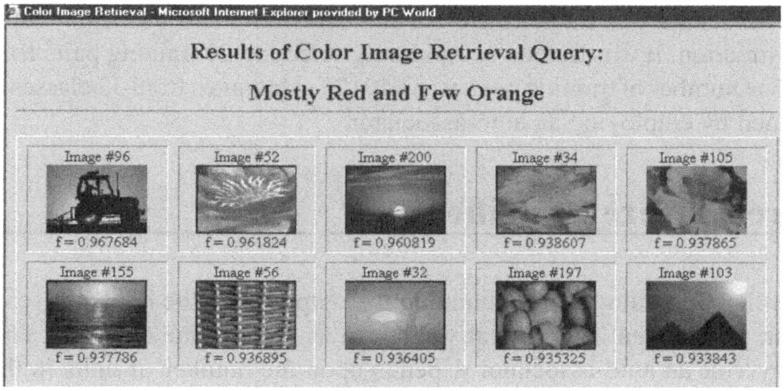

A. Query: Mostly and Few Contents

If the user wants to retrieve images from a database such as mostly green and few yellow, the two colors and content types are selected from the color and content pallets. Figure 4 shows the result of the query along with the colors and the content types specified for the query. The images are displayed along with their image number and confidence factor. In Figure 4, Image #96 contains a good combination of mostly red and few orange from all the images in the database, so this image has a maximum confidence factor of 0.967684. All other images are obtained in descending order with their confidence factor.

Texture Feature Extraction

The objective of these experiments is to illustrate that the proposed neural-based texture feature extraction provides a powerful tool to aid in image retrieval. The experiments were conducted in two stages: first, the training of the auto-associator, and second, the training of the classifier. Before training of an auto-associator, it is necessary to provide different parameters to a multi layer perceptron (MLP). Table 6 shows the values of parameters used for training the auto-associator. The number of training and testing subimages was 1,152 (96×12) and 384 (96×4). The mean and standard deviation was calculated for each row of the subimage. In the case of the auto-associator, the same patterns were applied to the output,

Table 6. Training parameters for the auto-associator

Number of inputs	Number of outputs	Number of training pairs	Learning rate η	Momentum α
256	256	1152	0.8	0.7

giving the number of outputs as 256. The values of the learning rate (η) and momentum (α) were 0.8 and 0.7, respectively.

The auto-associator was trained different numbers of hidden units and iterations to improve feature extraction. It was trained by randomly selecting 20 training pairs from each texture pattern. The number of training pairs was 640 (20 subimages from 32 classes). These images were trained by employing an auto-associator.

Texture Feature Classifier

The classifier was trained after obtaining the output from the hidden layer from the auto-associator. The hidden layer output was given as input to the classifier. The output of the hidden layer of an auto-associator depends upon the number of units in the hidden layer and the number of training pairs. The number of inputs to the classifier is the same as the number of hidden units used to train the auto-associator.

A. Classification Results for Reduced Training and Testing Sets

The training and testing sets were prepared by randomly selecting subimages from the Brodatz album. For 32 classes, 20 training pairs from each subimage were chosen for training the classifier, and five testing pairs were chosen for testing. The classifier was trained by keeping the number of hidden units constant. The number of hidden units used was 16. The classifier was trained and tested for different hidden layer outputs of the auto-associator.

The classifier was trained with the output of the hidden layer of the auto-associator. The hidden layer output was obtained by training it with 32 hidden units and 25,000 iterations. Table 7 shows the classification results obtained for training and testing sets. Initially, experiments were conducted for 1,000 iterations and increased gradually to 40,000 iterations. For 1,000 iterations, the classification rate was 87.96% for the training set and 81.87% for the testing

Table 7. Classification results for training and testing sets using the output of an auto-associator with 25,000 iterations

Iterations	RMS Error	Classification Rate Training		Classification Rate Testing	
		640	[%]	160	[%]
1000	0.00895	563	87.96	131	81.87
10000	0.00551	616	96.25	147	91.87
5000	0.00647	601	93.90	148	92.50
34000	0.00459	627	97.96	151	94.37
30000	0.00434	631	98.59	155	96.87
35000	0.00445	630	98.43	156	97.50

set. The maximum classification rate of 98.59% was obtained with 30,000 iterations on the training set. The same classification rate also was obtained for 40,000 iterations. For the testing set, the maximum classification rate of 97.5% was achieved for 35,000 iterations. For the same number of iterations, the classification rate for training set was 98.43%.

Comparitive Analysis of the Results

The results obtained using the proposed neuro-fuzzy-based technique for color image retrieval and the auto-associator- and classifier-based technique for the texture feature extractor are analyzed and compared with other CBIR techniques. The results obtained for color image retrieval are discussed in overall findings, neural network architecture and confidence values, and so forth. For texture image retrieval, the results are compared with other traditional- and neural-based texture feature extractions and classifications of these features.

Neural Network Architecture and Confidence Values

A unique feature of this research was that the confidence factor was provided for each image in the database for a specific query. The confidence factor shows the confidence of each image for the particular query. It was the actual output of the neural network on testing a set of images. The neural network was trained only once on the queries and not on the database; therefore, it was not necessary to retrain it if there was any change in the database.

Fusion of Multiple Queries

The neuro-fuzzy technique was used to fuse the multiple queries, and the results were discussed and compared with fuzzy AND and binary AND techniques for fusion. For the fuzzy AND technique, the confidence value was selected, which was the lowest among the two values for the two queries. This technique worked satisfactorily if there was not a significant difference between the two confidence factors. By proposing the neuro-fuzzy technique, it was seen that there was improvement in the confidence factors for the images of that particular query.

MLP as a Classifier

The classification results obtained by the proposed techniques were compared with other techniques. Other researchers also used the Brodatz album texture database. Jones and Jackway (1999) used the granold technique for texture classification. A granold is a texture representation technique that uses two parameterized monotonic mappings to transform an input image into a two-and-a-half-dimensional surface. The granold spectrum for each image then was calculated, and the grey level and size marginals were formed. The confu-

Table 8. Comparison of classification results of the proposed techniques with other techniques

Granold Classification	Classification with NLC	Classification with NN	Auto-associator-Classifier	Statistical-Neural Classifier
76.9 %	95.48%	93.47%	97.5%	96.06%

sion matrix was used to calculate the classification rate, which was 76.9% on the Brodatz texture database. Wang and Lui (1999) compared Nearest Linear Combination (NLC) with Nearest Neighbor (NN) Classification. The testing set was formed from selecting random images from the training set with various sizes (100 x 100 to 200 x 200) to verify the classification rate. The highest classification rate was 95.48%; NLC and 93.47 % were obtained with NN. Table 8 shows the comparison of classification results of the proposed techniques with other techniques.

Gabor features achieved a performance close to 74% retrieval (Ma &Manjunath, 1996). This was followed by the MR-SAR features at 73%. The TWT features (64.4%) performed better than PWT features (68.7%). Chang and Kuo (1993) examined the classification algorithm with respect to a set of eight visually similar textures of size 256×256. The maximum classification rate of 90% was achieved. The proposed technique also worked well for real-world texture images. For real-world texture images, the maximum classification rate of 94.04% was obtained with 18 hidden units and 23,000 iterations. In order to achieve this classification rate, the auto-associator was trained for 25,000 iterations with 32 hidden units.

Conclusion

Neural networks have been applied successfully by various researchers to solve the problem of CBIR. Self-Organizing Map (SOM), Radial Basis Function (RBF), and Error Back-Propagation (EBP) have been proposed for feature extraction and classification of these features. Neural networks have the ability to calculate the similarity between the two images, and they also have been used as a relevance feedback technique to refine queries. In this chapter, we have reviewed traditional- as well as neural-based techniques for content-based image retrieval and have proposed a neural approach in conjunction with fuzzy logic for retrieving the images based on color and texture.

For color image retrieval, the single color approach extended to pose the query in terms of multiple colors and multiple content types. The new technique for fusion of the queries was introduced, and it was implemented for two colors and their content types. The results of the neuro-fuzzy technique were compared with fuzzy AND and binary AND techniques. It was observed from the results that the neuro-fuzzy AND technique has improved the confidence factor compared to fuzzy AND technique.

For texture image retrieval, an auto-associator was proposed to extract the texture features from the set of images and used to retrieve the images from the database on the basis of texture query. In order to verify the extracted texture features for different classes, the features were fed to the MLP classifier. The results obtained from the classifier showed that the auto-associator seems to be a promising feature extractor using only a single hidden layer. For the classifier, the highest classification rate obtained was 97.5%. The highest classification rate was obtained using the proposed autoassociator-classifier technique.

The neural network-based technique for CBIR proposed in this chapter achieved better results than some of the existing techniques, and the results are very promising. Based on the review of existing techniques and experimental results presented in this chapter, it is suggested that the use of neural networks in CBIR systems can significantly improve the overall performance of the system.

References

Brodatz, P. (1966). *Texture: A photographic album for artists and designers*. New York: Dover Publications.

Carson, C., Belongie, S., Greenspan, H., & Malik, J. (1998). Blobworld: Image segmentation using expectation-maximization and its application to image querying. *Journal of Pattern Analysis and Machine Intelligence, 24*(8), 1026–1038.

Carson, C., & Ogle, V. (1996). Storage and retrieval of feature data for a very large online image collections. *Bulletin of the IEEE Computer Society Technical Committee on Data Engineering, 19*(4), 19–27.

Chang, T., & Kuo, C. (1993). Texture analysis and classification with tree-structured wavelet transform. *IEEE Transactions on Image Processing, 2*(4), 429–441.

Datta, R., Li, J., & Wang, J. (2005). Content-based image retrieval: A survey on the approaches and trends of the new age. In *Proceedings of the ACM International Workshop on Multimedia Information Retrieval, ACM Multimedia*, Singapore (pp. 77–82).

Duan, L., Gao, W., & Ma, J. (2000). An adaptive refining approach for content-based image retrieval. In *Proceedings of First IEEE Pacific-Rim Conference on Multimedia*, Sydney, Australia (pp. 188–191).

Faloutsos, C., et al. (1994). Efficient and effective querying by image content., *Journal of Intelligent Information Systems, 3*, 231–262.

Flickner, M., et al. (1995). Query by image and video content: The QBIC system. *IEEE Computer, 28*(9), 23–32.

Fournier, J., Cord, M., & Philipp-Foliguet, S. (2001). Back-propagation algorithm for relevance feedback in image retrieval. In *Proceedings of the International Conference on Image Processing* (Vol. 1, pp. 686–689).

Fukushima, S., & Ralescu, A. (1995). Improved retrieval in a fuzzy database from adjusted user input. *Journal of Intelligent Information Systems, 5*(3), 249–274.

Gupta, A. (1996). *Visual information retrieval: A virage perspective* (Tech. Rep. Revision No. 4]. San Diego, CA: Virage Inc. Retrieved from http://www.virage.com/wpaper

Ikeda, T., & Haiwara, M. (2000). Content-based image retrieval using neural networks. *International Journal of Neural Systems, 10*(5), 417–424.

Jones, D., & Jackway, P. (1999) Using granold for texture feature classification. In *Proceedings of the International Conference on Digital Image Computing, Techniques and Applications*, Perth (pp. 270–274).

Koskela, M., & Laaksonen, J. (2005). Semantic annotation of image groups with self-organizing maps. In *Proceedings of the 4th International Conference on Image and Video Retrieval,* Singapore (pp. 518–527).

Koskela, M., Laaksonen, J., & Oja, E. (2002). Implementing relevance feedback as convolutions of local neighbourhoods on self-organizing maps. In *Proceedings of the International Conference on Artificial Neural Networks,* Spain.

Kreyss, J., Roper, M., Alshuthe, P., Hermes, T., & Herzog, O. (1997). Video retrieval by still image analysis with imageminer. In *Proceedings of the IS&T/SPIE Symposium on Electronic Imaging: Science and Technology.*

Kulkarni, S., & Verma, B. (2001). An auto-associator for automatic texture feature extraction. In *Proceedings of the 4th International Conference on Computational Intelligence and Multimedia Applications*, Yokosuka, Japan (pp. 328–332).

Kulkarni, S., & Verma, B. (2002). An intelligent hybrid approach for content based image retrieval. *International Journal of Computational Intelligence and Applications, 2*(2), 173–184.

Kulkarni, S., Verma, B., Sharma, P., & Selvaraj, H. (1999). Content based image retrieval using a neuro-fuzzy technique. In *Proceedings of IEEE International Joint Conference on Neural Networks,* Washington (pp. 846-850).

Laaksonen, J., Koskela, M., & Oja, E. (1999a). PicSOM: A framework for content based image database retrieval using self-organising maps. In *Proceedings of the SCIA,* Kangerlussuaq, Greenland (pp. 151–156).

Laaksonen, J., Koskela, M., & Oja, E. (1999b). Content-based image retrieval using self-organising maps. In *Proceedings of VISual,* The Netherlands (Vol. 1614, pp. 541–548).

Lee, D., Barber, R., Biblack, W., Flickner, M., Hafner, J., & Petkovic, D. (1994). Indexing for complex queries on a query-by-content image database. In *Proceedings of the IEEE International Conference on Image Processing* (pp. 213–222).

Lee, H., & Yoo, S. (2000). A neural network based flexible image retrieval. In *Proceedings of the 4th World Multi-Conference on Circuits, Systems, Communications & Computers* (pp. 4591–4596).

Lee, H., & Yoo, S. (2001). Applying neural network to combining the heterogeneous features in content-based image retrieval. *SPIE Applications of Artificial Neural Networks in Image Processing, 4305*(13), 81–89.

Lee, H., & Yoo, S. (2001). A neural network based image retrieval using nonlinear combination of heterogeneous features. *International Journal of Computational Intelligence and Applications, 1*(2), 137–149.

Ma, W., & Manjunath, B. (1996). Texture features and learning similarity. In *Proceedings of the IEEE Conference on Computer Vision and Pattern Recognition* (pp. 425–430).

Ma, W., & Manjunath, B. (1999). NETRA: A toolbox for navigating large image databases. *Journal of ACM Multimedia Systems, 7*(3), 184–198.

Manjunath, B., & Ma, W. (1996). Browsing large satellite and ariel photographs. In *Proceedings of the 3rd IEEE Conference on Image Processing* (Vol. 2, pp. 765–768).

Muneesawang, P., & Guan, L. (2001). Interactive CBIR using RBF-based relevance feedback for WT/VQ coded images. In *Proceedings of the IEEE Conference on Acoustics, Speech and Signal Processing* (Vol. 3, pp. 1641–1644).

Muneesawang, P., & Guan, L. (2002a). Automatic machine interactions for CBIR using self organized tree map architecture. *IEEE Transactions on Neural Networks, 13*(4), 821–834.

Muneesawang, P., & Guan, L. (2002b). Automatic machine interactions for CBIR using self organized tree map architecture, *IEEE Transactions on Neural Networks, 13*(4), 821–834.

Naqa, I., Wernick, M., Yang, Y., & Galatsanos, N. (2000). Image retrieval based on similarity learning. In *Proceedings of the IEEE Conference on Image Processing*, British Columbia, Canada (Vol. 3, pp. 722–725).

Oja, E., Laaksonen, J., Koskela, M., & Brandt, S. (1997). Self organising maps for content based image database retrieval. In E. Oja & S. Kaski (Eds.), *Kohonen maps* (pp. 349–362). Amsterdam.

Pala, P., & Santini, S. (1999). Image retrieval by shape and texture. *Journal of Pattern Recognition, 32*(3), 517–527.

Parulekar, A., Datta, R., Li, J., & Wang, J. (2005). Large-scale satellite image browsing using automatic semantic categorization and content-based retrieval. In *Proceedings of the IEEE International Workshop on Semantic Knowledge in Computer Vision* (pp. 130–136).

Santini, S., & Jain, R. (1996). Similarity queries in image databases. In *Proceedings of the IEEE International Conference on Computer Vision and Pattern Recognition,* San Francisco (pp. 646–651).

Tsai, C. (2003). A novel solution to bridge the semantic gap for content-based image retrieval. *Journal of Online Information Review, 27*(3), 442-445.

Tsai, C., McGarry, K., & Tait, J. (2003). Using neuro-fuzzy technique based on a two stage mapping model for concept-based image database indexing. In *Proceedings of the Fifth International Symposium on Multimedia Software Engineering*, Taiwan (pp. 10–12).

Yang, Z., & Laaksonen, J. (2005a). Partial relevance in interactive facial image retrieval. In *Proceedings of the 3rd International Conference on Advances in Pattern Recognition*, Bath, UK (pp. 216–225).

Yang, Z., & Laaksonen, J. (2005b). Interactive retrieval in facial image database using self-organizing maps. In *Proceedings of IAPR Conference on Machine Vision Applications (MVA 2005),* Tsukuba Science City, Japan (pp. 112–115).

Yu, C, & Aslandogan, Y. (2000). Multiple evidence combination in image retrieval: Dio-
genes searches for people on the Web. In *Proceedings of the 23rd Annual International
Conference on Research and Development of Information Retrieval* (pp. 88–95).

Verma, B., & Kulkarni, S. (2004). Fuzzy logic based interpretation and fusion of colour
queries. *Journal of Fuzzy Sets and Systems, 147*(1), 99–118.

Wang, L., & Lui, J. (1999). texture classification using multiresolution Markov random field
models. *Journal of Pattern Recognition Letters, 20*(2), 171–182.

Wood, M., Campbell, N., & Thomas, B. (1998). Iterative refinement by relevance feedback
in content-based image retrieval. In *Proceedings of the ACM International Conference*, Bristol, UK (pp. 13–20).

Zhang, C., & Chen, X. (2005). Region-based image clustering and retrieval using multiple
instance learning. In *Proceedings of the 4th International Conference on Image and
Video Retrieval,* Singapore (pp. 194–204).

Zhuang, Y., Liu, X., & Pan, Y. (2000). Apply semantic template to support content-based
image retrieval. In *Proceedings of the SPIE Storage Retrieval Multimedia Database,*
San Jose, CA (pp. 422–449).

Chapter XIII

Semantic-Based Video Scene Retrieval Using Evolutionary Computing

Hun-Woo Yoo, Yonsei University, Korea

Abstract

A new emotion-based video scene retrieval method is proposed in this chapter. Five video features extracted from a video are represented in a genetic chromosome, and target videos that the user has in mind are retrieved by the interactive genetic algorithm through the feedback iteration. After the proposed algorithm selects the videos that contain the corresponding emotion from the initial population of videos, the feature vectors from them are regarded as chromosomes, and a genetic crossover is applied to those feature vectors. Next, new chromosomes after crossover and feature vectors in the database videos are compared, based on a similarity function to obtain the most similar videos as solutions of the next generation. By iterating this process, a new population of videos that a user has in mind are retrieved. In order to show the validity of the proposed method, six example categories—action, excitement, suspense, quietness, relaxation, and happiness—are used as emotions for experiments. This method of retrieval shows 70% of effectiveness on the average over 300 commercial videos.

Introduction

A variety of data available in image and video format are being generated, stored, transmitted, analyzed, and accessed with advances in computer technology and communication networks. To make use of these data, an efficient and effective technique needs to be developed for retrieval purposes. This chapter proposes a video scene retrieval method based on its emotion.

In general, earlier video retrieval systems extract video shot boundaries, select key frame(s) from each shot, and cluster similar shots with low-level features such as color, texture, motion of object, shot time, and so forth, to construct scenes (Ahanger & Little, 1996; Brunelli, Mich, & Modena, 1999; Gargi, Kasturi, & Strayer, 2000; Jain, Vailaya, & Xiong, 1999; Meng, Juan, & Chang, 1995; Yeo & Liu, 1995; Zabih, Miller, & Mai, 1999; Zhang, Kankanhalli, Smoliar, & Tan, 1993). However, these low-level features are not enough to describe the richer content of a video, because a human feels semantics rather than low-level features from a video. Thus, a semantic-based retrieval method has become an active research topic lately. Video genre classification (Roach, Mason, Xu, & Stentiford, 2002; Roach, Mason, & Pawlewski, 2001; Truong, Venkatesh, & Dorai, 2000) or event-based video retrieval (Chen, Shyu, Chen, & Zhang, 2004; Haering, Qian, & Sezan, 2000; Shearer, Dorai, & Venkatesh, 2000; Zhang & Chang, 2002) falls into this category.

Representation of emotions with a visual medium is very important. Many posters and movie previews are designed to appeal to the potential customer by containing specific moods or emotions. Hence, emotion-based retrieval would be one of the essential applications in the near future. Yet related studies associated with computer vision are still in infancy.

Several researchers have studied retrieval methods based on emotion (Cho, 2002; Colombo, Bimbo, & Pala, 1999; Colombo, Bimbo, & Pala, 2001; Takagi, Noda, & Cho, 1999; Um, Eum, & Lee, 2002). In Cho (2002), based on wavelet coefficients, gloomy images were retrieved through feedbacks called Interactive Genetic Algorithm (IGA). This method, however, is limited to differentiating only two categories: gloomy images or not. In a similar approach, Takagi et al. (1999) designed psychology space (or factor space) that captured human emotions and mapped those onto physical features extracted from images. Also, IGA was applied to emotion-based image retrieval. Based on Soen, Shimada, & Akita's (1987) psychological evaluation of color patterns, Um et al. (2002) proposed an emotional model to define a relationship between physical values of color image patterns and emotions. It extracted color, gray, and texture information from an image and input into the model. Then, the model returned the degree of strength with respect to each of 13 emotions. It, however, has a generalization problem because it experimented on only five images and could not be applied to the image retrieval directly, since, for image retrieval, an emotion keyword should be presented first as a query, and then images with the presence of the associated emotion are retrieved (Um's model has a reverse procedure). Moreover, these methods (Cho, 2002; Takagi et al, 1999; Um et al., 2002) deal only with images, not with videos that would be handled here in this chapter.

Based on the color theory of Itten (1961), Colombo et al. (1999) retrieved art paintings from the Renaissance and contemporary eras by mapping expressive and perceptual features onto emotions. It segmented the image into homogeneous regions, extracted features such as color, warmth, hue, luminance, saturation, position, and size from each region, and then used its contrast and harmony relationship to another region to capture emotions. However, it was designed for art painting retrieval only. Authors also proposed emotion-based retrieval method for video case. Based on the semiotic category, commercials are classified into four emotions: utopic, critical, practical, and playful (Colombo et al., 1999; Colombo et al., 2001). Hanjalic and Xu (2005) designed two-axis affective space in which one axis is called the arousal that describes an emotional strength, and the other is called the valence that describes an emotional type. They performed video retrieval according to affective curve (how a video traces over this space). Also, they claimed that it is possible for videos to be classified according to where a video is located over the space.

Audio-based retrieval with relation to emotions also is found in several studies (Cees, Snoek, & Worring, 2005; Pickens et al., 2002; Toivanen & Seppänen, 2002). In Toivanen and Seppänen (2002), audio information in the form of prosody features is extracted to describe the emotional content of audio files. In Pickens et al. (2002), music retrieval is performed by a so-called query-by-humming approach. Most recently, in Cees et al. (2005), audio information as a multimodal approach was surveyed for video indexing purposes.

So far, the researches on multimedia retrieval in terms of emotion are very few (in particular, for video case). In this chapter, we propose a new emotion-based video retrieval method that uses emotion information that is extracted from a video scene. The proposed method retrieves video scenes that a user has in mind through the evolutionary computation called interactive genetic algorithm (IGA). In the preprocessing step, for each database video, video shots are detected, and a key frame is selected from each shot. Next, the features that describe emotional content, such as the average color histogram, average brightness, average-edge histogram, average shot duration, and the gradual change rate, are extracted from the key frames among all the shots. Then, these features are encoded as a chromosome and indexed into the database. For all the database videos, this procedure is repeated. For retrieval, an interactive genetic algorithm is exploited that performs optimization with the human evaluation. The system displays 15 videos, obtains a relevance feedback of the videos from a human, and selects the candidates based on the relevance. A genetic crossover operator is applied to the selected candidates. In order to find the next 15 videos, the stored video information is evaluated by each criterion. Fifteen videos that have higher similarity to the candidates are provided as a result of the retrieval.

The IGA is different from the well-known genetic algorithm (GA) in that a human being is involved as a fitness function. Since retrieving videos that a user has in mind depends on personal preference, IGA is a technique that is suitable to emotion-based video retrieval. The overall retrieval procedure is described in Figure 1.

Figure 1. Proposed method

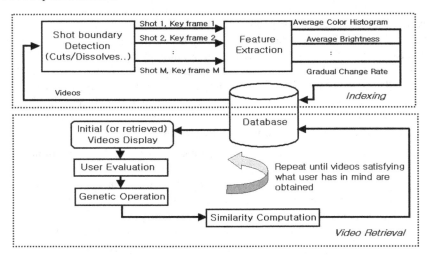

Main Thrust of the Chapter

Video Segmentation

A video consists of many still frames, which are compressed by spatial-temporal information. For the emotion-based scene retrieval, we have to segment a video into shots and choose key frames and extract meaningful features from the shots, which capture a specific emotion (mood) of the video.

In general, shot boundaries are detected for two cases: the abrupt transition called *cut* and the gradual transitions called *fade-ins, fade-outs,* or *dissolves*. For shot boundary detection, the correlation coefficient of gray-level histogram and edge histogram are used (Yoo & Jang, 2004).

Correlation Sequence Using Gray-Level Histogram

Let m_{lk} and σ_{lk} be the average and variance of the gray-level information in the kth frame. The gray-level interframe correlation between two consecutive frames k and $(k+1)$ is obtained as follows:

$$S_{LIFC}(k, k+1) = \frac{\sum_{i=0}^{H-1}\sum_{j=0}^{W-1}(X_{lk}[i][j] - m_{lk})(X_{l(k+1)}[i][j] - m_{l(k+1)})}{\sigma_{lk}\sigma_{l(k+1)}} \qquad (1)$$

$-1 \le S_{LIFC}(k, k+1) \le 1$

in which W and H are the width and height of a frame, and $X_{lk}[i][j]$ is the gray-level of (i, j) coordinate in the kth frame.

Correlation Sequence Using Edge Histogram

The number of edges is acquired in each 3×3 uniform rectangular region of the frame. Take the number of edges in each region as histogram bin values in that frame. Let m_{ek} and σ_{ek} be the average and variance of the number of edges in the kth frame. The edge interframe correlation between two consecutive frames, k and $(k+1)$, is obtained as follows:

$$S_{EIFC}(k, k+1) = \frac{\sum_{i=0}^{8}(X_{ek}[i] - m_{ek})(X_{e(k+1)}[i] - m_{e(k+1)})}{\sigma_{ek}\sigma_{e(k+1)}} \tag{2}$$

$-1 \le S_{EIFC}(k, k+1) \le 1$

in which $X_{ek}[i]$ is the number of edges of the ith region in the kth frame.

Integrated Correlation

Finally, two correlations are integrated for detection purposes.

$$S_{IFC}(k, k+1) = w_l S_{LIFC}(k, k+1) + w_e S_{EIFC}(k, k+1) \tag{3}$$

in which w_l and w_e are the weights for relative importance between the use of gray-levels and edges, and are subject to $w_l + w_e = 1$.

Abrupt and Gradual Detection

We consider a current frame as an abrupt shot boundary frame when it is sufficiently different from a previous frame, and a current frame is considered as a gradual shot boundary frame when it is sufficiently different from a previous shot boundary frame. The criterion of deciding the sufficient difference is based on the interframe correlation.

Detection processes are as follows: The correlation between the previous frame (p) and the current frame (c) is computed using Equation (3) and compared with the predetermined threshold t_l for abrupt detection. If this correlation is lower than t_l, an abrupt shot change is declared. If not, the current frame (c) is compared with the previous shot boundary frame (*b*) (if previous shot boundary frame does not exist, the first frame of video is considered a default shot boundary frame). If the correlation between (*c*) and (*b*) is lower than a predetermined threshold t_3, a gradual shot change is declared. Otherwise, the shot boundary is not declared, and the current frame (*c*) is set to the previous frame (*p*). Then, the aforementioned process with the next frame is repeated until the end of the video frames.

It requires so much computational costs that this procedure is applied to every consecutive frame. In order to reduce the cost, we have reduced temporal resolution by sampling every 15 frames instead of every frame. Since videos are used to have temporal redundancies, it is reasonable to reduce the temporal resolution for detection purpose.

Feature Extraction

In order to retrieve video scenes with the presence of specific emotion, the features for representing scene content effectively should be extracted. For that purpose, the shot boundaries in the scene are detected, the key frames in each of shots are chosen, and the features from each key frame are extracted to represent the scene content. In this research, visual content such as color, gray intensity, and edge information and visual rhythm such as shot duration in time and the ratio of the number of gradual boundaries among total number of shot boundaries is used for describing features.

Average Color Histogram $\Im_1$

Color has been used successfully in content-based image/video retrieval for a long time. This attribute could be used effectively to describe a specific emotion. For example, the images with "warm" sensation tend to have much red colors. The images with "cold" sensation tend to have much blue or cyan colors. Also, the "action" videos contain red or purple colors, and the videos with "quietness" sensation contain blue, green, and white colors (Colombo et al., 1999).

In our method, for describing the color in a shot, we first extract an RGB joint histogram from a key frame and multiply the value that each RGB joint histogram bin has by the number of frames within the shot. Then, by applying the previous procedure to every shot within a scene and dividing it by the total number of frames, we obtain the average color histogram in the scene as follows.

$$\Im_1 = H^{color}[i]$$
$$= \frac{\sum_k H_k^{color}[i] \times SL_k}{N_T}, \quad i = 0,1,2,...26 \tag{4}$$

in which $H_k^{color}[i]$ is the ith bin value of 27-dimensional joint RGB histogram (each of R, G, and B channels is equally quantized [i.e., 3×3×3]) in the key frame of the kth shot, SL_k is the length of the kth shot, which is denoted in the number of frames, and N_T is the total number of frames within the scene.

Average Brightness Average Color Histogram $\Im_2$

In general, people feel light and happy from bright images. Contrarily, people tend to feel hard, heavy, and gloomy from dark images. In videos, the scenes with quietness tend to have bright tones. Therefore, we include an average brightness within a scene for describing a feature.

$$\Im_2 = \frac{\sum_k Bright_k \times SL_k}{N_T} \tag{5}$$

in which $Bright_k$ is the average brightness of a key frame in the kth shot, SL_k is the length of the kth shot, which is denoted in the number of frames, and N_T is the total number of frames within the scene.

Average-Edge Histogram $\Im_3$

It is easily noticeable that in the images with a "gloomy" sensation, the number of edges are relatively few, and vice-versa in the images with a "joyful" sensation (Cho, 2002). We extract dominant edges through the Canny edge detector (Yoo, Jang, Jung, Park, & Song, 2002) and use the number of edges with respect to 72 edge directions as a feature.

$$\Im_3 = H^{edge}[i]$$
$$= \frac{\sum_k H_k^{edge}[i] \times SL_k}{N_T}, \quad i = 0,1,2,...,71 \tag{6}$$

in which $H_k^{edge}[i]$ is the ith bin value of 72-dimensional direction edge histogram in the key frame of the kth shot, SL_k is the length of the kth shot, which is denoted in the number of frames, and N_T is the total number of frames within the scene.

Average Shot Duration $\Im_4$

In general, the scenes such as "action" and "excitement" tend to have short shots in order to increase tension. On the other hand, "quietness," "relaxation," and "happiness" scenes tend to have long shots and do not have much change in the shot content. Here, we compute shot durations within the scene and sum them up. Then, by dividing the summation by the total number of shots, we obtain an average shot duration as a feature.

$$\Im_4 = \frac{\sum_k SD_k}{N_S} \tag{7}$$

in which SD_k is the duration time of the kth shot, which is denoted in seconds, and N_S is the total number of shots within the scene.

Gradual Change Rate $\Im_5$

Often, gradual shot boundaries in videos induce specific emotions. Many "quietness" scenes contain gradual shot boundaries such as dissolve. Hence, the ratio of the number of gradual boundaries over the total number of shot boundaries is used.

$$\Im_5 = \frac{N_G}{N_S}$$

(8)

in which N_G and N_S are the total number of gradual shot boundaries and the total number of shots within the scene, respectively.

Emotion-Based Video Retrieval

Interactive Genetic Algorithm

Genetic algorithm (GA) provides a powerful technique for searching large problem spaces. A GA solves problems by evolving a population of potential solutions to a problem using standard genetic operations such as crossover and mutation until an acceptable solution emerges (Goldberg, 1989). In other words, solutions that are appropriate only to a certain fitness function survive for every evolving step. However, most of the conventional applications of GA lack the capability to utilize human intuition or emotion appropriately. Consider retrieving the most favorite videos from human-machine interaction systems. Such systems must be evaluated subjectively, and it is hard or even impossible to design a fitness function.

Interactive genetic algorithm is a technique that searches a possible solution based on human evaluation. A human can obtain what he or she has in mind through repeated interactions with the method when the fitness function cannot be defined explicitly. It has been applied to a variety of areas such as graphic art, industrial design, musical composing, voice processing, virtual reality, information retrieval, education, games, and so forth (Banzhaf, 1997; Biles, 1994; Caldwell & Johnston, 1991; Cho, 2002; Lee & Cho, 1998; Takagi et al., 1999). An extensive survey on IGA can be found in Takagi (2001).

When the videos are presented to a user for emotion-based video scene retrieval, the user has to decide whether each video agrees with what he or she has in mind or not. Therefore, here, we try to achieve video scene retrieval by formulating it to the problem of the interactive genetic algorithm. For that purpose, five features are extracted from a video as in the previous section and encoded as a chromosome. Then, the retrieval is achieved by evolving the chromosomes of initially selected videos until satisfactory videos are obtained. Figure 2 shows how a chromosome is encoded from a video.

For scene-based retrieval, videos must be divided into many meaningful scenes, and those scenes must be indexed into a database, or short videos composed of one scene (e.g., commercial videos such as CF) must be indexed into a database. Here, we have used the short videos of a single scene for experimental convenience.

Retrieval Method

Step 1: Extract features from videos, represent them as chromosomes, and index them into the database.

Select a video from a database and input it to the shot boundary detection module. Detect abrupt shot boundaries (cuts) and gradual shot boundaries (dissolves, etc.) and take the first frame of each shot boundary as a key frame of the corresponding shot. Next, input each key frame to the feature extraction module in order to extract five features: average color histogram, average brightness, average edge histogram, average shot duration, and gradual change rate. Then, index them into the database and search table. Remember that it is represented as a chromosome in Figure 2. Iterate this procedure for all database videos (see the indexing box at the top of Figure 1).

Figure 2. Chromosome representation from a video

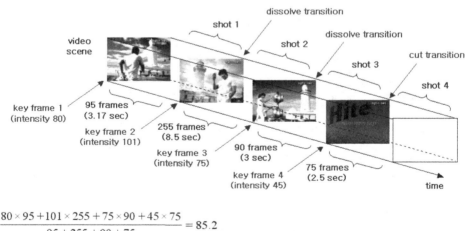

$$\frac{80 \times 95 + 101 \times 255 + 75 \times 90 + 45 \times 75}{95 + 255 + 90 + 75} = 85.2$$

$$\frac{3.17 + 8.5 + 3 + 2.5}{4} = 4.29 \qquad \frac{2}{4} = 0.5$$

24.2 32.1 . . . . .1.6	85.2	16.2 2.1 . . . . .18.3	4.29	0.5
$\mathfrak{I}_1$	$\mathfrak{I}_2$	$\mathfrak{I}_3$	$\mathfrak{I}_4$	$\mathfrak{I}_5$

Step 2: Present randomly initial videos to the user and select videos similar to what a user has in mind.

The system presents randomly selected 15 videos to the user. Only the first frames from each video are displayed (see the system GUI in Figure 3). Then, a user examines whether his or her preference videos exist among them through watching each video sequence by pressing the play button (Figure 4a). If a video is similar to what he or she has in mind (we call it target emotion here), select that video by checking. If not, do not check. Sometimes, since a user can feel that it is burdensome to check each video sequence because of fatigue or lack of time, we also include the function of examining only the key frames of every video (Figure 4b).

Step 3: Obtain target chromosomes by applying the crossover genetic operator to checked videos.

Suppose that M videos are checked in Step 2. Extract M-associated chromosomes of checked videos from the database. Produce 15 chromosomes by listing M-associated chromosomes plus 15-M chromosomes by selecting randomly some chromosomes more than once (if M is lower than 15). Finally, obtain 15 target chromosomes by applying a crossover operator among some of the 15 chromosomes as follows: (1) select random pairs among some of the 15 chromosomes; (2) select crossover points randomly in each pair; (3) swap a part of the chromosome on the basis of those points. In our research, one-point crossover of four possible points is used, as shown in Figure 5.

Figure 3. System GUI (graphic user interface)

Figure 4. Interface for examining a video: (a) the way of seeing the whole frame sequence (by pressing the ▶ button under key frame image); (b) the way of seeing only key frames (by pressing ⓘ under key frame image)

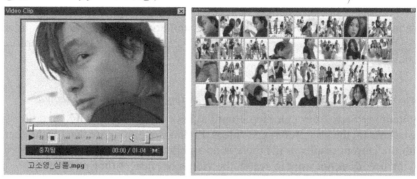

Figure 5. Crossover application: (a) possible crossover points; (b) example of one crossover point

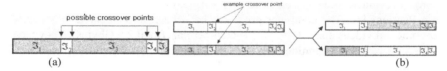

Step 4: Obtain result videos by computing the similarity between each of the target chromosomes and each of the chromosomes of the database videos.

The 15 new videos in which each of their chromosomes is most similar to each of the target chromosomes are obtained, based on the similarity function $S(Q,D)$ between the presented target chromosome Q and the chromosome D of the database video. The degree of similarity is computed by integrating the distances between Q and D for the five features:

$$S_i(Q, D), \ i=1,...5.$$

$$S(Q,D) = \sum_{i=1}^{5} w_i \times S_i(Q,D)$$

$$S_i(Q,D) = \sqrt{\mathfrak{I}_i(Q) - \mathfrak{I}_i(D)} \qquad \textit{for all i except for } i = 1,3 \qquad (9)$$

in which $S_i(Q,D)$ is the difference between i^{th} feature; that is, target chromosome $\mathfrak{I}_i(Q)$ and database chromosome $\mathfrak{I}_i(D)$. However, for the average color histogram feature $\mathfrak{I}_1$, we summed 31 differences because it has 31 entries. Also, for the average edge histogram feature $\mathfrak{I}_3$, we summed 71 differences because it is 71-dimensional. Here, w_i, $i = 1,2,...5$ are weights for each of the differences subject to:

$$\sum_{i=1}^{5} w_i = 1.$$

The lower value $S(Q, D)$ has, the higher the similarity between Q and D is. In our research, $w_{color} = 0.2$, $w_{Bright} = 0.2$, $w_{edge} = 0.1$, $w_{ShotT} = 0.25$, and $w_{GradR} = 0.25$ are empirically chosen as default weights.

Step 5: Iterate Step 2 through Step 4 until ending conditions are satisfied.

By iterating Step 2 through Step 4 until at least one of the ending conditions is satisfied, we can obtain the user preference videos. Ending conditions are that (1) the user is satisfied with the current results or (2) there is no significant change even after many iterations or (3) the user does not want to search the videos because of time or fatigue.

Experiments

In order to evaluate the proposed video retrieval method, several experiments were performed on a Pentium PC. The computer programs have been implemented in Visual C++. The experimental videos consist of 300 TV commercials (total 2.5 Gbytes). They have a single scene for a very short running time (less than one minute). We chose the crossover rate as 0.54 to take eight chromosomes from 15 chromosomes for crossover candidates.

The system presents an initial population that consists of 15 videos selected randomly, as shown in Figure 3. A user watches each video by playing it or examining key frames, and then selects all videos similar to what he or she has in mind. Based on the selected videos, 15 chromosomes are produced. Then, among 15 chromosomes, take eight chromosomes randomly for crossover candidates and also take a crossover point randomly. Then, list 15 target chromosomes by taking eight crossovered chromosomes plus seven noncrossovered chromosomes. Next, the system presents 15 database videos that are similar to the 15 target chromosomes through the similarity function (9). This procedure is repeated until at least one of the ending conditions is met. In case the results of the next iteration are not satisfactory, the system allows the user to increase or decrease the weights of each feature.

For what he or she has in mind, six emotions of "action," "excitement," "suspense," "quietness," "relaxation," and "happiness" were used for experiments. According to Colombo et al. (1999), the "action" scene has the presence of red and purple, short sequences joined by cuts, and a high degree of motion. The "excitement" scene has short sequences joined through cuts. The "suspense" scene that is associated with "action" has both long and short sequences joined through frequent cuts. The "quietness" scene has the presence of blue, orange, green, and white colors, and long sequences joined through dissolves. The "relaxation" scene does not have relevant motion components. The "happiness" scene shares "happiness" features and has relevant motion components. These characteristics can be represented as chromosomes with the five features proposed in this chapter. We have asked 10 subjects to categorize 300 videos into the six aforementioned emotions. Resultant categories are 36 videos for "action," 56 videos for "excitement," 22 videos for "suspense," 72 videos for "quietness," 71 videos for "relaxation," and 82 videos for "happiness." Since emotion is subject greatly to individuals, all the 300 videos used in the experiments are what all participants agree with for each of the emotions. The example key frames of "action" and "happiness" are listed in Figure 6.

In order to evaluate the proposed method, we performed the retrieval process up to a maximum 10 iterations and check the user's satisfaction (effectiveness) on the resulting videos. Ten graduate students who had been trained about how the system works and what the six features mean participated in the experiment. For each emotion, the user's satisfaction D_{sat} (we call it system effectiveness here) is computed using Equation (10). Final results were obtained by averaging 10 results from the 10 graduates.

$$D_{sat} = \frac{N_{correct}}{N_{total}}$$

(10)

in which N_{total} is the number of displayed videos (i.e., 15), and $N_{correct}$ is the number of videos that the user satisfies among 15 videos.

For the six emotions, the system effectiveness up to the 10th feedback (it is called the 10th generation in genetic algorithm) is depicted as the best, worst, and average cases, as shown in Figure 7. As we expect, it increases according to the feedback. In the 10th generation, it shows about 0.7 on the average (0.77 for "action," 0.71 for "excitement," 0.49 for "suspense," 0.77 for "quietness," 0.61 for "relaxation," and 0.85 for "happiness"). This means that among

Figure 6. Example key frames with the presence of six emotions: (a) action and (b) happiness

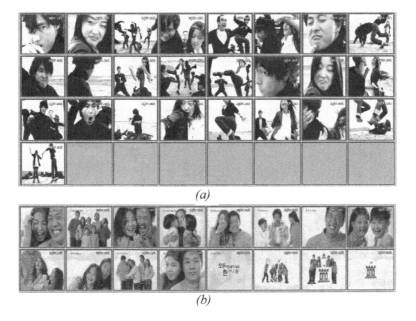

(a)

(b)

the 15 retrieved videos after the 10th feedback, the user found it among 10 videos that he or she had in mind. Except for "suspense" and "relaxation," the effectiveness is more than 0.7. "Happiness" shows the best result, 0.85. "Suspense" shows the worst result, 0.49.

For "action," time information such as average shot duration is very important, since they consist of many cuts for a short time. For "excitement," the low satisfaction of 0.2 is attributed to the fact that some videos have a long shot duration and contain dissolve transitions, which are not common in "excitement" videos. For "suspense," the lowest effectiveness is shown among the results of the six emotions. For "quietness," time information such as the average shot duration and the gradual change rate are important, since they consist of many cuts and dissolve for a long time. Similarly, for "relaxation," time information such as average shot duration is important, since they consist of many cuts and dissolve for a long time. For "happiness," the highest effectiveness is shown among the results of the six

Figure 7. User's satisfaction (system effectiveness) according to six emotions: (a) action; (b) excitement; (c) suspense; (d) quietness; (e) relaxation; and (f) happiness

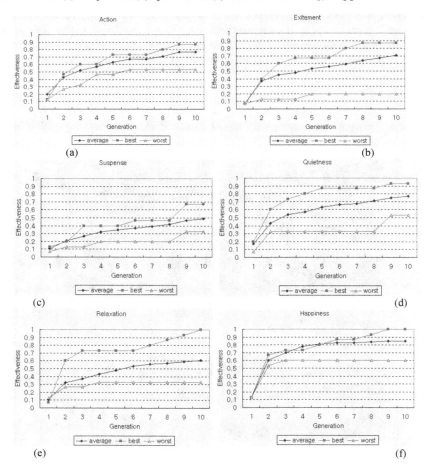

emotions. Among 300 videos, 82 videos are related to "happiness" and most of them have white, green, and blue colors and long average shot duration.

As you can see in Figure 7, it is interesting to note that some cases have room for improvement after more than 10 feedbacks. Thus, the user can perform more feedbacks to obtain better results (if time is allowed, and if the user does not feel fatigued).

Commercials used in these experiments have numerous content variations, because they have to deliver many pieces of information in a short time. On the other hand, videos such as news, drama, and movies do not. This observation has led us to believe that our method will yield better effectiveness for the scene retrieval of general videos (e.g., scenes of an anchor person in the news video, conversation scenes, etc.). Also, in the GUI, we give two levels for user satisfaction on each video just by checking a box. The checked video is considered to be the one that the user has in mind, whereas a nonchecked video is not. In Cho (2002), a slider bar is provided to give a variety of levels for the user's degree of satisfaction. This method will have a chance that in the next generation, more videos with which the user is satisfied will be retrieved, since more satisfactory videos likely are to be chosen for crossover. However, in this chapter, we did not use that approach, because the use of a slider bar can cause greater fatigue for the user than the use of a checkbox.

Conclusion

In this chapter, a new emotion-based video scene retrieval method has been presented. The videos that a user has in mind can be retrieved by mapping his or her emotion space onto physical feature space through interactive genetic algorithm. Experiments on 300 commercial videos have yielded an average of 70% satisfaction for six selected emotions—action, excitement, suspense, quietness, relaxation, and happiness—after 10 generations. In addition, it has provided room for better results if the user tries to perform more than 10 generations.

However, although the performance of the proposed method could be a solution for emotional video retrieval application, it has some limitations to overcome. One of them is that the five features (average color histogram, average brightness, average edge histogram, average shot duration, and gradual change rate) do not describe emotions induced from moving objects in the video. Hence, it would be necessary to add related features to the chromosomes by using a motion vector or optical flow technique. Also, one of the major problems of IGA is that it cannot evolve over many generations because of user fatigue. To deal with the problem, we presented a quick overview of key frames as well as playing whole videos. However, evolving over many generations cannot eliminate the user's fatigue completely. Interface that reduces the burden of the user and fast convergence methodology of IGA remains the main subject for further research. To convince more of the usefulness of the method, subjects more than 10 and subjects naïve to the five features should participate in the experiments. Currently, Asian commercials were used for experiments. The use of well-known popular movies in the world also would show the performance of the method in a more reliable manner.

References

Ahanger, G., & Little, T. D. C. (1996). A survey of technologies for parsing and indexing digital video. *Journal of Visual Communication and Image Representation, 7*(1), 28–43.

Banzhaf, W. (1997). Interactive evolution. In *Handbook of Evolutionary Computation.*

Biles, J. A. (1994). GenJam: A genetic algorithm for generating jazz solos. In *Proceedings of the International Computer Music Conference* (pp. 131–137).

Brunelli, R., Mich, O., & Modena, C. M. (1999). A survey on the automatic indexing of video data. *Journal of Visual Communication and Image Representation, 10*(1), 78–112.

Caldwell, C., & Johnston, V. S. (1991). Tracking a criminal suspect through face-space with a genetic algorithm. In *Proceedings of the International Conference on Genetic Algorithm* (pp. 416–421).

Cees, G., Snoek, M., & Worring, M. (2005). Multimodal video indexing: A review of the state-of-the art. *Multimedia Tools and Applications, 25*(1), 5–35.

Chen, S. C., Shyu, M. L., Chen, M., & Zhang, C. (2004). A decision tree-based multimodal data mining framework for soccer goal detection. In *Proceedings of the IEEE International Conference on Multimedia and Expo* (pp. 265–268).

Cho, S.-B. (2002). Towards creative evolutionary systems with interactive genetic algorithm. *Applied Intelligence, 16*(2), 129–138.

Colombo, C., Bimbo, A., & Pala, P. (1999). Semantics in visual information retrieval. *IEEE Multimedia, 6*(3), 38–53.

Colombo, C., Bimbo, A., & Pala, P. (2001). Retrieval of commercials by semantic content: The semiotic perspective. *Multimedia Tools and Applications, 13*(1), 93–118.

Gargi, U., Kasturi, R., & Strayer, S. H. (2000). Performance characterization of video-shot-change detection methods. *IEEE Transactions on Circuits and Systems for Video Technology, 10*(1), 1–13.

Goldberg, D. E. (1989). *Genetic algorithms in search, optimization, and machine learning.* Addison-Wesley.

Haering, N. C., Qian, R., & Sezan, M. (2000). A semantic event detection approach and its application to detecting hunts in wildlife video. *IEEE Transactions on Circuits and Systems for Video Technology, 10*(6), 857–868.

Hanjalic, A., & Xu, L.-Q. (2005). Affective video content representation and modeling. *IEEE Transactions on Multimedia, 7*(1), 143–154.

Itten, J. (1961). Art of color (kunst der farbe) (in German). Ravensburg, Germany: Otto Maier Verlag.

Jain, A. K., Vailaya, A., & Xiong, W. (1999). Query by video clip. *Multimedia systems: Special issue on video libraries, 7*(5), 369–384.

Lee, J.-Y., & Cho, S.-B. (1998). Interactive genetic algorithm for content-based image retrieval. In *Proceedings of the Asia Fuzzy Systems Symposium* (pp. 479–484).

Meng, J., Juan, Y., & Chang, S. F. (1995). Scene change detection in a MPEG compressed video sequence. In *Proceedings of the SPIE/IS&T Symposium on Electronic Imaging Science and Technology: Digital Video Compression: Algorithms and Technologies* (Vol. 2419, pp. 14–25).

Pickens, J., et al. (2002). Polyphonic score retrieval using polyphonic audio queries: A harmonic modeling approach. In *Proceedings of the ISMIR* (pp. 13–17).

Roach, M., Mason, J., & Pawlewski, M. (2001). Video genre classification using dynamics. In *Proceedings of the International Conference on Acoustics, Speech and Signal Processing* (pp. 1557–1560).

Roach, M., Mason, J., Xu, L.-Q., & Stentiford, F. (2002). Recent trends in video analysis: A taxonomy of video classification problems. In *Proceedings of the International Conference on Internet and Multimedia Systems and Applications* (pp. 348–353).

Shearer, K., Dorai, C., & Venkatesh, S. (2000). Incorporating domain knowledge with video and voice data analysis in news broadcasts. In *Proceedings of the International Workshop on Multimedia Data Mining* (pp. 46–53).

Soen, T., Shimada, T., & Akita, M. (1987). Objective evaluation of color design. *Color Research and Application, 12*(4), 184–194.

Takagi, H. (2001). Interactive evolutionary computation: Fusion of the capabilities of EC optimization and human evaluation. *Proceedings of the IEEE, 89*(9), 1275–1296.

Takagi, H., Noda, T., & Cho, S.-B. (1999). Psychological space to hold impression among media in common for media database retrieval system. In *Proceedings of the IEEE International Conference on System, Man, and Cybernetics* (pp. 263–268).

Toivanen, J., & Seppänen, T. (2002). *Prosody-based search features in information retrieval* (vol. 44). TMH-QPSR, Fonetik.

Truong, B. T., Venkatesh, S., & Dorai, C. (2000). Automatic genre identification for content-based video categorization. In *Proceedings of the International Conference on Pattern Recognition* (pp. 230–233).

Um, J.-S., Eum, K.-B., & Lee, J.-W. (2002). A study of the emotional evaluation models of color patterns based on the adaptive fuzzy system and the neural network. *Color Research and Application, 27*(3), 208–216.

Yeo, B. L. & Liu, B. (1995). Rapid scene analysis on compressed video. *IEEE Transactions on Circuit and Systems for Video Technology, 5*(6), 533–544.

Yoo, H.-W., & Jang, D.-S. (2004). Automated video segmentation using computer vision technique. *International Journal of Information Technology and Decision Making, 3*(1), 129–143.

Yoo, H.-W., Jang, D.-S., Jung, S.-H., Park, J.-H., & Song, K.-S. (2002). Visual information retrieval system via content-based approach. *Pattern Recognition, 35*(3), 749–769.

Zabih, R., Miller, J., & Mai, K. (1999). A feature-based algorithm for detecting and classifying production effects. *Multimedia Systems, 7*(2), 119–128.

Zhang, D., & Chang, S. F. (2002). Event detection in baseball video using superimposed caption recognition. In *Proceedings of the ACM International Conference on Multimedia* (pp. 315–318).

Zhang, H. J., Kankanhalli, A., Smoliar, S. W., & Tan, S. Y. (1993). Automatic partitioning of full motion video. *ACM Multimedia Systems, 1*(1), 10–28.

Section VI

Miscellaneous Techniques
in Applications

Chapter XIV

Managing Uncertainties in Image Databases

Antonio Picariello, University of Napoli, Italy

Maria Luisa Sapino, University of Torino, Italy

Abstract

In this chapter, we focus on those functionalities of multimedia databases that are not present in traditional databases but are needed when dealing with multimedia information. Multimedia data are inherently subjective; for example, the association of a meaning and the corresponding content description of an image as well as the evaluation of the differences between two images or two pieces of music usually depend on the user who is involved in the evaluation process. For retrieval, such subjective information needs to be combined with objective information, such as image color histograms or sound frequencies, that is obtained through (generally imprecise) data analysis processes. Therefore, the inherently fuzzy nature of multimedia data, both at subjective and objective levels, may lead to multiple, possibly inconsistent, interpretations of data. Here, we present the FNF^2 data model, a Non-First Normal Form extension of the relational model, which takes into account subjectivity and fuzziness while being intuitive and enabling user-friendly information access and manipulation mechanisms.

Introduction

In the multimedia age, which is characterized by new emergent kinds of data such as images, sounds, texts, and video objects, the need for information storage and retrieval requirements cannot be satisfied simply by relying on traditional databases. The various properties of these objects cannot be captured properly by relational or object-oriented models. Therefore, multimedia databases have to provide new functionalities, depending on the type of —possibly heterogeneous—multimedia data being stored. Within this context, new challenges ranging from problems related to data representation to challenges related to the indexing and retrieval of such complex information, have to be addressed.

In this chapter, we focus on those functionalities of multimedia databases that are not present in traditional databases but are needed when dealing with multimedia information. Multimedia data are inherently subjective; for example, the association of a meaning and the corresponding content description of an image as well as the evaluation of the differences between two images or two pieces of music usually depend on the user who is involved in the evaluation process. Furthermore, such subjective information usually needs to be combined with objective information, such as image color histograms or sound frequencies, that are obtained through data analysis. Data analysis processes generally are imprecise. Therefore, the inherently *fuzzy* nature of multimedia data, both at the subjective and objective levels, may lead to multiple, possibly inconsistent interpretations of data. Thus, providing a data model that can take into account subjectivity and fuzziness, while being intuitive and enabling user-friendly information access and manipulation mechanisms, is not trivial.

Although most of the content presented in this chapter also applies to diverse multimedia information management scenarios, for the purposes of illustration, we focus on image data that illustrate the subjectivity and fuzziness aspects that are common to all such scenarios.

In order to store a collection of images properly in a database, the system must offer appropriate capabilities to explore the relationships among the different images, to recognize the relevant image features, to provide methods and techniques to express those relationships and features, and to query on them. As opposed to the classical relational data model in which queries are posed textually (or through some visual interface that does not increase the expressive power of the textual format), in image databases, queries usually are expressed in nontextual forms. This is the case, for example, when using *Query By Example* or *Query by Content* forms, in which a query may include an image as part of it, and the returned result does not rely on a crisp evaluation process but rather on a notion of similarity between the query and the images in the database. In particular, each returned image has a degree of satisfaction relative to the query, which represents to which extent the result image can be considered similar to the query image according to the chosen notion of similarity.

Fuzziness and uncertainty related to image query processing cannot be represented directly in the relational data model. Therefore, several approaches (Raju & Majumdar, 1998; Takashi, 1993; Yang, Zhang, Wu, Nakajima, & Rishe, 2001) have been proposed to extend the relational data model to include these aspects appropriately. Zaniolo et al. (1997) extend the relational model to incorporate uncertainty at tuple as well as attribute levels. In the *tuple-level* approaches, the schema of the relations can include attributes representing uncertainty values. Thus, each tuple may contain one or more uncertainty attributes, each one representing the fuzziness degree associated with a different interpretation of the data values stored in the remaining attributes of the tuple. The uncertainty attributes usually have real values or are expressed in terms of intervals on real numbers. In the *attribute-level* approaches, on the other hand, instead of associating a value representing the uncertainty of the data to the tuple as a whole, a degree of uncertainty is associated directly to every single attribute value. In image databases (in which images are represented in terms of their various feature values that are extracted from the image using appropriate image processing and analysis processes) attribute-level approaches are more applicable; it is easier to store and maintain detailed information about the various relevant aspects of a given image using an attribute-based approach instead of associating a unique, global value to the overall image tuple.

In the next section, we present the background on modeling and accessing image data collections and the state-of-the-art on the problem of dealing with fuzzy information in image databases. After discussing the problem of image representation, we present related work in the area of image retrieval and comment on relevant fuzzy models for image databases. We then build on this background to introduce the FNF^2 data model and describe its suitability for image retrieval. Concluding remarks are given in the final section.

Background and Related Work

Problems related to modeling and accessing image databases have been approached from different perspectives in various scientific communities. Some aspects, such as issues related to feature extraction and representation, have been studied in both computer vision and image processing communities (Del Bimbo, 1999), while those aspects related to storage, indexing, and query processing have received great interest in the database community (Buche, Dervin, Haemmerlé, & Thomopoulos, 2005; Grosky, 1997; Krishnapuram, Medasani, Jung, Choi, & Balasubramaniam, 2005; Subrahmanian, 1998).

Given the high dimensionality of the problem, researchers tend to concentrate on specific problems while making simplifying assumptions on other aspects. For example, a simplifying assumption widely adopted in the database community is that textual description and representation of images are available; that is, data are annotated. On the other hand, approaching the problem from a different perspective, most researchers from the computer vision and the image processing communities make simplifying hypotheses about data modeling and data retrieval methods and work in the context of image repositories or image directories instead of image databases.

A successful image database system should benefit from the integration of methods and results from the various communities. In this section, we introduce the main components of an image database system in order to motivate the role of fuzziness in image databases, and we present alternative approaches for fuzzy data modeling for image retrieval (Petry, 1986; Raju et al., 1998; Subrahmanian, 1998; Takahashi, 1993; Yang et al., 2001; Yazici, Buckles, & Petry, 1999).

Image Representation

Images are represented in the database as collections of low-level features. In order to detect the prominent features in the stored images, image databases use feature extraction modules that are implemented on the basis of computer vision techniques. The most frequently used image features are colors, textures, shapes, and spatial descriptors (Del Bimbo, 1999).

Colors usually are described through color histograms (Swain & Ballard, 1991), which associate a bin to each distinct color. Each bin contains those pixels that have the corresponding color in the image; thus, the size of the bin reflects how much that color is present in the image. Color histograms capture the color distribution in a given image from a quantitative point of view, but they are sufficient neither to describe spatial color distribution nor to handle color correlations in the images. These aspects of color and, more specifically, spatial relationships between colored pixels in a certain image region, can be represented better by analyzing textures and shapes (Del Bimbo, 1999; Smeulders, Worring, Santini, Gupta, & Jain, 2000; Van de Wouver, Scheunders, Livens, & Van Dyck, 1999; Wang Jing & Dana, 2003).

According to the feature contrast model (Jiang, Li, Zhan, & Gu, 2004), a visual stimulus may be characterized as a set of binary features (i.e., a stimulus is represented by means of the set of feature properties that it satisfies). For every feature, the set of possible values is fixed. Consequently, any visual object can be represented as a set of binary values that denote the fact that the corresponding feature can be considered as having or not having that particular value. Equivalently, the feature set for a given stimulus can be characterized as a set of logic predicates that the stimulus does satisfy (Santini , Gupta, & Jain, 2001).

On the other hand, binary (Boolean) logic is not always suitable for modeling image features. With the goal of taking into account the noise of visual stimuli perception and representation, fuzzy theory is recognized as a natural modeling framework. In this framework, any image I is characterized in terms of a number of fuzzy measurements v_i on the image features and properties. For example, let us assume that the shape of an object in an image is the interesting feature to be represented. It can be the case that different observers (or different shape-extraction algorithms) provide different shape characterizations for the same object. One observer might say that a given shape is "highly" oval or that it is "almost" rectangular. Expressions such as *highly* or *almost* recall this notion of fuzziness, which is embedded implicitly in the similarity evaluation of visual stimuli.

Each image can be characterized in terms of its physical representation and the information extracted from the image or provided by the user. More formally, let I be the set of all the images in the given image database. Given a set of classes of membership M and a set of possible memberships values P, the analysis process is a function computed by a *system V* (either a human, or a computer vision system) which associates to any feature in I a set of elements from M, each one with its specific grade of membership $p \in P$, that is:

$$V : I \rightarrow (M \rightarrow P)$$

If $V(i) = \{(c_1, p_1),...,(c_k, p_k)\}$, then for any class c_i, the value p_i is a measure of the membership of the image i to the class c_i. The function V abstracts the process of analyzing the

Figure 1. An example visual system for extracting descriptions from images (Chianese, Picariello, & Sansone, 2001)

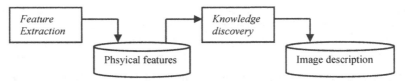

physical representations of the images to produce a description for each image in terms of its features and the corresponding grades of *uncertainty*. In the following discussion and examples, when representing the set $V(i)$, we do not list the pairs whose second components (i.e., grades) are 0; these correspond to feature properties that are classified as missing in the considered object. Thus, we read the set $V(i) = \{(c_1, p_1),...,(c_k, p_k)\}$ as the specification of all the properties that are satisfied by the object i, with their degrees of satisfaction. For retrieval, given a set of low-level features, a notion of *similarity/dissimilarity* measure is defined over the features. Naturally, similarity and dissimilarity are opposite each other. The definition of the dis/similarity between two stimuli depends on the formalisms in which the stimuli are represented. The retrieval process is performed using this dis/similarity concept in the corresponding feature space. In many cases, dissimilarity measures characterize the difference between a given pair of images as a distance in some suitable, mostly metric, feature space (Ashby & Perrin, 1988; Santini et al., 2001).

Figure 1 schematically represents a system and the underlying workflow designed to extract the image representation presented above (Chianese et al., 2001). In this system, the structural—or physical—features are extracted by a feature extraction module. These features describe colors, textures, and shapes in the image. The system also includes a knowledge discovery module to analyze the structural features of an image in order to produce a description of the image itself. In other words, the knowledge discovery module classifies the images in the database based on their structural properties. The module associates a reference class and a related membership grade to each discovered image descriptor through the use of a collection of genetic algorithms. These algorithms have been trained by using supervised learning, by associating a symbolic description (determined by a human expert) to each extracted feature vector (Chianese et al., 2001).

The Image Retrieval Problem

Intuitively, the image retrieval problem is deciding whether and to which extent any image stored in the database matches a criterion that is specified by means of a descriptive query or an example. Even when low-level features of an image are not fuzzy, image information content (needed for measuring the degree of match between a query and the image) cannot always be considered univocal. In general, what the image represents and what is important in the image depends on the observer as well as on the context in which the image is inserted into the collection (Itti & Koch, 2001). Therefore, there is not a unique notion of similarity. To account for this, similarity models can be classified in a way to distinguish between perceived similarity and judged similarity. If a and b are any two stimuli, and if A

and B are their representations in the feature space, then the perceived similarity between the two stimuli is a function $s(A;B)$ of their representations, while the judged similarity is a function of the perceived similarity; that is, a function $\sigma(A,B) = g[s(A;B)]$ in which g is a suitable, monotonically nondecreasing function (Itti & Koch, 2001). Monotonicity ensures that it cannot be the case that two images are seen subjectively as more similar than another pair, while being less similar according to perceived similarity. Specific similarity functions have been defined to capture the uncertainty related to similarity concepts (Wang & Simoncelli, 2005).

In some cases, the image description comes from human-provided annotations. In these cases, query processing requires the evaluation of a semantic similarity between the query terms describing the target image and the annotations describing the images in the database. This task usually is done by means of pattern recognition techniques (Boccignone, Chianese, Moscato, & Picariello, 2005), which return high-level features (as opposed to low-level features like color histograms) with associated grades of confidence. A grade of confidence denotes the degree of membership of the given feature to the discovered pattern.

Uncertainty-Based Models for Image Databases

In the following, we present the literature that focuses on the role and importance of uncertainty, which is necessary for defining and implementing an effective image database management system.

Buckles and Petry (1982) were among the first authors to propose a formal treatment of fuzziness in relational databases. Their objective was to address fuzziness in a general sense, not specifically focussing on images or other multimedia data. In their model, fuzziness is associated to data by means of linguistic terms (e.g., terms like *bad* and *good*), and they define a notion of similarity on such linguistic terms. More recently, the same authors classified databases in various groups that are characterized as dealing with precise data, imprecise data, or vague data (Buckles & Petry, 1995). The approach is applied best to enterprises (or parts of enterprises) in which the linguistic sets are finite and may be extended to continuous domains. The case of precise data virtually includes all of the database systems in widespread use. On the other hand, an imprecise data model is the basis of the studies on uncertainty in databases. The key notion is that while only one value applies to the enterprise, the database extension may contain a set, and each database object is *surely*, *maybe*, or *surely not* a response to the query (possibility theory). The vague database refers to the databases that deal with attribute values for which it is assumed that no precise value exists. They are represented as linguistic terms that are related to each other by similarity relationships.

A natural extension of the previous works is provided by Yazici, Buckles, and Petry (1999). This work motivates the need for nonfirst-normal form models in order to model and manage uncertainty in the relational data model. It introduces the ExIFO model for conceptual design of fuzzy databases and for a logical design based on non-first normal form logical database models. In particular, the ExIFO model deals with incompleteness, null data, and fuzziness. It represents uncertainty at the attribute level by means of new constructors at the conceptual and logical levels, thus defining fuzzy entity types. Raju and Majumdar (1988) represent ambiguities in data values as well as impreciseness in the association among them through a fuzzy relational data model. The authors describe and formalize the treatment of the integrity constraints for a fuzzy data model; they define relational operators for fuzzy relations and investigate the applicability of fuzzy logic in order to capture integrity constraints. Moreover, they address the problem of lossless join decomposition of fuzzy relations for a given set of fuzzy functional dependencies.

In a collection of works edited by Zaniolo et al. (1997), Subrahmanian introduces a foundational theory for managing uncertainty in databases and knowledge bases in general. In particular, uncertainty is represented at the tuple level; that is, each tuple is extended to include one or more uncertainty attributes that represent the likelihood of the overall information associated with the tuple. Given this model, a probabilistic extension of classical relational algebra operators also is provided. These take into account the uncertainty aspects included in the relations. The model is general enough to be applied to various fuzzy knowledge management scenarios, including image databases. However, in this case, it fully relies on an annotated scenario and does not consider the numerous low-level or intermediate level descriptions (or features) that can be extracted independently from image data by means of image processing techniques.

More recently, Atnafu, Chbeir, Coquil, and Brunie (2004) proposed an integration of similarity-based queries into image DBMS. In particular, the authors propose an image data repository model, a formal algebra for content-based image operations, and several image processing and retrieval techniques. Similarity-based operators adapted to image data are discussed within an Object Relational DBMS framework. Uncertainty is incorporated in the definition of similarity- (distance-) based operators; the authors consider metric space computations returning similar images based on the value of an uncertainty threshold that can be chosen appropriately based on the peculiarities of an application. Starting from this concept, they define a content-based algebra with selection and join operators.

Krishnapuram et al. (2004) propose a content-based image retrieval system based on fuzzy logic, FIRST (fuzzy image retrieval system). In particular, the authors propose a data model based on fuzzy attributed relational graph (FARG), in which each image is represented as a graph whose nodes represent objects, and edges represent relations (e.g., spatial relations) between them. A given query is converted to a FARG, and the query processing is reduced to a subgraph matching problem. In order to reduce the NP complexity of the subgraph matching problem, Krishnapuram et al. (2004) also propose indexing schemes based on a leader-clustering algorithm. The underlying data structure, which extends the relational model, is more expressive than the previously existing approaches. As a tradeoff, creating and maintaining such data structure is more complex.

Buche et al. (2005) describe a relational model extended with the so-called multi-views fuzzy querying. The model is well-suited specifically for biological processes, but it has interesting applications to image databases as well. As an innovative contribution, fuzzy data are integrated and dealt with by referring to ontologies and semantic rules. The model captures incompleteness and impreciseness, and it expresses user preferences with a fuzzy model in which an ontology is used to express fuzziness; the values of a domain are connected by using the "a kind of" semantic link.

In the next section, we discuss an attribute-level fuzzy extension of the relational model. Unlike the other approaches already presented, this model captures the uncertainty in the description of images from multiple angles.

The *FNF²* Model

In this section, we introduce the *FNF²* (fuzzy non-first normal form) model for image databases. We first introduce motivating examples and then present a formal definition of the model.

Motivating Examples

Let us assume that we are given a database that contains a collection of digital pictures, and let us consider a set of queries that we may want to pose on this collection:

Query 1. *Find those images that contain a high quantity of sunlight and also a mountain.*

In order to process this query, the system first needs a color-based retrieval module in order to find all the images in which those colors that usually are associated to the light (i.e., red and white) appear, in such a quantity that the system would classify their amount as high. This process will involve a fuzzy description of the color content of the digital image. Second, the system also needs semantic information (extracted automatically or obtained from a repository of human annotations) about the pictures in the database in order to tell whether they contain any mountains.

As described earlier, an automatic vision system analyzes a given image and extracts information about colors, textures, shapes, and spatial descriptors, each associated with a degree of certainty. For example, the analysis of the Ves_pict.gif image in Figure 2 by means of an automatic vision system developed at the University of Napoli (Chianese et al., 2001), returns fuzzy feature descriptors. In order to illustrate the model, let us consider the various color descriptors. In this image, multiple colors appear, and the system associates to each color attribute value a certainty degree: grey appears with certainty degree 0.6, red with certainty 0.75, white with certainty 0.7, blue with certainty 0.411, and yellow with certainty 0.417. Texture attribute is recognized as thin with certainty degree 0.6, mixed thin and coarse

Figure 2. Ves_pict.gif, a picture of the Vesuvio Vulcan at sunset

Figure 3. Coast_pict.gif, a coast landscape

with certainty 0.9, net with certainty 0.571, and crisp with certainty 0.322. The recognized semantic attribute is mountain with a confidence 0.871.

Table 1 contains the color, texture, and semantic content information extracted by this system from the image in Figure 2 as well as from the images in Figures 3, 4, and 5. Figure 3 presents a picture of a coast landscape in which blue is the dominant color. Figure 4 presents a picture of a mountain covered with snow in which white is the dominant color, and the image in Figure 5 corresponds to a sunset in a saguaro desert in which purple and burgundy dominate. To answer Query 1, this information has to be fetched from the database and combined during the query answering process in order to identify images with colors that usually are associated with light and that contain objects identified as mountains.

Query 2. *Find those images that are similar to Ves_pict.jpg.*

This second type of query is referred to as *query by example*; images in the database must be retrieved based on their similarities to the given example image. The major difference from

Figure 4. Snow_pict.gif, a snow-mountain picture

Figure 5. Sunset_pic.gif, a saguaro-mountain picture

Query 1 is that, in this case, the aspects of interest in the image are not enumerated explicitly, but the relevant information on which similarity is to be verified has to be detected automatically. Therefore, knowing that our system has extracted from Ves_pict.gif, color, texture, and semantic information, the query can be rephrased as "Find all the images that have red and white as predominant colors, that have a thin texture, and that contain a mountain."

In order to move toward a model that can handle queries like Query 1 and Query 2, we have to address two main issues. First, we have to represent uncertainty in data at the attribute level (e.g., uncertainties associated with colors, textures, contents, etc.). Second, we have to develop mechanisms to process queries when uncertainties are associated with the available data at the attribute level.

The *FNF²* Data Model

The FNF² fuzzy relational model is an extension of the standard relational model. In this extension, the considered attribute value domains are fuzzy. Fuzzy data models can be in-

Table 1. Fuzzy attribute values extracted from images in Figures 2, 3, 4, and 5

File	Color	Texture	Content
Ves_pict.gif	<grey,0.600>, <red,0.750>, <blue,0.750>, <yellow,0.417>, <white, 0.700>	<thin,0.600>, <thin-Coar,0.900>, <crisp,0.322>, <net,0.571>	< mountain,0.871>
Coast_pict.gif	<blue,0.827>, <beige,0.765> <green,0.816>	<thin, 0.704>	<SeaCoastPicture,0.939>
Snow_pict.gif	<white,0.939>, <gray, 0.569>	<mixt_texture, 0.674	<SnowMountain,0.918>
Sunset_pict.gif	<purple,0.866> <yellow,0.786> <burgundy,0.856>	<thin,0.704> <crisp,0.346>	<Sunset, 0.839>

terpreted as extensions of traditional data models using fuzzy set theory (Zadeh, 1971) and the possibility theory (Zadeh, 1978). We first define a fuzzy tuple.

***FNF²* fuzzy tuple.** Let $D_1, ..., D_n$ be n domains. A fuzzy n-tuple is any element of the Cartesian product $2^{D1} \times ... \times 2^{Dn}$ in which 2^{Di} is the fuzzy powerset of D_i; that is, the set of all fuzzy subsets of D_i.

According to the definition, any n-tuple is a sequence $< vs_1, ..., vs_n >$ in which each vs_i is a set of elements of the form $<v_j, f_j>$. Here, v_j is a value from the corresponding fuzzy domain D_i, and f_j is its corresponding fuzzy membership value.

For the sake of simplicity and readability, we denote those attribute value sets vs_i that are singletons and where the only member of the set has fuzzy membership degree equal to *1* (which represents full, certain membership) by means of the only domain value. This is the case, for example, for the Ves_pict.gif, Coast_pict.gif, Snow_pict.gif, and Sunset_pict.gif values of the File field in the data presented in Table 1.

As we already have mentioned, we also consider as a special case the presence of a membership degree of *0*. This represents the certain nonmembership. In our model, the presence of pairs $< v, 0>$ in an attribute value does not provide any information in addition to what we could have if we did remove that pair from the set, since domain values that do not appear in an attribute value implicitly are associated with the membership degree *0*. For example, the attribute values $\{< red, 0.5 >\}$ and $\{<red, 0.5 >, < green, 0.0>\}$ provide the same information. Thus, we assume that our attribute value sets do not contain any such pair. Since the fuzzy values are derived from data returned by some automatic vision systems, and since data provided by visual systems are only nonzero values (systems only give information about feature values they find in the considered image), then this also is not a restriction from an implementation point of view.

***FNF²* fuzzy relation schema.** A fuzzy relation schema is used to associate attribute names to domains. In the following, we also use the term *fuzzy attribute* to denote an attribute A whose domain *dom(A)* is a fuzzy set.

A fuzzy relational schema is defined as a symbol R, which is the name of the fuzzy relation, and a set $X = \{ A_1, ..., A_n \}$ of (names of) fuzzy attributes. The schema is denoted as $R(X)$.

***FNF²* fuzzy relation.** A fuzzy relation is an instance of a fuzzy relation schema; that is, a fuzzy relation is a set of fuzzy tuples, as stated in the following definition.

Let $R(\{ A_1, A_2 ... , A_n\})$ be a relational schema. A fuzzy relation, defined over R, is a set of fuzzy tuples $t = < vs_1, ..., vs_n >$ such that each vs_i is a fuzzy subset of *dom(A_i)*.

An example of a fuzzy relation has been given in Table 1. The schema of this example relation has four attributes: File, Color, Texture, and Content. Each attribute value is a set of <domain_value, fuzzy_value> pairs.

Manipulation of *FNF²* Fuzzy Relations

FNF² relations are accessed and manipulated by means of a corresponding *FNF²* algebra. The usual set theoretic operators can be extended to fuzzy sets in different ways, depending on the specific semantics associated with the fuzzy logical connectives. In fact, several semantics have been proposed for fuzzy logical connectives, *and* ($\wedge$) and *or* ($\vee$), but it has been proved that the min-max semantics is the only semantics for conjunction and disjunction that preserves logical equivalences and satisfies the idempotency property (Yager, 1982). This motivates our choice to adopt the following definitions for fuzzy logical operators.

-Fuzzy intersection (and): $A \wedge B = \{< u, min(\mu_A(u), \mu_B(u))> \mid u \in U\}$

-Fuzzy union (or): $A \vee B = \{<u, max(\mu_A(u), \mu_B(u)) > \mid u \in U\}$

- Fuzzy complementation (not): $\neg A = \{<u, 1 - \mu_A(u)) > \mid u \in U\}$

In these definitions, U represents the universe (i.e., the domain against which fuzzy membership is evaluated).

In this chapter, we do not present the details of the algebra. Instead, we concentrate on the primary aspects on which the *FNF²* algebra is founded: (i) tuple comparison, which is the basis of all operations that would require some logical predicate on tuples to be evaluated; (ii) tuple combination, on which most algebraic operations rely for putting information together after identifying the relevant tuples to be combined; and (iii) tuple ordering, the basis of all operations having to do with ranking based on a given criterion.

Tuple comparison. Tuple comparison is the basis of several algebraic operations, including the set oriented ones. In *FNF²* data model, attribute values are sets of pairs. Depending on the query, comparison either can involve the complete information represented by the tuples or can be restricted to a particular component. Indeed, it might be necessary to recognize and combine those tuples that contain, for every attribute in their schema, the same sets of values; these values might differ at most because of membership degrees.

This notion is formalized in the definition of ***data identical* tuples**. Let $R(\{A_1, A_2, ..., A_n\})$ be a relation schema and r be an instance of R. Let $t_1 = <vs_1^1, ..., vs_n^1>$ and $t_2 = <vs_1^2, ..., vs_n^2>$ denote two tuples of r. The tuples t_1 and t_2 are *data identical* iff for every index i, it holds that $data_1 = \{u \mid <u, \mu_{dom(Ai)}(u)> \in vs_i^1\} = data_2 = \{u \mid <u, \mu_{dom(Ai)}(u)> \in vs_i^2\}$. As an example, consider the tuples t_1 and t_2 in Table 2, which are data identical.

As the comparison operator, depending on the intended meaning of the query processing operations, either more traditional equality or data identicalness can be used. Consider, for example, the *set union* operation. Union merges the tuples appearing in the given sets and removes duplicates. If equality is used as the comparison operator, then the resulting relation can contain multiple instances of data-identical tuples, which are not eliminated because of the different degrees of certainty of their attribute values. This could be the intended meaning if the operation has the goal of keeping track of all the data retrieved by the feature extraction modules. On the other hand, for applications that need to return a single certainty information for every feature data value in the result, the existence of data-identical tuples cannot

Table 2. Example of data identical tuples

	File	Color	Texture	Content
t1	Ves_pict.gif	<black,0.7>, <red,0.815>, <beige,0.414>, <brown,0.311>, <white, 0.628>	<thin,0.715>, <mixed,0.715>, <net,0.511>, <crisp, 0.121>	< human, 0.95>
t2	Ves_pict.gif	<black,0.75>, <red,0.8>, <beige,0.6>, <brown,0.3>, <white, 0.7>	<thin,0.6>, <mixed,0.715>, <net,0.7>, <crisp,0.1>	< human, 0.9>

be acceptable. Therefore, data-identical tuples should be combined. Similar considerations apply to the other set operators as well. In particular, *intersection* returns the tuples that are in common. Therefore, it depends on the meaning associated to "being in common," and on the treatment chosen for data-identical (but distinct) tuples occurring in the relations being intersected. Similarly, *set difference* returns the tuples appearing in the first relation and not appearing in the second one. Therefore, it also depends on the comparison operator chosen and on the treatment selected for data-identical tuples.

Tuple combination. Different alternative tuple combination functions can be defined, depending on the intended meaning of the union operation. In particular, if we want to adopt a skeptical treatment toward the combination results, we use conjunction as the combination function for data-identical tuples, while an optimistic treatment of combination would make disjunction preferable:

- **Optimistic combination** ($\oplus_o$): Let $t_1 = <vs_1^1, ..., vs_n^1>$ and $t_2 = <vs_1^2, ..., vs_n^2>$ be two data-identical tuples. $t_1 \oplus_o t_2 = <vs_1, ..., vs_n>$, where for each i, $vs_i = \{<u, \mu_1(u)> \lor <u, \mu_2(u)> \mid <u, \mu_1(u)> \in vs_i^1$ and $<u, \mu_2(u)> \in vs_i^2\}$.
- **Skeptical combination** ($\oplus_s$) is defined in the analogous way by applying the *fuzzy and* operator on the values instead of the *fuzzy or*: $t_1 \oplus_s t_2 = <vs_1, ..., vs_n>$, where for each i, $vs_i = \{<u, \mu_1(u)> \land <u, \mu_2(u)> \mid <u, \mu_1(u)> \in vs_i^1$ and $<u, \mu_2(u)> \in vs_i^2\}$.

As an example, we consider the two data-identical tuples, t_1 and t_2, introduced previously. Their optimistic and skeptical combinations are as shown in Table 3.

Both optimistic and skeptical combinations are commutative and associative. They inherit the properties from commutativity and associativity of *fuzzy and* and *fuzzy or* and from the standard set union. Therefore, combination operators can be extended straightforwardly to sets of data-identical tuples.

The comparison and combination operators defined so far are needed mainly for defining and implementing set-oriented operations. In order to move toward a complete algebra on the fuzzy data model, we need to be able to express more general conditions on the tuple and on their components. The fuzzy attribute-based model is very flexible and allows us to express many interesting relationships and conditions. A representative sample of conditions is provided next.

Table 3. Optimistic and skeptical combinations

	File	Color	Texture	Content
$t_1 \oplus_o t_2$	Ves_pict.gif	<black, 0.75>, <red, 0.815>, <beige, 0.6>, <brown, 0.311>, <white, 0.7>	<thin, 0.715>, <mixed, 0.715>, <net, 0.7>, <crisp, 0.121>	< human, 0.95>
$t_1 \oplus_s t_2$	Ves_pict.gif	<black, 0.7>, <red, 0.8>, <beige, 0.414>, <brown, 0.3>, <white, 0.628>	<thin, 0.6>, <mixed, 0.715>, <net, 0.511>, <crisp, 0.1>	< human, 0.9>

Selection conditions on tuples: Selection conditions are either basic or complex. Basic conditions are those that are *atomic*. Complex conditions are defined by means of conjunction, disjunction, and negation of atomic conditions. The FNF^2 model admits different classes of conditions. We list some illustrative examples as starting points to help to define selection conditions.

- **Comparing with constants:** These conditions are defined on individual attribute values (i.e., a single set of value/degree pairs).

    1. *Fuzzy pair membership condition* tests whether a given pair belongs to a given set. This condition generally is used to retrieve images with a specific feature value; for example, in any query, such as "Find all images in the relation *r* in which the color red appears with certainty 0.8." Syntactically, this selection condition can be expressed as <red,0.8> $\in$ r.Color.

    2. *Fuzzy pair membership condition, restricted to data value*, tests whether a pair with a specific data value belongs to a given attribute value. For example, this condition allows the retrieval of images in which some quantity of red is identified by the feature extraction module. This query could be expressed as <red, _> $\in$ *r.Color*, where "_" is a wildcard.

    3. *Fuzzy pair membership condition, restricted to certainty-value*, tests whether a pair with specific certainty value belongs to a given attribute value. For example, we might be interested in knowing if there is any color that has been recognized with certainty 0.9 in a given collection of pictures. This query can be written as < _, 0.9 > $\in$ *r.Color*.

    4. *Fuzzy pair membership condition, restricted to certainty value thresholds*, tests whether there is any pair with a certainty value above a given threshold (or, similarly, below the threshold). Conditions of this sort allow users to retrieve those images whose data certainty values are above the threshold of interest for the specific application. For example, the condition < _, 0.5> $\in_{<=}$ *r.Color* would retrieve those images that have at least one color with a certainty value of at least 0.5.

    When we need to combine different tuples (as is the case for join operations in the relational algebra), we need to be able to express conditions that relate those attribute values relevant to the query in the given tuples. The following is a representative but nonexhaustive list of conditions that can be expressed in our FNF^2 model in order to compare attribute values.

- **Comparing two attribute values** (i.e., comparing two sets of fuzzy pairs):

    1. *Equality of the sets.* This condition can be used for testing whether two tuples have exactly the same value for a given attribute. For example, the condition $r_1.Color = r_2.Color$ could be used to retrieve those image pairs in relations r_1 and r_2, which are described as having the same colors with the same certainty values.

    2. *Equality of the two sets, restricted to the data component.* If we are interested in image pairs described as having the same colors, but we do not care about the certainty values associated with their descriptions, we can restrict the attribute value comparison to data values as follows: $r_1.Color =_d r_2.Color$.

    3. *Equality of the two sets, restricted to the certainty component.* This case is similar to the previous one but applies when we are interested in comparing the degrees of certainty without checking the corresponding data values: $r_1.Color =_c r_2.Color$.

    4. *Set inclusion.* If we are interested, for example, in checking whether the color descriptions of an image is included, with the same certainty degrees, in the description of another image, we could express this condition as $r_1.Color \subseteq r_2.Color$.

    5. *Set inclusion, restricted to the data component.* We can use this type of condition, for instance, if we want to check whether all the colors that describe a given image also appear in the description of another image (possibly with different certainty values). A restricted set inclusion type condition that expresses this request is as follows: $r_1.Color \subseteq_d r_2.Color$.

    6. *Set inclusion restricted to the certainty component.* In some cases, we might want to check whether the data about two images have been collected with the same degrees of certainty, no matter what the corresponding data values are. We would express this as $r_1.Color \subseteq_c r_2.Color$.

    7. *Overlapping of the two sets.* This condition allows us, for instance, to test whether two images have some color in common and described with the same certainty; that is, if their Color attributes have a nonempty intersection. Syntactically, this condition would be expressed as $(r_1.Color \cap r_2.Color) \neq \emptyset$.

    8. *Overlapping of the two sets, restricted to the data component.* If we are interested in the presence of common colors, but we can ignore their degrees of certainty, then we can restrict the test to the data component of the common pairs: $(r_1.Color \cap_d r_2.Color) \neq \emptyset$.

    9. *Overlapping of the two sets, restricted to the certainty component.* Analogously, if we are interested in the presence of any color described with the same certainty, we can use $(r_1.Color \cap_c r_2.Color) \neq \emptyset$.

    10. *Relative ordering of the two sets.* In many cases, different value sets might result from the same data processed by different vision systems. The FNF^2 model allows us to test whether data in common are more reliable (i.e., more certain) according to one set or the other. The condition $r_1.Color \leq r_2.Color$ checks, for example, if all the common values in the Color attributes in the two tuples (one from r_1 and the other from r_2) have higher certainty values in the second image than in the first one.

These conditions, as well as other conditions that we could express in FNF^2, can be seen as atomic conditions that can be combined in a complex condition by means of conjunction, disjunction, and negation connectives. Thus any combination of predicates is applicable either on the data elements or on the certainty elements. For example, FNF^2 allows us to check whether two tuples "have exactly the same color information, but the granularity of texture in the first one is finer than the second one." Naturally, for the execution of this query, a partial order over granularity of texture description must be defined a priori. Thus, the FNF^2 model enables a natural extension (suitable for managing image data) of the relational algebra with both standard (atomic and complex) selection conditions and ad hoc (atomic and complex) conditions described in this section.

Application of FNF^2 to Image Retrieval

The richness and flexibility of the FNF^2 data model enable its use in many image retrieval systems. Naturally, the performance of the resulting system depends on the chosen feature extraction, indexing, and clustering mechanisms. Yet, in order to experience directly the expressive power of our model and to verify its applicability for retrieval of visual data, we developed a prototype system called FIB (Fuzzy Image Database). FIB extends relational databases both in terms of the data model (fuzzy relations are used instead of standard ones) and in terms of the implemented algebraic operators, which include selection conditions of all the forms we listed in the previous section. As for the tuple combination functions, in the existing prototype only the skeptical version is used.

Figure 6 illustrates the architecture of the system, which consists of the following major components:

- With the goal of supporting visual, textual, and visual/textual queries, *a graphical user interface* is available for users' query expressions.

- An image *query processing engine* implements the various feature extraction algorithms described in Chianese et al. (2001). This module provides the fuzzy information to be stored in the fuzzy relations.

- An *algebraic optimizer* rewrites the queries, taking into account a number of algebraic equivalences. These equivalences are basically extensions (to the FNF^2 model) of the well-known equivalences of the standard relational algebra. They allow significant cost reduction in the query evaluation process by properly choosing the ordering of the operators (e.g., by anticipating projections and selections over Cartesian products and joins).

- A *query translator* transforms algebraic queries in a sequence of PL/SQL statements; that is, multimedia queries are translated into queries over an object-relational database.

- An *object relational database* system contains information about the images and their contents. The database engine is written in PL/SQL code in an Oracle 9i environment.

Figure 6. System architecture

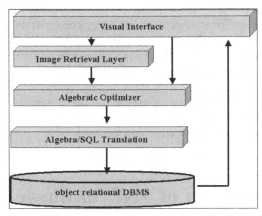

Conclusion

The problem of managing uncertainty in image databases, with the ultimate goal of implementing image retrieval methods that match users' expectations as well as possible, is becoming more and more important. Unfortunately, this challenging problem is yet to be addressed in a way that can be acceptable for the needs of the database and image processing communities. The major current trends in the image database research include the following:

- Development of a theoretical framework to manage uncertainty using several theories (probability, fuzziness).
- Development of a unified data model based on relational or object-oriented databases.
- Development of a model to measure the complexity of image database content and image database retrieval (similarity) algorithms.

The goal of this chapter was to highlight uncertainty-related challenges inherent in image databases and to discuss several cutting-edge solutions, illustrating the theory, tools, and technologies available to support various types of uncertainty. In particular, we have introduced a new and powerful model, *FNF²*, developed by the authors, and we have described the main features and technical contributions of this model. The model is well-suited for the definition of an extension of the relational algebra for managing image data. A preliminary version of the extended algebra already has been defined by the authors (Chianese, Picariello, Sansone, & Sapino, 2004) under some simplifying hypotheses about comparison and combination operators. In particular, in the algebra, the tuple combination operators have been limited to their *skeptical* versions. A prototype system based on the *FNF²* model and including the extended relational algebra operators also has been implemented at the University of Napoli "Federico II." An extended version of the algebra, with the presence of both skeptical and optimistic versions of the operators, is a work in progress.

References

Ashby, F.G., & Perrin, N.A. (1988). Toward a unified theory of similarity and recognition. *Psychological Review, 95*(1), 124–150.

Atnafu, S., Chbeir, R., Coquil, D., & Brunie, L. (2004). Integrating similarity-based queries in image DBMSs. In *Proceedings of the ACM Symposium on Applied Computing* (pp. 735–739).

Boccignone, G., Chianese, A., Moscato, V., & Picariello, A. (2005). Foveated shot detection for video segmentation. *IEEE Transactions on Circuits and Systems for Video Technology, 15*(3), 365–377.

Buche, P., Dervin, C., Haemmerlé, O., & Thomopoulos, R. (2005). Fuzzy querying of incomplete, imprecise, heterogeneously structured data in the relational model using ontologies and rules. *IEEE Transactions on Fuzzy Systems, 13*(3), 373–383.

Buckles, B. P., & Petry, F. E. (1982). A fuzzy representation of data for relational databases. *Fuzzy Sets and Systems, 7*, 213–226.

Buckles, B. P., & Petry, F. E. (1995). Fuzzy databases in the new era. In *Proceedings of the 1995 ACM Symposium on Applied Computing* (pp. 497–502).

Chianese, A., Picariello, A., & Sansone, L. (2001). Query by examples in image database using a fuzzy knowledge base. In *Proceedings of the SCI 2001* (Vol. 16, pp. 362–367).

Chianese, A., Picariello, A., Sansone, L, & Sapino, M. L. (2004). Managing uncertainties in image databases: A fuzzy approach. *Multimedia Tools and Applications, 23*(3), 237–252.

Del Bimbo, A. (1999). *Visual information retrieval*. Morgan Kaufmann.

Grosky, W. I. (1997). Managing multimedia information in database systems. *Communications of the ACM, 40*(12), 72–80.

Itti, L., & Koch, C. (2001). Computational modelling of visual attention. *Nature Reviews— Neuroscience, 2*, 1–11.

Jiang, W., Li, M., Zhan, H., & Gu, J. (2004). *Online feature selection based on generalized feature contrast model*. In *Proceedings of the IEEE International Conference on Multimedia and Expo* (pp. 1995–1998).

Krishnapuram, R., Medasani, S., Jung, S. H., Choi, Y.S., & Balasubramaniam, R. (2004). Content-based image retrieval based on a fuzzy approach. *IEEE Transactions on Knowledge and Data Engineering, 16*(10), 1185–1199.

Petry, F.E. (1986). *Fuzzy databases: Principles and applications*. Norwell, MA: Kluwer Academic Publishers.

Raju, K. V. S. V. N., & Majumdar, A. K. (1998). Fuzzy functional dependencies and loss-less join decomposition of fuzzy relational database systems. *ACM Transactions on Database Systems (TODS), 13*(2), 129–166.

Santini, S, Gupta, A., & Jain, R. (2001). Emergent semantics though interactions in image databases. *IEEE Transactions on Knowledge and Data Engineering, 13*, 337–351.

Smeulders, A. W., Worring, M., Santini, S., Gupta, A., & Jain, R. (2000). Content-based image retrieval at the end of the early years. *IEEE Transaction on Pattern Analysis and Machine Intelligence, 22*(12), 1349–1380.

Subrahmanian, V. S. (1998). *Principles of multimedia database systems.* Morgan Kaufmann.

Swain, M. J., & Ballard, D. H. (1991). Color indexing. *International Journal of Computer Vision, 7*(1), 11-32.

Takahashi, Y. (1993). Fuzzy database query languages and their relational completeness theorem. *IEEE Transactions on Knowledge and Data Engineering, 5*(1), 122–125.

Van de Wouwer, G., Scheunders, P., Livens, S., & Van Dyck, D. (1999). Wavelet correlation signature for color texture characterization. *Pattern Recognition, 32*, 175–183.

Wang, J., & Dana, K. J., (2003). A novel approach for texture shape recovery. *Proceedings of the 9th International Conference on Computer Vision* (pp. 1374–1380).

Wang Z., & Simoncelli, E. P. (2005). Translation insensitive image similarity in complex wavelet domain. In *Proceedings of the IEEE International Conference on Acoustic, Speech and Signal Processing* (pp. 573–576).

Yager, R. R. (1982). Some procedures for selecting fuzzy set-theoretic operators. *International Journal of General Systems, 8*, 235–242.

Yang, Q., Zhang, C., Wu, J., Nakajima, C., & Rishe, H. (2001). Efficient processing of nested fuzzy SQL queries in a fuzzy database. *IEEE Transactions on Knowledge and Data Engineering, 13*(6), 884–901.

Yazici, A., Buckles, B. P., & Petry, F. E. (1999). Handling complex and uncertain information in the Exifo and nf2. *IEEE Transactions on Fuzzy Systems, 7*(1), 659–676.

Zadeh, L. (1971). Quantitative fuzzy semantics. *Information Systems, 3*, 159–176.

Zadeh, L. (1978). Fuzzy sets as a basis for a theory of possibility. *Fuzzy Sets and Systems, 1*, 3–28.

Zaniolo, C., et al. (1997). *Advanced database systems.* Morgan Kaufmann Publishers.

Chapter XV

A Hierarchical Classification Technique for Semantics-Based Image Retrieval

Mohammed Lamine Kherfi, Université du Québec à Trois-Rivières, Canada

Djemel Ziou, Université de Sherbrooke, Canada

Abstract

We present a new approach for improving image retrieval accuracy by integrating semantic concepts. First, images are represented according to various abstraction levels. At the lowest level, they are represented with visual features. At the upper level, they are represented with a set of very specific keywords. At subsequent levels, they are represented with more general keywords. Second, visual content together with keywords are used to create a hierarchical index. A probabilistic classification approach is proposed, which allows the grouping of similar images into the same class. Finally, this index is exploited in order to define three retrieval mechanisms: the first is text-based, the second is content-based, and the third is a combination of both. Experiments show that our combination allows one to nicely narrow the semantic gap encountered by most current image retrieval systems. Furthermore, we show that the proposed method helps to reduce retrieval time and improve retrieval accuracy.

Introduction

The number of images available on the World Wide Web and in the electronic collections is becoming very great, and it continues to grow every day. Image retrieval engines have proved to be very useful tools that allow people to easily access this information and to benefit from it. We can distinguish two main techniques in images retrieval. The first technique, known as *text-based image retrieval*, dates back to the late 1970s and is due to the database management community (Kherfi, Ziou, & Bernardi, 2004b). In this approach, images first are annotated with text, then text retrieval techniques can be applied. Many commercial systems have adopted this technique; however, it suffers from two main drawbacks. First, images are not always annotated, and their manual annotation may prove very expensive and time-consuming. Second, human annotation is subjective; the same image may be annotated differently by different observers. Furthermore, relying exclusively on text may prove insufficient, especially when the user is interested in visual components of the image that hardly can be described by words. The second approach involves using image content such as color and texture. This approach, known as *content-based image retrieval* (CBIR), was proposed in the early 1990s and comes from the computer vision community (Kherfi et al., 2004b). The main drawback of current CBIR systems is what is called the *semantic gap*. This drawback comes from the lack of connection between the visual description extracted from an image and the interpretation that a user assigns to the same image in a given situation. Indeed, people associate a multitude of high-level concepts, such as sensations and moods, to images. However, low-level features (such as color) that can be extracted automatically from images are still unable to derive high-level concepts.

In this chapter, we attempt to overcome the lack of the two mentioned approaches. We propose a new method based on a hierarchical indexing of images. We apply this method to combine visual features with semantics-based ones in order to narrow the semantic gap. We will start the chapter with a literature review in which we show the importance of each of the approached issues (indexing and semantics retrieval) and review the main existing techniques. In subsequent sections, we respectively explain how we perform image annotation, indexing, and retrieval. In the experimentation section, we show that combining image content with text considerably helps to narrow the semantic gap. We finish the chapter with some conclusions and a discussion on future trends.

Background

Semantic Gap and Combining Visual Features with Text

In addition to low-level features such as color and shape, people use high-level concepts in order to categorize and identify images. This has led to the emergence of two main levels in image retrieval: low-level retrieval and high-level retrieval. Low-level retrieval comprises retrieval by primitive features such as color, texture, shape, and spatial location of image elements. High-level retrieval comprises retrieval of named objects and persons and retrieval by abstract attributes such as emotions that can be associated with a given image. Current computer vision techniques allow for the automatic extraction of low-level features from images with a good degree of efficiency. However, except for some simple cases, it is still difficult to extract high-level concepts from images without resorting to text. The lack of coincidence between the information that can be extracted from the visual data and the interpretation that a user assigns to the same data in a given situation is known as the semantic gap (Smeulders, Worring, Santini, Gupta, & Jain, 2000). In order to automatically derive high-level concepts, some researchers have investigated the possibility of combining text-based with content-based retrieval, since the text encodes high-level semantics better than the image content. Barnard and Forsyth (2001) cluster images by modeling the distribution of words and image regions. Grosky and Zhao (2001) apply Latent Semantic Indexing (LSI) to image retrieval. They concatenate visual features with keyword features into a unique vector that they use for retrieval. Benitez and Chang (2002) extract perceptual knowledge from collections of annotated images by clustering images based on visual and text descriptors. A number of other researchers have been interested in how to link terms used by the user to express his or her queries and those used by the indexer when annotating images. Lu, Hu, Zhu, Zhang, and Yang (2000) use a semantic network to relate keywords to images in a seamlessly retrieval framework. Zhou and Huang (2002) propose a seamless joint querying and relevance feedback scheme based both on keywords and visual features. They develop a Word Association via Relevance Feedback algorithm that they apply in learning the semantic relations between keywords. Kherfi, Brahmi, and Ziou (2004a) propose an improvement of CBIR by integrating semantic features extracted from keywords. They try to model the user's preference by using a relevance feedback schema. In the experimentation section, we will show that the model we propose in the current chapter is more general and yields better results.

Indexing

The way people look at images and how they try to locate them varies from user to user, and even for the same user at different times. According to Fidal (1997), the use of images lies between the data pole and the objects pole. At the data pole, images are used as sources of information. At the objects pole, images are defined in terms of some task (to be used in the creation of an advertisement, book jacket, or brochure). At the data pole, users want the smallest set that can provide the information needed. At the objects pole, users want to be able to browse larger sets of retrieved items. This allows us to identify two main functions that a retrieval system should ensure: query-based retrieval and browsing.

First, in query-based retrieval, the user can formulate his or her queries by using example images and/or text keywords; then, the system goes through the collection looking for all the images that correspond to this query. Most current image retrieval engines offer this functionality. For example, we can cite Kherfi, Ziou, and Bernardi (2003), Kherfi and Ziou (2006), and Li and Wang (2003). Indexing can be very useful for query-based retrieval since it helps to reduce retrieval time and increase accuracy. Indeed, the great number of images, the size of each image feature, and the use of a storage device make it necessary to use effective indexing techniques in order to facilitate access and thereby reduce retrieval time. Using an index can help to minimize the retrieval time dramatically, because the system will not have to go through the whole database of features when it looks for images but simply can retrieve the most similar images according to the index. These images generally belong to the same class as the query or to neighboring classes.

A browsing catalog is important, since it summarizes the whole content of a collection. It is very handy, especially for users who have a broad query in mind. By going more deeply into the catalog's levels, users can narrow their queries and identify what they are looking for. Among the rare systems that propose a browsing catalog, we can cite Smith and Chang (1997). In order to create a catalog, indexing techniques should be adopted in order to create an index with several abstraction levels. In higher levels, images are grouped into big sets according to more general concepts, such as the class Animals, which regroups images of all kinds of animals. By going more deeply into the index, the classes of images are subdivided according to more specific concepts. The class Animals, for example can be subdivided into Birds, Mammals, and Reptiles. This subdivision can go further until we reach very specific classes that constitute the leaves of the catalog.

Many text retrieval engines comprise indexes, such as those specialized in retrieval from the Web (e.g., Google, Yahoo!, etc.). However, few image retrieval engines offer this functionality, and those that offer it have adopted tree techniques developed originally for text databases, which are not suitable for image databases (Smeulders et al., 2000). Techniques used include k-d trees, priority k-d trees, quad-trees, K-D-B trees, hB trees, R-trees, R+ trees, R* trees, and X-trees (Smeulders et al., 2000; Rui, Huang, & Chang, 1999; Berchtold, Keim, & Kriegel, 1996). K-d-trees are a generalization of binary trees in k dimensions (Friedman, Bentley, & Finkel, 1997). Examples of image retrieval engines that use k-d trees include ImageRover (Sclaroff, Taycher, & La Cascia, 1997) and ImageScape (Lew, 2000). K-D-B-trees are a further improvement as a multidimensional generalization of standard B-trees with splitting capacity of k-d-trees (Robinson, 1981). R-trees first were proposed to index data items of nonzero size in high-dimensional spaces (Guttman, 1984), and then they were adapted to index multidimensional data. R*-trees are one of the most successful variants of R-trees (Beckmann, Kriegel, Schnieder, & Seeger, 1990). They have been used in QBIC. SS-trees are similar to R-trees with the major difference being that SS-trees use minimum bounding spheres instead of minimum bounding rectangles (White and Jain, 1996). SR-trees are a combination of SS-trees and R*-trees (Katamaya & Satoh, 1997). The PicToSeek image retrieval engine uses this technique (Gevers & Smeulders, 1999). The major drawback of tree-based techniques is that they were designed for traditional database queries (point queries and range queries) and not for similarity queries used in image databases (Lew, 2000). Indeed, these techniques make two assumptions: (1) the dissimilarity between two objects is an Euclidian distance of the points in the feature space, and (2) the dimensionality of the feature space is low. However, both of these assumptions may not be held in image retrieval context (Perrin, 1988). Furthermore, in general, these index schemes do not maintain the neighborhood information between points within a partition, especially the partition's boundary points; thus, they are not suitable for nearest-neighbor search encountered in image retrieval. Tree-based techniques hardly can be incorporated with probabilistic models developed recently for image retrieval, such as Cox, Miller, Minka, and Papathomas (2000) and Kherfi and Ziou (2006), and their combination may prove inefficient.

Some authors tried to develop organization techniques that were specific for image collection. Charikar, Chekur, Feder, and Motwani (1997) proposed an incremental clustering technique for dynamic information retrieval that handles high-dimension data and supports non-Euclidian similarity measures. Barnard et al. (2001, 2003) proposed an indexing algorithm in which they cluster images by hierarchically modeling the distribution of words and image regions. This, then, is applied for associating words with pictures, and unsupervised learning is used for object recognition.

The method we propose in this chapter is specific for image collections and more general than existing methods. It models various aspects of the collection by estimating the probability of observing an image and a keyword together in the context of different clusters with several abstraction levels.

Summary of the Proposed Method

In this chapter, we propose an indexing mechanism that we apply for semantics-based retrieval. Our method can be summarized as follows. First, images are annotated with keywords by using a multi-level thesaurus that we establish beforehand. Second, images are grouped hierarchically in order to create an index. Finally, this index is used to develop three retrieval mechanisms: content-based, text-based, and a combination of both. Compared with existing methods, our model has the following characteristics: (1) it is developed specifically for image databases; (2) it is applicable for various purposes, including indexing, retrieval, and browsing; (3) it supports different similarity measures; (4) it is probabilistic and, thus, suitable for retrieval of the nearest-neighbor-based queries encountered in image databases; (5) it is hierarchical, making it suitable for indexing; and (6) it combines image content with keywords in an attempt to narrow the semantic gap.

Thesaurus Establishment and Image Annotation

In order to be able to apply our model to an image collection, images must be annotated with keywords. In case images are found embedded in documents such as Web pages, related keywords can be found in the same document as the image. In case images are found as stand-alone objects, techniques for manual or automatic annotation should be adopted. Manual annotation may be very time-consuming, while entirely automatic annotation may not be efficient. In this work, we opt for a compromise consisting of semi-automatic annotation. We use the tool developed in Chaouch (2005) for keyword propagation. A description of that technique will be given in the section "How We Annotate Images." Before we can perform annotation, another issue must be resolved. It concerns the thesaurus establishment. This consists of identifying the set of keywords that will be used when annotating images. Depending on the image collection that will be annotated, the set of keywords may not be the same. If we have a collection made up of bird images, then keywords like Eagle may be relevant and should be present in the thesaurus, while keywords like Chair may be absent. Furthermore, the use of a thesaurus is indispensable because we want to have different abstraction levels, as explained before. Details on our thesaurus are given in the next section.

How We Establish the Thesaurus

Our thesaurus has the following characteristics:

1. It is multilevel. Semantic concepts that may be associated with images belong to various levels ranging from very broad concepts (e.g., animal) to very specific ones (e.g., hawk).
2. For each level, we scoured all the images of our collection, identifying the various concepts (e.g., object names) that could be associated with them.
3. Terms in our thesaurus focus essentially on image content (e.g., object name) instead of external attributes (e.g., image author).
4. Keywords of the lowest level may describe object name (bus), name of individual object or person (CN-Tower), set name (flock), manufacture material (bronze), and object's characteristics (sharp). Keywords of the mid-level may describe object family (bird), material family (metal), manufacturing technique (arabesque), action (flight), place (Maghreb), time (spring), use (inquiry), and general features (futurist). Finally, keywords of the highest level may describe the big family of objects (animal), global characteristics of the image (wildlife, mythology) sensations and moods (love), civilization (Greek), and era (Iron Age).

How We Annotate Images

After we established the thesaurus, we annotated images using the semi-automatic tool of Chaouch (2005), which works as follows:

First, we mention that each image is represented with an annotation vector. Each component of this vector represents the probability that relates our image to a given keyword. When there is no relation between the keyword and the image, then the corresponding component should be set to zero. When, on the other hand, the image and the keyword are related, this probability should be different from zero; it should reflect the importance of the keyword in representing the image content. Annotation is performed in two phases:

First Phase: Manual Annotation of the Training Set

First, some images that represent the general content of the collection are selected. Those images constitute the training set.

The training images are annotated manually by using the thesaurus that we described in the previous paragraph. This is done by associating each image with a set of keywords together with the probabilities that relate the keywords to it.

Second Phase: Keyword Propagation to the Remaining Collection

For each of the images not yet annotated, we initialize its annotation vector with zeros.

We use each of the training set images to create a query that we submit to the retrieval tool that we developed in Kherfi, Ziou, and Bernardi (2003). Some details about this tool will be given at the end of the current section.

Once the retrieval results are obtained, keywords of the query are automatically propagated to each image retrieved. This is done by increasing their importance in the resulting images' annotations. This importance depends on factors such as the relevance of the keyword to the query image itself and the rank of each image retrieved.

Finally, we perform a manual validation of the annotation; we can choose to keep or to remove an association that the system created between an image and a keyword. We also can choose to approve the probability it calculated or to modify it.

The annotation mechanism was based on our image retrieval engine of Kherfi et al. (2003). In this tool, image retrieval is performed as follows:

The user formulates a query by choosing one or more images as a positive example, and eventually, one or more images as a negative example.

The system uses an optimization model to calculate the parameters that correspond to the query. Those parameters are the ideal query that is equal to the average of the query images and the feature weights. The feature weighting mechanism of this tool is based on the following principle: important features are those for which positive example images are close to each other, negative example images are close to each other, and positive example is well-separated from a negative example. Such features are given more importance than other features.

The query parameters then are used to calculate a kind of distance between each image in the collection and the query.

The collection images are sorted in an increasing order according to their distance from the query.

Finally, the top-ranked ones are returned to the user.

Indexing Mechanism

Our objective now is to organize the collection into an index with various abstraction levels. Let us first explain how we represent images of the collection. Visually, we represent each image with a set of features that describe color, texture, and shape (Figure 1) in a way similar to Kherfi and Ziou (2006). For color, we use the *CIE-L*a*b** values of the image pixels. For texture, we calculate nine subfeatures derived from the co-occurence matrix: Mean, Variance, Energy, Correlation, Entropy, Contrast, Homogeneity, Cluster Shade, and Cluster Prominence. For shape, we calculate the normalized dominant orientation of edge pixels. Semantically, we represent each image with a set of keywords as in the Vector Space Model (VSM) used in text retrieval. We use a vector to represent the semantics of each im-

Figure 1. An image and its visual and semantic representation according to various abstraction levels

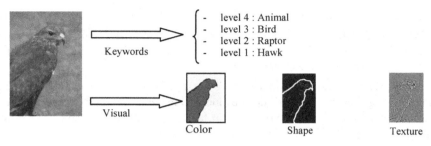

Keywords
- level 4 : Animal
- level 3 : Bird
- level 2 : Raptor
- level 1 : Hawk

Visual

Color   Shape   Texture

age. Each component of this vector gives the probability that relates a given keyword to this image. Furthermore, keywords should describe our images at various abstraction levels. If we consider the image of Figure 1, then at the lowest level, it can be annotated with the term *Hawk*, at the following level with *Bird*, and finally *Animal* at the highest level.

Our collection can be seen as one or more clouds of points in the multidimensional feature space (Figure 2). Since then, indexing consists in a hierarchical grouping of these data according to each level. In the first level (lowest level), data are grouped into classes on the basis of visual features and, eventually, very specific concepts (e.g., keywords describing objects names). In the second level, first-level classes are grouped into big classes according to more general concepts (e.g., keywords describing objects family). This process is continued until the highest level in which all the classes are grouped into a unique class (see Figure 2). Furthermore, we want our index to be soft. This means that instead of assigning each image to a unique class, the image can belong to several classes with different membership probabilities.

Figure 2. Hierarchical grouping of images

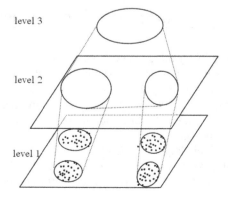

level 3

level 2

level 1

Now, consider the image r and the keyword w. We want to calculate the probability of observing the pair (r,w). First, this probability can be decomposed into a weighted sum (or mixture) of the same observation conditioned on different level contexts:

$$P(r,w) = \sum_l P(l) P(r,w \mid l)$$

in which l indexes levels. Second, we can decompose further our sum by introducing the classes constituting each level:

$$P(r,w) = \sum_l P(l) \sum_{c_l} P(r,w,c \mid l)$$

$$= \sum_l P(l) \sum_{c_l} P(c \mid l) P(r,w \mid l,c)$$

$$= \sum_l P(l) \sum_{c_l} P(c \mid l) P(r \mid l,c) P(w \mid r,l,c) \qquad (1)$$

We will call a *node* each pair (l,c), because it constitutes a node in our hierarchical index. We use the aspect model introduced by Hofmann (1999), which is based on a conditional independence assumption stipulating that, conditioned on latent classes, keywords are generated independently of the specific document identity (Hofmann, 1999). In our case, this translates to the fact that conditioned on nodes (l,c), keywords are generated independently of the specific image identity r. This particularly allows us to write (1) as:

$$P(r,w) = \sum_l P(l) \sum_{c_l} P(c \mid l) P(r \mid l,c) P(w \mid l,c)$$

or also

$$P(r,w) = \sum_l \sum_{c_l} P(l) P(c \mid l) P(r \mid l,c) P(w \mid l,c)$$

$$= \sum_l \sum_{c_l} P(l,c) P(r \mid l,c) P(w \mid l,c)$$

Following the likelihood principle, we can calculate the parameters that characterize classes and levels by maximizing the likelihood function, which is nothing but the product of all the pairs of images and keywords (r,w) of the probability of observing these pairs, given by equation (2):

$$L = \prod_r \prod_w P(r,w) = \prod_r \prod_w \sum_l \sum_{c_l} P(l,c)P(r\mid l,c)P(w\mid l,c) \tag{2}$$

in which:

$P(l,c)$, *is the a priori probability of the node (l,c).*
$P(r\mid l,c)$ *is the probability the image r is generated by the node (l,c).*

In order to calculate $P(r\mid l,c)$, we first need to endow each node with a visual representation Θ_{lc}. This representation is a mixture of several components, which enables us to decompose $P(r\mid l,c)$ as follows:

$$P(r\mid l,c) = \sum_{k_{lc}} P(l,c,k)P(r\mid \Theta_{lck}) \tag{3}$$

Each component is described with a set of I features. This allows further decomposing of $P(r\mid \Theta_{lck})$ into a sum over all the features we consider:

$$P(r\mid \Theta_{lck}) = \sum_i P(r_i\mid \Theta_{lcki}) \tag{4}$$

The model we propose (equation (2)) is general and enables one to choose different probability distributions to model $P(r\mid l,c)$. The only difference will be in the calculation of optimal parameters. In this chapter, we give the results of derivation that corresponds to the choice of Gaussian mixtures. They easily can be replaced with other distributions, however.

$P(w\mid l,c)$ is the probability that the keyword w is generated by the node (l,c). In order to calculate it, we need each cluster to be provided with a vector that contains the appearance frequency of the keywords assigned to its images. If such an information is available, then a possible estimation of $P(w\mid l,c)$ would be to consider these probabilities equal to the fraction of times the keywords are observed, that is:

$$P(w\mid l,c) = \frac{n_{lcw}}{\mid \vec{n}_{lc}\mid},$$

in which n_{lcw} is the frequency of the keyword w in the node (l,c), and $\mid \vec{n}_{lc}\mid$ is the sum of frequencies for all the keywords appearing in the node (l,c). However, if we use this method, we make an error, because we can give some keywords a zero probability of occurring, while this should not be the case. To understand this problem well, let us consider the example of assessing the fairness of a coin. A coin is said to be fair if the probability to get heads or tails is equal. If we toss a coin 10 times, and each time it comes up heads, we might think that the coin is biased. But if we toss the coin 200 times among which only 100 tosses come up heads, then we conclude that 10 heads in a row was just a statistical fluke and will not change our *a priori* assumption that the coin is fair. Hence, in order to avoid an over-fitted estimation of our probabilities, we have to introduce a Bayesian framework into the construction of this statistical model, which combines term frequency vectors $\vec{n}_{lc}$ with the prior belief we have about term occurrence.

Since we want to model keywords' frequency vectors, we opt for multinomial Dirichlet distributions, because it has been shown that they achieve a good modeling of count vectors (Kherfi et al., 2004a). Assume that the observed frequency vector $\vec{n}_{lc}$ is generated according to a multinomial distribution with parameters $\vec{p}_{lc}$ (distribution vector of term frequencies) chosen according to a Dirichlet density with parameters $\vec{\alpha}_{lc}$. The hidden process that estimates the probability $P(w \mid l,c)$ of a given term, considering a Dirichlet density with parameters $\vec{\alpha}_{lc}$ and observed frequency vector $\vec{n}_{lc}$, then is modeled by:

$$P(w \mid l,c) = P(w \mid \vec{\alpha}_{lc}, \vec{n}_{lc}) = \int_{\vec{p}_{lc}} P(w \mid \vec{p}_{lc}) P(\vec{p}_{lc} \mid \vec{\alpha}_{lc}, \vec{n}_{lc}) d\vec{p}_{lc} \qquad (5)$$

$P(w \mid \vec{p}_{lc})$ is simply p_{lcw}, the wth element of the distribution vector $\vec{p}_{lc}$. Concerning $P(\vec{p}_{lc} \mid \vec{\alpha}_{lc}, \vec{n}_{lc})$, it represents the posterior probability of the distribution $\vec{p}_{lc}$ under the Dirichlet parameters $\vec{\alpha}_{lc}$, given that we have observed term frequencies $\vec{n}_{lc}$. We obtain:

$$P(w \mid l,c) = \frac{n_{lcw} + \alpha_{lcw}}{\mid \vec{n}_{lc} \mid + \mid \vec{\alpha}_{lc} \mid} \qquad (6)$$

such that $\mid \vec{n}_{lc} \mid = \sum_w n_{lcw}$ and $\mid \vec{\alpha}_{lc} \mid = \sum_w \alpha_{lcw}$. The probability $P(w \mid l,c)$ models the distribution of terms in the node (l,c). According to equation (6), computing $P(w \mid l,c)$ requires having the values of $\vec{n}_{lc}$ and $\vec{\alpha}_{lc}$. Concerning $\vec{n}_{lc}$, it is given by the frequency of terms used to annotate the collection images. As for $\vec{\alpha}_{lc}$, it is given by equation (11).

Optimal Parameters

We maximize the log likelihood of equation (2) in order to calculate the optimal parameters of our model. Hereafter is a summarization of the obtained results.

A nodes *a priori* probability:

$$P(l,c) = \frac{\sum_r \sum_w P(l,c \mid r,w)}{N_r N_w} \qquad (7)$$

in which N_r is the total number of images in the collection, and N_w is the total number of keywords in the collection.

A component's average:

$$\mu_{lcki} = \frac{\sum_r \sum_w \dfrac{P(l,c \mid r,w)P(r_i \mid \Theta_{lcki})}{P(r \mid l,c)} r_i}{\sum_r \sum_w \dfrac{P(l,c \mid r,w)P(r_i \mid \Theta_{lcki})}{P(r \mid l,c)}} \qquad (8)$$

A component's covariance:

$$\Sigma_{lcki} = \frac{\sum_r \sum_w \dfrac{P(l,c\,|\,r,w)P(r_i\,|\,\Theta_{lcki})}{P(r\,|\,l,c)}(r_i - \mu_{lcki})(r_i - \mu_{lcki})^T}{\sum_r \sum_w \dfrac{P(l,c\,|\,r,w)P(r_i\,|\,\Theta_{lcki})}{P(r\,|\,l,c)}} \tag{9}$$

A component's *a priori* probability:

$$P(l,c,k) = \frac{\sum_r \sum_w P(l,c\,|\,r,w)P(k\,|\,r,l,c)P(l,c)}{N_r N_w} \tag{10}$$

Dirichlet parameter:

$$\alpha_{lcw} = (|\,\vec{n}_{lc}\,| + |\,\vec{\alpha}_{lc}\,|)\frac{\sum_r P(l,c\,|\,r,w)}{\sum_r \sum_{w'} P(l,c\,|\,r,w')} - n_{lcw} \tag{11}$$

Retrieval

After annotating images and creating the index, in the current section we show how we apply it for retrieval. We will present three mechanisms for image retrieval: content-based, text-based, and the combination of both. We will be showing that this combination improves retrieval accuracy, thereby narrowing the semantic gap. Furthermore, all of the retrieval mechanisms we created are very quick. This is due to the fact that they make use of the index. This allows them to limit search space to the most probable nodes, which enables one to considerably reduce retrieval time and improve retrieval precision.

Content-Based Image Retrieval

The principle of our content-based retrieval mechanism is to consider retrieval as:

1. Estimating the probability that each image is emitted by the query q.
2. Sorting images according to this probability.
3. Returning the top-ranked images to the user.

Consider that we are performing retrieval in a collection B. In order to alleviate writing, we refer to a node (l,c) with m. The probability that image r is emitted by the query q can be written as a summation over all the index nodes m of the probability that this image has been generated by this node, weighted by the probability the node emits the query, that is:

$$P(r \mid q, B) = \sum_m P(r \mid m) P(m \mid q).$$

Concerning $P(r \mid m)$, it is calculated as in the section "Indexing Mechanism." As for $P(m \mid q)$, it is estimated as follows: we first represent the node m with its visual average and then take $P(m \mid q)$ as the probability of this average, given the query parameters, in the same way as Kherfi and Ziou (2006).

Text-Based Image Retrieval

Typically, text-based retrieval for a query composed of one keyword w consists of computing $P(r \mid w)$ for each image r. As in content-based retrieval, we can write this probability as a summation, over all the nodes, of the probability that this image has been generated by this node and weighted by the probability that the node emits the keyword, that is:

$$P(r \mid w, B) = \sum_m P(r \mid m) P(m \mid w).$$

$P(r \mid m)$ is calculated as in the section "Indexing Mechanism." As for $P(m \mid w)$, it can be written:

$$P(m \mid w) = \frac{P(w \mid m) P(m)}{P(w)} = \frac{P(w \mid m) P(m)}{\sum_r P(r, w)},$$

in which $P(w \mid m) = \dfrac{n_{mw} + \alpha_{mw}}{\mid \vec{n}_m \mid + \mid \vec{\alpha}_m \mid}.$

Composed and Combined Retrieval

This section deals with the case of combined queries (i.e., constituted of a combination of keywords and images). We notice that composite queries (i.e., visual-only queries with several images or text-only queries with several keywords) are special cases of combined queries. Consider that we have a query Q made up of several keywords and several images: $Q = \{w_1, .., w_g, r_1, ..., r_h\}$. Retrieval can be performed in a way similar to the last two cases (content-based retrieval and text-based retrieval). In effect, we can write:

Figure 3. A comparison of the three retrieval modes for different image categories: Content-based=C, Keyword-based=S, Combination=CS

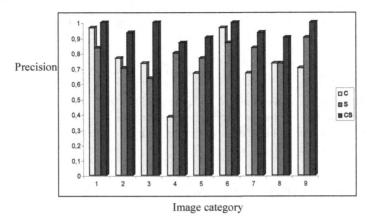

Image category

$$P(r \mid Q,B) = \sum_{m} P(r \mid m)P(m \mid Q).$$

Concerning $P(r \mid m)$, it is calculated as in the section "Indexing Mechanism." As for $P(m \mid q)$, we decompose it as a product for all the query components q, that is:

$$P(m \mid Q) = \prod_{q \in Q} P(m \mid q).$$

If q is an image, then $P(m \mid q)$ is computed as in the section "Content-Based Image Retrieval," and if q is a keyword, then it is computed as in the section "Text-Based Image Retrieval."

Experiments

Narrowing the Semantic Gap

First Experiment: Narrowing the Semantic Gap for Various Image Categories

Four users participated in this experiment, which attempts to measure the improvement achieved, thanks to combining visual features with semantics. We asked our users to perform several retrieval sessions, trying each time to locate images that belong to one of nine categories in as many feedback steps as they want, then to assign a degree of satisfaction

Figure 4. A full comparison of the three retrieval modes of our engine

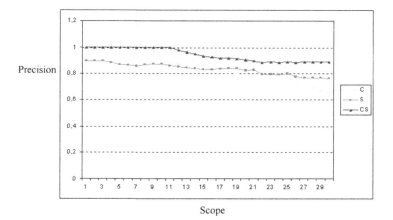

to each of the retrieved images. The user can formulate a query by selecting one or more example images. Each query then is processed in three ways:

- **C (i.e., Content).** Only the query images are submitted to the retrieval module.
- **S (i.e., Semantics).** Only the query keywords (annotation of the selected images) are submitted to the retrieval module.
- **CS (i.e., Content and Semantics).** Both the query images and keywords are submitted to the retrieval module.

Figure 3 shows the average precision obtained, in which C is represented with yellow sticks, S with green sticks, and CS with blue sticks. We first notice that for all of our experiments, combining visual features with semantics performs better than relying exclusively on any of them alone. This is completely natural, because each image possesses some visual characteristics and depicts some semantic events, and hence, both visual descriptors and keywords are important in describing its content and context. Second, as a comparison between C and S, we notice that each of them performs better than the other for some categories of images. C performs better with images that can be described easily with visual features such as those of the category Sunsets. On the other side, S performs better with images that embed complex high-level semantics that the visual features are unable to catch and that require textual keywords.

Second Experiment: Narrowing the Semantic Gap, Average

The second experiment attempts to measure the average precision in the three cases. In order to do so, each of our four users was asked to formulate 50 queries of his or her choice and then to assign a degree of satisfaction to each image retrieved. In Figure 4, we draw up

Figure 5. A comparison between our retrieval engine and that of 30 in terms of retrieval precision

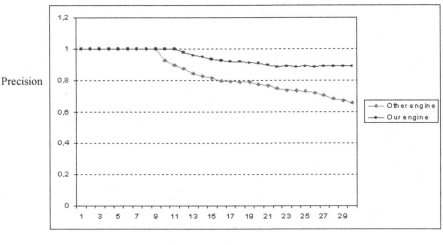

Scope

the curves of the average precision as a function of the scope in the three cases (scope is the number of images returned to the user). C is represented with a yellow curve, S with a green curve, and CS with a blue one. From Figure 4, we can see that our curves have three shapes. For C, the precision starts high for low values of scope but drops for higher values. This is due to the fact that visual features allow one to retrieve images that are very similar to the query in the top positions and then yield images that visually resemble the query but that are completely irrelevant. For S, we notice that the precision always is acceptable but never really high, whatever the value of the scope. This can be explained by the fact that all the images returned on the basis of semantic features contain the same concepts as the query, but many of them may not resemble it visually; also, they will not necessarily be ranked correctly according to their visual resemblance to it. Finally, the precision of CS starts higher than that of S and drops less quickly than that of C. This is natural because the semantic features limit the retrieved images to those that contain the same concepts as the query, while the visual ones enable one to sort these images according to their resemblance to the query. This experiment shows again the importance of combining visual features with semantics.

Third Experiment: Comparison With Another Engine

The aim of this experiment was to compare the performance of our system with the system of Kherfi et al. (2004a) in terms of retrieval accuracy. Note that both of those two systems are based on a probabilistic framework and enable one to combine visual features with key-words. The 200 queries of the last experiment were submitted to the two engines. Figure 5 shows the average precision that each engine yields as a function of scope. We notice that for low values of the scope, both of the engines achieve a good score. This corresponds

Table 1. Average retrieval time (in seconds) for several image categories

	category 1	category 2	category 3	category 4	category 5	category 6	category 7	category 8	category 9
Sequential	5.70	5.72	5.72	5.72	5.72	5.72	5.72	5.72	5.72
With Index	0.20	0.14	0.12	0.18	0.20	0.16	0.12	0.14	0.16

to the images that considerably resemble the query and, consequently, have been located easily by each of them. For high values of the scope, however, our model achieves a noticeable improvement over the other one. This may be explained by various facts. First, the probabilistic framework proposed in this chapter is more general than the one of Kherfi et al. (2004a); while Kherfi et al. (2004a) model the co-occurence of images and keywords in the whole image collection, here we model them but without forgetting the context of each cluster in which they appear. This experiment confirms that this decomposition allows one to achieve a better modeling of images and keywords in a more effective way. Second, the goodness of the results returned by our engine is due in part to the fact that it limits the search space to the most likely classes. The engine of Kherfi et al. (2004a) performs a sequential search by visiting all of the images of the collection, which increases the risk of retrieving noise images. On the other hand, our engine makes use of the index, which restricts the set of images to be visited to those belonging to the most likely clusters, which increases the probability of retrieving relevant images and limits the risk of retrieving irrelevant ones.

Improving Retrieval by Using the Index

We implemented two versions of the retrieval system. The first one performs a sequential retrieval by traversing the entire database and comparing each image with the query. The second one makes use of the index by limiting the comparison process to the clusters that are most likely related to the query. Two potential gains may be achieved, thanks to the use of the index by retrieval module, reducing retrieval time and improving retrieval accuracy. Each of the following experiments attempts to measure one of those aspects. We used our database of 5,000 images, which our tool organized into an index in which the lowest level constituted 38 classes.

Fourth Experiment: Gain in Retrieval Time

We performed nine retrieval sessions, each time using images that belong to a given category. The same query was submitted to two modules of retrieval: a sequential one and one that uses the index. Table 1 compares the average retrieval time of images that belong to different categories. We see that in all cases, the use of the index renders retrieval very quickly. Without using the index, retrieval time turns around 5.72 seconds. By using the index, however, this

Table 2. Average precision for different values of the scope

	10 Top-Ranked Images	
	Sequential	**With Index**
class 1	100%	100%
class 2	90%	100%
class 3	100%	100%
class 4	80%	100%
class 5	90%	90%
class 6	100%	100%
class 7	90%	100%
class 8	80%	100%
class 9	100%	100%
average	92%	99%

	20 Top-Ranked Images	
	Sequential	**With Index**
class 1	90%	100%
class 2	85%	95%
class 3	100%	100%
class 4	90%	95%
class 5	80%	95%
class 6	95%	100%
class 7	85%	85%
class 8	75%	100%
class 9	90%	100%
average	88%	97%

	30 Top-Ranked Images	
	Sequential	**With Index**
class 1	86.7%	96.7%
class 2	80%	96.7%
class 3	83.3%	93.3%
class 4	80%	96.7%
class 5	73.3%	90%
class 6	93.3%	100%
class 7	70%	86.7%
class 8	66.7%	96.7%
class 9	76.7%	100%
average	79%	95%

time drops down for all the clusters, reaching 0.12 seconds in some cases. In the sequential case, retrieval time is high, because retrieval involves comparing a high number of images (corresponding to the whole database) with the query. This time decreases when we use the index, because rather than going through all the images, the retrieval module compares the query with those belonging to the most likely clusters, which decreases traversing time dramatically. The second notice we made is that while retrieval time is almost constant in the sequential case, it varies relatively when we use the index. This is natural because in the sequential case, the query always is compared with the same number of images, while retrieval time in the index case depends on the size of the clusters nearest to our query. For big clusters, this time increases, while for small clusters, it decreases; but in all cases, it remains very low compared to sequential retrieval. Finally, we notice that, on average, retrieval time is reduced nearly 37 times, thanks to the index.

Fifth Experiment: Gain in Precision

The same experiment was repeated with the two versions of retrieval (sequential and index-based), but this time we were interested in computing the number of relevant images that each module retrieves. Table 2 gives the percentage of relevant images in the two cases for different values of the scope (scope is the number of images returned to the user). The first notice we make from Table 2 is that for the small values of the scope, both retrieval modules achieve a good performance. The reason is that the top-ranked images are those that highly resemble the query and, consequently, cannot be missed by any of the retrieval modules. The second notice is that for high values of the scope, while the precision of images retrieved by the index-based module remains appreciably high, it drops more quickly for the sequential module. The reason is that the sequential module visits all the images of the database, which increases the risk of making an error by retrieving noise images. On the other hand, the use of the index restricts the set of images to be visited to those that belong to the most likely clusters, which limits the risk of retrieving irrelevant images. The last two experiments confirm our expectation that the use of the index improves retrieval effectiveness and accuracy.

Table 3. Confusion matrix (cluster × cluster) of classified/misclassified images

	c1	c2	c3	c4	c5	c6	c7	c8	c9	c10
c1	100	0	0	0	0	0	0	0	0	0
c2	0	97	0	0	3	0	0	0	0	0
c3	0	0	99	0	0	0	0	0	1	0
c4	0	0	0	95	0	0	3	2	0	0
c5	0	4	0	0	96	0	0	0	0	0
c6	0	0	0	0	0	100	0	0	0	0
c7	0	0	0	3	0	0	97	0	0	0
c8	0	0	0	0	0	0	0	100	0	0
c9	0	0	0	0	0	0	0	0	98	02
c10	0	0	0	0	1	0	0	0	0	99

Browsing

In order to validate our image cataloguing process, we indexed our database with the proposed mechanism into an index with three abstraction levels. We performed a set of experiments, each of which aims to measure a given aspect of our cataloguing process.

Sixth Experiment: Homogeneity of Clusters

This experiment aims to test whether the clusters our model provides are sufficiently coherent. This is particularly important because coherent clusters make sense to the user, while incoherent clusters do not. First, we took the 38 clusters our algorithm generated (the classes of the lowest level). Then, we randomly generated a set of 38 random clusters from the same database. After that, the two sets of clusters were mixed randomly into a whole set of 76 clusters containing index clusters and random clusters. Finally, we asked a user to visit each cluster and to say whether it was coherent enough to be considered a true cluster. Our user was aware that a few noise images that may appear in a given cluster should not affect his decision on the cluster coherence. Among the 38 clusters that our program generated, the user was able to identify 37 clusters, which is a success rate of 97%. This allows us to say that most of the clusters that our program identified make sense to the user.

Seventh Experiment: Precision of Classification

This experiment aims to show the rate of correctly classified images as well as misclassified ones. With this intention, and in order to be able to visualize our confusion matrix, we limited ourselves to 1,000 images that we clustered into the following 10 clusters: Oenochoe, Sheep, Hawks, Eagles, Horses, Knives, Planes, Seagulls, Resplendent Leaves, and Pink Roses. Table 3 gives confusion matrix for the 10 clusters in which, due to limited space, we refer to our clusters respectively as $c1, c2 \ldots c10$, and omit the percentage sign (%). The value in the entry corresponding to the row $c1$ and the column $c2$, for example, represents the percentage of images that should be classified as members of $c1$ that have been assigned to $c2$. First, we notice that values on the diagonal are very high, which means a high number of correctly classified images. We notice also that most of the remaining values of this matrix are zero, which indicates a low percentage of misclassified images. Finally, we notice that our matrix presents a kind of symmetry for some clusters. For example, some images that should be assigned to the cluster Sheep have been misclassified and assigned to Horses, and symmetrically, some of the images that belong to Horses have been misclassified and assigned to Sheep. This can be explained by the fact that those two clusters are near each other (visually and semantically), and hence, it is completely natural that each of them tries to attract images that belong to its neighbor. However, confusion rate is very low and does not exceed 4% in the worst case, which means that our classification is very precise.

Future Trends and Conclusion

In this chapter, we presented a new mechanism for combining visual features with textual annotation in an attempt to narrow the semantic gap encountered by most of the current image retrieval engines. First, we annotate images with keywords according to different abstraction levels. This is done using a thesaurus that we created beforehand. Second, we create an index by hierarchically grouping images. Finally, we use this index to develop three retrieval mechanisms as well as some other applications. Experimental results show that our method of combining visual features with keywords always achieves better results than relying on each of them alone. This is completely natural, because each image possesses some visual characteristics that are described better with visual features but simultaneously depicts some semantic events that are described better with keywords.

We are aware that considerable work needs to be done before we become able to retrieve images according to both low-level and high-level concepts. Since the text describes high-level concepts better than image content, there is first a vital need to develop good algorithms for automatic image annotation. Indeed, most current techniques for semantics-based image retrieval rely on keywords. Those keywords either are added manually to images or extracted from the text and titles surrounding them. Manual annotation of images is fraught with a lot of problems; it depends greatly on the annotator, because two people may associate different keywords with the same image. As for keyword extraction from text surrounding the images, existing algorithms are still unable to do it with enough precision.

A second promising direction for semantics-based retrieval is based on automatic concept extraction from images. Rather than representing images with keywords, this technique consists of trying to find a combination of visual feature values that better represent each high-level concept. For example, a lemon can be represented as an object with a round shape and a yellow color. This description then is translated into visual feature values and after that used to perform high-level retrieval. In Li and Wang (2003), for example, the authors train the system with hundreds of statistical models, each representing a given concept. In Chaouch (2005), the authors create some concepts based on color and shape and apply them to perform concept-based retrieval. Even though it seems to be an objective and very promising approach, automatic concept extraction cannot be applied to extracting complex concepts. It may be applied to detect simple concepts such as the presence of some objects in the image. However, it is still unclear how it could be applied to extracting more complex concepts such as emotions.

References

Barnard, K., & Forsyth, D. (2001). Learning the semantics of words and pictures. In *Proceedings of the IEEE ICCV*, Vancouver, Canada.

Barnard, K., et al. (2003). Matching words and pictures. *Journal of Machine Learning Research, 3*, 1107–1135.

Beckmann, N., Kriegel, H. P., Schneider, R., & Seeger, B. (1990). The R*-tree: An efficient and robust access method for points and rectangles. In *Proceedings of the ACM SIGMOD*, Atlantic City, NJ.

Benitez, A. B., & Chang, S. F. (2002). Perceptual knowledge construction from annotated image collections. In *Proceedings of the IEEE ICME*, Lausanne, Switzerland.

Berchtold, S., Keim, D. A., & Kriegel, H. P. (1996). The X-tree: An index structure for high-dimensional data. In *Proceedings of the International Conference on VLDB*, Bombay, India.

Chaouch, Z. (2005). *Annotation d'images* [master's thesis]. Canada: Université de Sherbrooke.

Charikar, M., Chekur, C., Feder, T., & Motwani, R. (1997). Incremental clustering and dynamic information retrieval. In *Proceedings of the ACM Symposium on Theory of Computing*.

Cox, I. J., Miller, M. L., Minka, T. P., & Papathomas, T. V. (2000). The Bayesian image retrieval system, PicHunter: Theory, implementation, and psychophysical experiments. *IEEE Transactions on Image Processing, 9*(1), 20–37.

Fidal, R. (1997). The image retrieval task: Implications for the design and evaluation of image databases. *The New Review of Hypermedia and Multimedia, 3*, 181–199.

Friedman, J. H., Bentley, J. L., & Finkel, R. A. (1997). An algorithm for finding best matches in logarithmic expected time. *ACM Transactions on Mathematical Software, 3*(3), 209–226.

Gevers, T., & Smeulders, A. W. M. (1999). The PicToSeek WWW image search system. In *Proceedings of the IEEE International Conference on Multimedia Computing and Systems*, Florence, Italy.

Grosky, W. I., & Zhao, R. (2001). Improved text-based Web document retrieval using visual features. In *Proceedings of the International Conference on Integration of Multimedia Contents*, Gwangju, Korea.

Guttman, A. (1984). *R-trees: A dynamic index structure for spatial searching*. In *Proceedings of the ACM SIGMOD*, Boston.

Hofmann, T. (1999). Probabilistic latent semantic indexing. In *Proceedings of the ACM Conference on Research and Development in Information Retrieval*, Berkeley, CA.

Lu, Y., Hu, C., Zhu, X., Zhang, H. J., & Yang, Q. (2000). A unified semantics and feature based image retrieval technique using relevance feedback. In *Proceedings of the ACM Multimedia*, Los Angeles.

Katamaya, N., & Satoh, S. (1997). *The SR-tree: An index structure for high-dimensional nearest neighbor queries*. In *Proceedings of the ACM SIGMOD*, Tuscon, AZ.

Kherfi, M. L., Brahmi, D., & Ziou, D. (2004a). Combining visual features with semantics for a more efficient image retrieval. In *Proceedings of the IEEE ICPR*, Cambridge, UK.

Kherfi, M. L., Ziou, D., & Bernardi, A. (2003). Combining positive and negative examples in relevance feedback for content-based image retrieval. *Journal of Visual Communication and Image Representation, 14*(4), 428–457.

Kherfi, M. L., Ziou, D., & Bernardi, A. (2004b). Image retrieval from the World Wide Web: Issues, techniques and systems. *ACM Computing Surveys, 36*(1), 35–67.

Kherfi, M. L., & Ziou, D. (2006). Relevance feedback for CBIR: A new approach based on probabilistic feature weighting with positive and negative examples. *IEEE Transactions on Image Processing, 15*(4), 1017–1030.

Lew, M. S. (2000). Next generation Web searches for visual content. *IEEE Computer, 33*(11), 46–53.

Li, J., & Wang, J. Z. (2003). Automatic linguistic indexing of pictures by a statistical modeling approach. *IEEE Transactions on PAMI, 25*(9), 1075–1088.

Perrin, A. (1988). Toward a unified theory of similarity and recognition. *Psychological Review, 95*(1), 124–150.

Robinson, J. T. (1981). The K-D-B-tree: A search structure for large multidimensional dynamic indexes. In *Proceedings of the ACM SIGMOD*, Ann Arbor, MI.

Rui, Y., Huang, T. S., & Chang, S. F. (1999). Image retrieval: Current techniques, promising directions, and open issues. *Journal of Visual Communication and Image Representation, 10*(1), 39–62.

Sclaroff, S., Taycher, L., & La Cascia, M. (1997). Image rover: A content-based image browser for the World Wide Web. In *Proceedings of the IEEE Workshop on Content-Based Access of Image and Video Libraries*, Puerto Rico.

Smeulders, A. W. M., Worring, M., Santini, S., Gupta, A., & Jain R. (2000). Content-based image retrieval at the end of the early years. *IEEE Transactions on PAMI, 22*(12), 1349–1380.

Smith, J. R., & Chang, S. F. (1997). An image and video search engine for the World-Wide Web. In *Proceedings of the IS&T/SPIE Conference on Storage and Retrieval for Image and Video Databases*, San Jose, CA.

White, D. A., & Jain, R. (1996). Similarity indexing with the SS-tree. In *Proceedings of the International Conference on Data Engineering*, New Orleans, LA.

Zhou, X. S., & Huang, T. S. (2002). Unifying keywords and visual contents in image retrieval. *IEEE Multimedia, 9*(2), 23–33.

Chapter XVI

Semantic Multimedia Information Analysis for Retrieval Applications

João Magalhães, Imperial College London, UK

Stefan Rüger, Imperial College London, UK

Abstract

Most of the research in multimedia retrieval applications has focused on retrieval by content or retrieval by example. Since the classical review by Smeulders, Worring, Santini, Gupta, and Jain (2000), a new interest has grown immensely in the multimedia information retrieval community: retrieval by semantics. This exciting new research area arises as a combination of multimedia understanding, information extraction, information retrieval, and digital libraries. This chapter presents a comprehensive review of analysis algorithms in order to extract semantic information from multimedia content. We discuss statistical approaches to analyze images and video content and conclude with a discussion regarding the described methods.

Introduction: Multimedia Analysis

The growing interest in managing multimedia collections effectively and efficiently has created new research interest that arises as a combination of multimedia understanding, information extraction, information retrieval, and digital libraries. This growing interest has resulted in the creation of a video retrieval track in TREC conference series in parallel with the text retrieval track (TRECVID, 2004).

Figure 1 illustrates a simplified multimedia information retrieval application composed by a multimedia database, analysis algorithms, a description database, and a user interface application. Analysis algorithms extract features from multimedia content and store them as descriptions of that content. A user then deploys these indexing descriptions in order to search the multimedia database. A semantic multimedia information retrieval application (Figure 1) differs eminently from traditional retrieval applications on the low-level analysis algorithms; its algorithms are responsible for extracting semantic information used to index multimedia content by its semantic. Multimedia content can be indexed in many ways, and each index can refer to different modalities and/or parts of the multimedia piece. Multimedia content is composed of the visual track, sound track, speech track, and text. All these modalities are arranged temporally to provide a meaningful way to transmit information and/or entertainment. The way video documents are temporally structured can be distinguished in two levels: semantic and syntactic structure (Figure 2).

At the syntactic level, the video is segmented into shots (visual or audio) that form a uniform segment (e.g., visually similar frames); representative key-frames are extracted from each shot, and scenes group neighboring similar shots into a single segment. The segmentation of video into its syntactic structure of video has been studied widely (Brunelli, Mich, & Modena, 1999; Wang, Liu, & Huang, 2000).

Figure 1. A typical multimedia information retrieval application

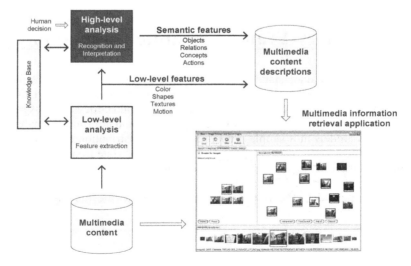

Figure 2. Syntactic and semantic structure of video

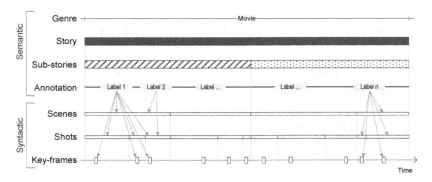

At the semantic level, annotations of the key-frames and shots with a set of labels indicate the presence of semantic entities, their relations, and attributes (agent, object, event, concept, state, place, and time (see Benitez et al., 2002, for details). Further analysis allows the discovery of logical sub-units (e.g., substory or subnarrative), logical units (e.g., a movie), and genres. A recent review of multimedia semantic indexing has been published by Snoek and Worring (2005).

The scope of this chapter is the family of semantic-multimedia analysis algorithms that automate the multimedia semantic annotation process. In the following sections, we will review papers on multimedia-semantic analysis: semantic annotation of key-frame images, shots, and scenes. The semantic analysis at the shot and scene level considers independently the audio and visual modalities and then the multi-modal semantic analysis. Due to the scope of this book, we will give more emphasis to the visual part than to the audio part of the multimedia analysis and will not cover the temporal analysis of logical substories, stories, and genres.

Key-Frame Semantic Annotation

Image analysis and understanding is one of the oldest fields in pattern recognition and artificial intelligence. A lot of research has been done since (Marr, 1983), culminating in the modern reference texts by Forsyth and Ponce (2003) and Hartley and Zisserman (2004). In the following sections we discuss different types of visual information analysis algorithms: single class models fit a simple probability density distribution to each label; translation models define a visual vocabulary and a method to translate from this vocabulary to keywords; hierarchical and network models explore the interdependence of image elements (regions or tiles) and model its structure; knowledge-based models improve the model's accuracy by including other sources of knowledge besides the training data (e.g., a linguistic database such as WordNet).

Single Class Models

A direct approach to the semantic analysis of multimedia is to learn a class-conditional probability distribution $p(w \mid x)$ of each single keyword w of the semantic vocabulary, given its training data x (see Figure 3). This distribution can be obtained by using Bayes' law

$$p(w \mid x) = \frac{p(x \mid w) p(w)}{p(x)}.$$

The data probability $p(x)$ and the keyword probability $p(w)$ can be computed straightforward, and the $p(x \mid w)$ can be computed with very different data density distribution models.

Several techniques to model the $p(x \mid w)$ with a simple density distribution have been proposed: Yavlinsky, Schofield, and Rüger (2005) used a nonparametric distribution, Carneiro and Vasconcelos (2005) a semi-parametric density estimation, Westerveld and de Vries (2003) a finite mixture of Gaussians, and Mori, Takahashi, and Oka (1999), Vailaya, Figueiredo, Jain, and Zhang (1999), and Vailaya, Figueiredo, Jain, and Zhang (2001) different flavors of vector quantization techniques.

Yavlinsky et al. (2005) modeled the probability density of images, given keywords as a nonparametric density smoothed by two kernels: a Gaussian kernel and an Earth Mover's Distance kernel. They used both global and 3×3 tile color and texture features. The best reported mean average precision (MAP) results with tiles achieved 28.6% MAP with the dataset of Duygulu, Barnard, de Freitas, and Forsyth (2002) and 9.2% with a Getty Images dataset.

Yavlinsky et al. (2005) showed that a simple nonparametric statistical distribution can perform as well or better than many more sophisticated techniques (e.g., translation models). However, the nonparametric density nature of their framework makes the task of running the

Figure 3. Inference of single class models

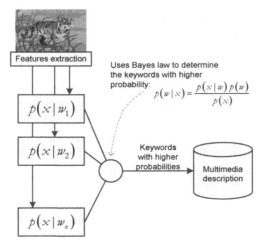

model on new data very complex. The model is the entire dataset meaning that the demands on CPU and memory increase with the training data.

Westerveld and de Vries (2003) used a finite-mixture density distribution with a fixed number of components to model a subset of the DCT coefficients:

$$p(x \mid \theta) = \sum_{m=1}^{k} \alpha_m p(x \mid \mu_m, \sigma_m^2),$$

in which k is the number of components, θ represents the complete set of model parameters with mean μ_m, covariance σ_m^2, and component prior α_m. The component priors have the constraints $\alpha_1, ..., \alpha_k \geq 0$ and $\sum_{m=1}^{k} \alpha_m = 1$. Westerveld (2003) tested several scenarios to evaluate the effect (a) of the number of mixture components, (b) of using different numbers of DCT coefficients (luminance and chrominance), and (c) of adding the coordinates of the DCT coefficients to the feature vectors. The two first factors produced varying results, and optimal points were found experimentally. The third tested aspect, the presence of the coefficients position information, did not modify the results.

Marrying the two previous approaches, Carneiro and Vasconcelos (2005) deployed a hierarchy of semi-parametric mixtures to model $p(x \mid w)$ using a subset of the DCT coefficients as low-level features. Vasconcelos and Lippman (2000) had already examined the same framework in a content-based retrieval system.

The hierarchy of mixtures proposed by Vasconcelos and Lippman (1998) can model data at different levels of granularity with a finite mixture of Gaussians. At each hierarchical level l, the number of each mixture component k^l differs by one from adjacent levels. The hierarchy of mixtures is expressed as:

$$p(x \mid w_i) = \frac{1}{D} \sum_{m=1}^{k^l} \alpha_{i,m}^l p(x \mid \theta_{i,m}^l).$$

The level $l=1$ corresponds to the coarsest characterization. The more detailed hierarchy level consists of a nonparametric distribution with a kernel placed on top of each sample. The only restriction on the model is that if node m of level $l+1$ is a child of node n of level l, then they are both children of node p of level $l-1$. The EM algorithm computes the mixture parameters at level l, given the knowledge of the parameters at level $l+1$, forcing the previous restriction.

Carneiro and Vasconcelos (2005) report the best published retrieval MAP of 31% with the dataset of Duygulu et al. (2002). Even though we cannot dissociate this result from the pair of features and statistical model, the hierarchy of mixtures appears to be a very powerful density distribution technique.

Even though the approaches by Carneiro and Vasconcelos (2005) and Westerveld and de Vries (2003) are similar, the differences make it difficult to do a fair comparison. The DCT features are used in a different way, and the semi-parametric hierarchy of mixtures can model classes with very few training examples.

The relationship between finite-mixture density modeling and vector quantization is a well-studied subject (see Hastie, Tibshirani, & Friedman, 2001). One of the applications of vector quantization to image retrieval and annotation was realized by Mori et al. (1999). Given the training data of a keyword, they divide the images into tiles and apply vector quantization to the image tiles in order to extract the codebook used to estimate the $p(x \mid w)$ density distribution. Later, they use a model of word co-occurrence on the image tiles in order to label the image. The words with the higher sum of probabilities across the different tiles are the ones assigned to that image.

Vailaya et al. (1999) and Vailaya et al. (2001) describe a Bayesian framework using a codebook to estimate the density distribution of each keyword. They show that the Minimum Description Length criterion selects the optimal size of the codebook extracted from the vector quantizer. The features are extracted from the global image, and there is no image tiling. The use of the MDL criterion makes this framework quite elegant and defines a statistical criterion to select every model parameter and without any user-defined parameters.

Translation Models

All of the previous approaches employ a direct model to estimate $p(x \mid w)$ with image global features and/or image tiles features. In contrast to this, the vector quantization (usually k-means) approach generates a codebook of image regions or image tiles (depending on the segmentation solution). The problem then is formulated as a translation problem between two representations of the same entity: English-Esperanto, word-blob codebook, or word-tile codebook.

Inspired by machine translation research, Duygulu et al. (2002) developed a method of annotating image regions with words. First, regions are created using a segmentation algorithm like normalized cuts (Shi & Malik, 2000). For each region, features are computed, and then blobs are generated by clustering the regional image features across an image collection. The problem then is formulated as learning the correspondence between the discrete vocabulary

Figure 4. Translation models

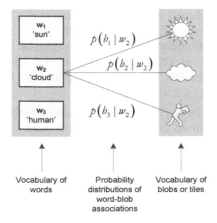

of blobs and the image keywords. The model consists of a mixture of correspondences for each word of each image in the collection:

$$p\left(w_j \mid I_n\right) = \sum_{i \in \{\text{blobs in } I_n\}} p\left(a_{nj} = i\right) p\left(w = w_{nj} \mid b = b_i\right),$$

in which $p(a_{nj} = i)$ expresses the probability of associating word j to blob i in image n, and $p(w = w_{nj} \mid b = b_i)$ is the probability of obtaining an instance of word w given an instance of blob b. These two probability distributions are estimated with the EM algorithm. The authors refined the lexicon by clustering indistinguishable words and ignoring the words with probabilities $p(w \mid b)$ below a given threshold.

The machine translation approach, the thorough experiments, and the dataset form strong points of this chapter (Duygulu et al., 2002). This dataset is nowadays a reference, and thorough experiments showed that (a) their method could predict numerous words with high accuracy, (b) increasing the probability threshold improved precision but reduced recall, and (c) the word clustering improved recall and precision.

Following a translation model, Jeon, Lavrenko, and Manmatha (2003), Lavrenko, Manmatha, and Jeon (2003), and Feng, Lavrenko, and Manmatha (2004) studied a model in which blob features $b_I^{(r)}$ of an image I are assumed to be conditionally independent of keywords w_i, that is:

$$p\left(w_i, b_I\right) = \sum_{J \in D} P(J) P\left(w_i \mid J\right) P\left(b_I \mid J\right)$$

$$= \sum_{J \in D} P(J) P\left(w_i \mid J\right) \prod_{r \in I} P\left(b_I^{(r)} \mid J\right)$$

Note that $b_I^{(r)}$ and w_i are conditionally independent, given the image collection D and that $J \in D$ act as the hidden variables that generated the two distinct representations of the same process (words and features).

Jeon et al. (2003) recast the image annotation as a cross-lingual information retrieval problem, applying a cross-media relevance model based on a discrete codebook of regions. Lavrenko et al. (2003) continued their previous work (Jeon et al., 2003) and used continuous probability density functions $p(b_I^{(r)} \mid J)$ to describe the process of generating blob features and to avoid the loss of information related to the generation of the codebook. Extending their previous work, Feng et al. (2004) replaced blobs with tiles and modeled image keywords with a Bernoulli distribution. This last work reports their best results, a MAP of 30%, with a Corel dataset (Duygulu et al., 2002).

Latent semantic analysis is another technique of text analysis and indexing; it looks at patterns of word distributions (specifically, word co-occurrence) across a set of documents (Deerwester, Dumais, Furmas, Landauer, & Harshman, 1990). A matrix M of word occurrences in documents is filled with each word frequency in each document. The singular value decomposition (SVD) of matrix M gives the transformation to a singular space in which projected documents can be compared efficiently.

Hierarchical Models

The aforementioned approaches assumed a minimal relation among the various elements of an image (blobs or tiles). This section and the following section will review methods that consider a hierarchical relation or an interdependence relation among the elements of an image (words and blobs or tiles).

Barnard and Forsyth (2001) studied a generative hierarchical aspect model, which was inspired by Hofmann and Puzicha's (1998) hierarchical clustering/aspect model. The data are assumed to be generated by a fixed hierarchy of nodes in which the leaves of the hierarchy correspond to soft clusters. Mathematically, the process for generating the set of observations O associated with an image I can be described by:

$$p(O \mid I) = \sum_{c} \left(p(c) \prod_{n \in O} \left(\sum_{l} p(o \mid l, c) p(l \mid c, I) \right) \right), \quad O = \{w_1, ..., w_n, b_1, ..., b_m\},$$

in which c indexes the clusters, o indexes words and blobs, and l indexes the levels of the hierarchy. The level and the cluster uniquely specify a node of the hierarchy. Hence, the probability of an observation $p(o \mid l, c)$ is conditionally independent given a node in the tree. In the case of words, $p(o \mid l, c)$ assumes a tabular form, and in the case of blobs, a single Gaussian models the regions' features. The model is estimated with the EM algorithm.

Blei and Jordan (2003) describe three hierarchical mixture models to annotate image data, culminating in the correspondence latent Dirichlet allocation model. It specifies the following joint distribution of regions, words, and latent variables (θ, z, y):

$$p(r, w, \theta, z, y) = p(\theta \mid \alpha) \left(\prod_{n=1}^{N} p(z_n \mid \theta) p(r_n \mid z_n, \mu, \sigma) \right)$$
$$\cdot \left(\prod_{m=1}^{M} p(y_m \mid N) p(w_m \mid y_m, z, \beta) \right).$$

This model assumes that a Dirichlet distribution θ (with α as its parameter) generates a mixture of latent factors: z and y. Image regions r_n are modeled with Gaussians with mean μ and covariance σ, in which words w_n follow a multinomial distribution with a β parameter.

This mixture of latent factors then is used to generate words (y variable) and regions (z variable). The EM algorithm estimates this model, and the inference of $p(w \mid r)$ is carried out by variational inference. The correspondence latent Dirichlet allocation model provides a clean probabilistic model for annotating images with multiple keywords. It combines the advantages of probabilistic clustering for dimensionality reduction with an explicit model of the conditional distribution from which image keywords are generated.

Li and Wang (2003) characterize the images with a hierarchical approach at multiple tiling granularities (i.e., each tile in each hierarchical level is subdivided into smaller sub-tiles). A color and texture feature vector represents each tile. The texture features represent the energy in high-frequency bands of wavelet transforms. They represent each keyword separately with two-dimensional, multi-resolution hidden Markov models. This method achieves a certain degree of scale invariance due to the hierarchical tiling process and the two-dimensional multiresolution hidden Markov model.

Network Models

In semantic-multimedia analysis, concepts are interdependent; for example, if a house is detected in a scene, then the probability of existing windows and doors in the scene are boosted, and vice-versa. In other words, when inferring the probability of a set of interdependent random variables, their probabilities are modified iteratively until an optimal point is reached (to avoid instability, the loops must exist over a large set of random variables [Pearl, 1988]). Most of the papers discussed next model keywords as a set of interdependent random variables connected in a probabilistic network.

Various graphical models have been implemented in computer vision to model the appearance, spatial relations, and co-occurrence of local parts. Markov random fields and hidden Markov models are the most common generative models that learn the joint probability of the observed data (X) and the corresponding labels (Y). These models divide the image into tiles or regions (other approaches use contour directions, but these are outside the scope of our discussion). A probabilistic network then models this low-level division in which each node corresponds to one of these tiles or regions and its label. The relation among nodes depends on the selected neighboring method. Markov random fields can be expressed as:

$$P(x,y) = \frac{1}{Z} \cdot \prod_i \left(\phi_i\left(x_i, y_i\right) \cdot \prod_{j \in N_i} \varphi_{i,j}\left(y_i, y_j\right) \right),$$

in which i indexes the image's tiles, j indexes the neighbors of the current i tile, ϕ_i is the potential function of the current tile x_i, and its possible labels y_i, and $\varphi_{i,j}$ are the interaction functions between the current tile label and its neighbors. Figure 5 illustrates the Markov random field framework.

The Markov condition implies that a given node only depends on its neighboring nodes. This condition constitutes a drawback for these models, because only local relationships are incorporated into the model. This makes it highly unsuitable for capturing long-range relations or global characteristics.

Figure 5. Two types of random fields

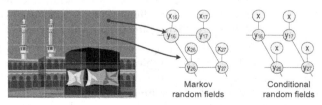

Markov random fields   Conditional random fields

In order to circumvent this limitation, Kumar and Herbert (2003a) propose a multi-scale random field (MSRF) as a prior model on the class labels on the image sites. This model implements a probabilistic network that can be approximated by a 2D hierarchical structure such as a 2D-tree. A multiscale feature vector captures the local dependencies in the data. The distribution of the multiscale feature vectors is modeled as a mixture of Gaussians. The features were selected specifically to detect human-made structures, which are the only types of objects that are detected.

Kumar and Herbert's (2003) second approach to this problem is based on discriminative random fields, an approach inspired on conditional random fields (CRF). CRFs, defined by Lafferty, McCallum, and Pereira (2001), are graphical models, initially for text information extraction, that are meant for visual information analysis in this approach. More generally, a CRF is a sequence-modeling framework based on the conditional probability of the entire sequence of labels (Y), given the all image (X). CRFs have the following mathematical form:

$$P(y|x) = \frac{1}{Z} \cdot \prod_i \left(\phi_i(y_i, x) \cdot \prod_{j \in N_i} \varphi_{i,j}(y_i, y_j; x) \right),$$

in which i indexes the image's tiles, j indexes the neighbors of the current i tile, ϕ_i is the association potential between the current tile and the image label, and $\varphi_{i,j}$ is the interaction potential between the current tile and its neighbors (note that it is also dependent on the image label). Figure 5 illustrates the conditional random field framework. The authors showed that this last approach outperformed their initial proposal of a multiscale random field as well as the more traditional MRF solution in the task of detecting human-made structures.

He, Zemel, and Carreira-Perpiñán (2004) combine the use of a conditional random field and data at multiple scales. Their multiscale conditional random field (mCRF) is a product of individual models, each model providing labeling information from various aspects of the image: a classifier that looks at local image statistics; regional label features that look at local label patterns; and global label features that look at large, coarse label patterns. The mCRF is shown to detect several types of concepts (i.e., sky, water, snow, vegetation, ground, hippopotamus, and bear) with classification rates better than a traditional Markov random field.

Quattoni, Collins, and Darrell (2004) extend the CRF framework to incorporate hidden variables and combine class-conditional CRFs into a unified framework for part-based object recognition. The features are extracted from special regions that are obtained with the scale-invariant feature transform or SIFT (Lowe, 1999). The SIFT detector finds points in locations at scales in which there is a significant amount of variation. Once a point of interest is found, the region around it is extracted at the appropriate scale. The features from this region then are computed and plugged into the CRF framework. The advantage of this method is that it needs fewer regions by eliminating redundant regions and selecting the ones with more energy on high-frequency bands.

One should note that all these approaches require a ground truth at the level of the image's tiles/regions as is common in computer vision. This is not what is found traditionally in multimedia information retrieval datasets in which the ground truth exists rather at a global level.

Knowledge-Based Models

The previous methods have only visual features as training data to create the statistical models in the form of a probabilistic network. Most of the time, these training data are limited, and the model's accuracy can be improved by other sources of knowledge. Prior knowledge can be added to a model either by a human expert who states the relations between concept variables (nodes in a probabilistic network) or by an external knowledge base in order to infer the concept relations (e.g., with a linguistic database such as WordNet) (Figure 6).

Tansley (2000) introduces a multimedia thesaurus in which media content is associated with appropriate concepts in a semantic layer composed by a network of concepts and their relations. The process of building the semantic layer uses Latent Semantic Indexing to connect images to their corresponding concepts, and a measure of each correspondence (image concept) is taken from this process. After that, unlabeled images (test images) are annotated by comparing them with the training images using a k-nearest-neighbor classifier. Since the concepts' interdependences are represented in the semantic layer, the concepts' probability computed by the classifier are modified by the others concepts.

Other authors have explored not only the statistical interdependence of context and objects but also have used other knowledge that is not present in multimedia data, which humans use to understand (or predict) new data. Srikanth, Varner, Bowden, and Moldovan (2005) incorporated linguistic knowledge from WordNet (Miller, 1995) in order to deduce a hierarchy of terms from the annotations. They generate a visual vocabulary based on the semantics of the annotation words and their hierarchical organization in the WordNet ontology.

Benitez and Chang (2002) and Benitez (2005) took this idea further and suggested a media ontology (MediaNet) to help to discover, summarize, and measure knowledge from annotated images in the form of image clusters, word senses, and relationships among them. MediaNet, a Bayesian network-based multimedia knowledge representation framework, is composed by a network of concepts, their relations, and media exemplifying concepts and relationships. The MediaNet integrates classifiers in order to discover statistical rela-

Figure 6. Knowledge-based models

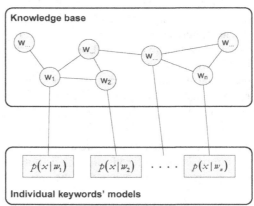

tionships among concepts. WordNet is used to process image annotations by stripping out unnecessary information. The summarization process implements a series of strategies to improve the images' description qualities, for example using WordNet and image clusters to disambiguate annotation terms (images in the same clusters tend to have similar textual descriptions). Benitez (2005) also proposes a set of measures to evaluate the knowledge consistency, completeness, and conciseness.

Tansley (2000) used a network at the concept level, and Benitez (2005) used the MediaNet network to capture the relations at both concept and feature levels. In addition, Benitez (2005) utilized WordNet, which captures human knowledge that is not entirely present in multimedia data.

Summary

The described algorithms vary in many different aspects such as in their low-level features, segmentation methods, feature representation, modeling complexity, or required data. While some concepts require a lot of data to estimate its model (e.g., a car), others are very simple and require just a few examples (e.g., sky). So, we advocate that different approaches should be used for different concept complexities.

Single-class models assume that concepts are independent and that each concept has its own model. These are the simplest models that can be used and the ones with better accuracy (e.g., Yavlinsky et al., 2005).

Translation models, hierarchical models, and network models capture a certain degree of the concept's interdependence (co-occurrence) from the information present in the training data. The difference between the models is linked to the degree of interdependence that can be represented by the model. In practice, when interdependencies information is incorporated in the model, it also inserts noise in the form of false interdependencies, which causes a decrease in performance. So, the theoretical advantage of these models is in practice reduced by this effect.

All these models rely exclusively on visual low-level features in order to capture complex human concepts and to correctly predict new unlabeled data. Most of the time, the training data are limited, and the model's accuracy can be improved by other sources of knowledge. Srikanth et al. (2005) and Benitez (2005) are two of the few proposals that exploit prior knowledge that is external to the training data in order to capture the interdependent (co-occurrence) nature of concepts.

At this time, knowledge-based models seem to be the most promising semantic analysis algorithms for information retrieval. Text information retrieval already has shown great improvement over exclusively statistical models when external linguistic knowledge was used (Harabagiu et al., 2000). Multimedia retrieval will go through a similar progress but at a slower pace, because there is no multimedia ontology that offers the same knowledge base as WordNet offers to linguistic text processing.

Shot and Scene Semantic Annotation

Shot and scene semantic analysis introduces the time dimension to the problem at hand. The time dimension adds temporal frames, resulting in more information to help the analysis. To take advantage of the sequential nature of the data, the natural choices of algorithms are based on hierarchical models or network models. The section is organized by modality, and within each modality, we don't detail the algorithms by technique due to space constraints. This way, we shed some light on multimodality shot and scene semantic analysis and keep the chapter's emphasis on visual information analysis.

Audio Analysis

Audio analysis becomes a very important part of the multimodal analysis task when processing TV news, movies, sport videos, and so forth. Various types of audio can populate the sound track of a multimedia document, the most common types being speech, music, and silence. Lu, Zhang, and Jiang (2002) propose methods to segment audio and to classify each segment as speech, music, silence, and environment sound. A k-nearest neighbor model is used at the frame level followed by vector quantization to discriminate between speech and nonspeech. A set of threshold-based rules is used in order to discriminate among silence, music, and environment sound. The authors also describe a speaker change detection algorithm based on Gaussian-mixture models (GMM); this algorithm continuously compares the model of the present speaker's speech with a model that is created dynamically from the current audio frame. After a speaker change has been detected, the new GMM replaces the current speaker's GMM.

In most TV programs and sport videos, sound events do not overlap, but in narratives (movies and soap operas), these events frequently occur simultaneously. To address this problem, Akutsu, Hamada, and Tonomura (1998) present an audio-based approach to video indexing by detecting speech and music independently, even when they occur simultaneously. Their framework is based on a set of heuristics over features histograms and corresponding thresholds. With a similar goal, Naphade and Huang (2000) define a generic statistical framework based on hidden Markov models (Rabiner, 1989) in order to classify audio segments into speech, silence, music, and miscellaneous and their co-occurrences. By creating an HMM for each class and every combination of classes, the authors achieved a generic framework that is capable of modeling various audio events with high accuracy.

Another important audio analysis task is the classification of the musical genre of a particular audio segment. This can capture the type of emotion that the director wants to communicate (e.g., stress, anxiety, happiness). Tzanetakis and Cook (2002) describe their work on categorizing music as rock, dance, pop, metal, classical, blues, country, hip-hop, reggae, or jazz (jazz and classical music had more subcategories). In addition to the traditional audio features, they also use special features to capture rhythmic characteristics and apply simple statistical models such as GMM and KNN to model each class' feature histogram. Interestingly, the best reported classification precision (61%) is in the same range as human performance for genre classification (70%).

All these approaches work as a single class model of individual classes/keywords. Note that the hidden Markov model is, in fact, a probabilistic network for modeling a single temporal event that corresponds to a given concept/keyword. So, even though it is a network model, it is used as a single class model.

Visual Analysis

Many of the visual video analysis methods are based on heuristics that are deduced empirically. Statistical methods are more common when considering multimodal analysis. Most of the following papers explore the temporal evolution of features to semantically analyze video content (e.g., shot classification, logical units, etc.). Video visual analysis algorithms are of two types: (a) heuristics-based, in which a set of threshold rules decides the content class, and (b) statistical algorithms that are similar to the ones described in Section 2.

Heuristic methods rely on deterministic rules that were defined in some empirical way. These methods monitor histograms, and events are detected if the histogram triggers a given rule (usually a threshold). They are particularly adequate for sport videos because broadcast TV follows a set of video production rules that result in well-defined semantic structures that ease the analysis of the sports videos. Several papers have been published on sports video analysis, such as football, basketball and tennis, in order to detect semantic events and to semantically classify each shot (Li & Sezan, 2003; Luo & Huang, 2003; Tan, Saur, Kulkarni, & Ramadge, 2000).

Tan et al. (2000) introduced a model for estimating camera movements (pan, tilt, and zoom) from the motion vectors of compressed video. The authors further showed how camera motion histograms could be used to discriminate various basketball shots. Prior to this, the video is segmented into shots based on the evolution of the intensity histogram across different frames. Shots are detected if the histogram exceeds a predefined threshold; then, they are discriminated based on (a) the accumulated histogram of camera motion direction (fast breaks and full-court advances), (b) the slope of this histogram (fast breaks or full-court advances), (c) sequence of camera movements (shots at the basket), and (d) persistence of camera motion (close-ups).

Other heuristic methods deploy color histograms, shot duration, and shot sequences to automatically analyze various types of sports such as football (Ekin, Tekalp, & Mehrotra, 2003) and American football (Li & Sezan, 2003).

The statistical approaches reviewed previously can be applied to the visual analysis of video content with the advantage that shapes obtained by segmentation are more accurate due to the time dimension. Also, analyzing several key-frames of the same shot and then combining the results facilitate the identification of semantic entities in a given shot.

Luo and Hwang's (2003) statistical framework tracks objects within a given shot with a dynamic Bayesian network and classifies that shot from a coarse-grain to a fine-grain level. At the course-grain level, a key-frame is extracted from a shot every 0.5 seconds. From these key-frames, motion and global features are extracted, and their temporal evolution is modeled with a hierarchical hidden Markov model (HHMM). Individual HHMMs (a single-class model approach) capture a given semantic shot category. At the fine-grain level analysis, Luo and Hwang (2003) employ object recognition and tracking techniques. After the coarse-grain level analysis, segmentation is performed on the shots to extract visual objects. Then, invariant points are detected in each shape to track the object movement. These points are fed to a dynamic Bayesian network to model detailed events occurring within the shot (e.g., human body movements in a golf game).

Souvannavong, Merialdo, and Huet (2003) used latent semantic analysis to analyze video content. Recall that latent semantic analysis algorithm builds a matrix M of word occurrences in documents, and then the SVD of this matrix is computed to obtain a singular space. The problem with multimedia content is that there is no text corpus (a vocabulary). A vector quantization technique (k-means) returns a codebook of blobs, the vocabulary of blobs from the shots' key-frames. In the singular feature space, a k-nearest-neighbor ($k=20$) and a Gaussian mixture model technique are used to classify new videos. The comparison of the two techniques shows that GMM performs better when there is enough data to correctly estimate the 10 components. The k-nn algorithm has the disadvantages of every nonparametric method—the model is the training data, and for the TRECVID dataset (75,000 key-frames), training can take considerable time.

Multimodal Analysis

In the previous analysis, the audio and visual modalities were considered independently in order to detect semantic entities. These semantic entities are represented in various modalities, capturing different aspects of that same reality. Those modalities contain co-occurring patterns that are synchronized in a given way because they represent the same reality. Thus, synchronization and the strategy to combine the multimodal patterns is the key issue in multimodal analysis. The approaches described in this section explore the multimodality statistics of semantic entities (e.g., pattern synchronization).

Sports video analysis can be greatly improved with multimodal features; for example, the level of excitement expressed by the crowd noise can be a strong indicator of certain events (foul, goal, goal miss, etc). Leonardi, Migliotari, and Prandini (2004) take this into account when designing a multimodal algorithm to detect goals in football videos. A set of visual features from each shot is fed to a Markov chain in order to evaluate their temporal evolution from one shot to the next. The Markov chain has two states that correspond to the goal state and to the nongoal state. The visual analysis returns the positive pair shots, and the shot audio loudness is the criterion to rank the pair shots. Thus, the two modalities never are combined but are used sequentially. Results show that audio and visual modalities together improve the average precision when compared only to the audio case (Leonardi et al., 2004).

In TV news videos, text is the fundamental modality with the most important information. Westerveld, et al. (2003) build on their previous work described previously to analyze the visual part and to add text provided by an Automatic Speech Recognition (ASR) system. The authors further propose a visual dynamic model to capture the visual temporal characteristics. This model is based on the Gaussian mixture model estimated from the DCT blocks of the frames around each key-frame in the range of 0.5 seconds. In this way, the most significant moving regions are represented by this model with an evident applicability to object tracking. The text retrieval model evaluates a given $Shot_i$ for the queried keywords $Q = q_1, q_2, q_3, ...\}$:

$$RSV\left(Shot_i\right) = \frac{1}{|Q|} \sum_{k=1}^{|Q|} \log\left(\lambda_{Shot}\, p\left(q_k \mid Shot_i\right) + {}_{Scene}\, p\left(q_k \mid Scene_i\right) + \lambda_{Coll}\, p\left(q_k\right)\right).$$

This measure evaluates the probability that one or more queried keywords appear in the evaluated shot, $p(q_k \mid Shot_i)$, or in the scene, $p(q_k \mid Scene_i)$, under the prior $p(q_k)$. The λ variables correspond to the probabilities of corresponding weights. This function, inspired by language models, creates the scene-shot structure of video content. The visual model and the text model are combined under the assumption that they are independent; thus, the probabilities are simply multiplied. The results with both modalities are reported to be better than using just one.

Naphade and Huang (2001) characterize single-modal concepts (e.g., indoor/outdoor, forest, sky, water) and multimodal concepts (e.g., explosions, rocket launches) with Bayesian networks. The visual part is segmented into shots (Naphade et al., 1998), and from each key-frame, a set of low-level features is extracted (color, texture, blobs, and motion). These features then are used to estimate a Gaussian mixture model of multimedia concepts at region level and then at frame level. The audio part is analyzed with the authors' algorithm described previously (Naphade & Huang, 2000). The outputs of these classifiers are then combined in a Bayesian network in order to improve concept detection. Their experiments show that the Bayesian network improves the detection performance over individual classifiers. IBM's research by Adams et al. (2003) extend the work of Naphade and Huang (2001) by including text from Automatic Speech Recognition as a third modality and by using Support Vector Machines to combine the classifiers' outputs. The comparison of these two combination strategies showed that SVMs (audio, visual, and text) and Bayesian networks (audio and visual) perform equally well. However, since in the latter case, speech information was ignored, one might expect that Bayesian networks can, in fact, perform better. More details about IBM's research work can be found in Naphade and Smith (2003), Natsev, Naphade, and Smith (2003), and Tseng, Lin, Naphade, Natsev, and Smith (2003).

The approach by Snoek and Worring (2005) is unique in the way synchronization and time relations between various patterns are modeled explicitly. They propose a multimedia semantic analysis framework based on Allen's (1983) temporal interval relations. Allen showed that in order to maintain temporal knowledge about any two events, only a small set of relations is needed to represent their temporal relations. These relations, now applied to audio and visual patterns, are the following: precedes, meets, overlaps, starts, during, finishes, equals, and no relation. The framework can include context and synchronization of heterogeneous information sources involved in multimodal analysis. Initially, the optimal pattern configuration of temporal relations of a given event is learned from training data by a standard statistical method (maximum entropy, decision trees, and SVMs). New data are classified with the learned model. The authors evaluate the event detection on a soccer video (goal, penalty, yellow card, red card and substitution) and TV news (reporting anchor, monologue, split-view and weather report). The differences among the various classifiers (maximum entropy, decision trees, and SVMs) appear to be not statistically significant.

Summary

When considering video content, a new, very important dimension is added: time. Time adds a lot of redundancy that can be explored effectively in order to achieve a better segmentation and semantic analysis. The most interesting approaches consider time either implicitly (Westerveld et al., 2003) or explicitly (Snoek & Worring, 2005).

Few papers show a deeper level of multimodal combination than Snoek and Worring (2005) and Naphade and Huang (2001). The first explicitly explores the multimodal co-occurrence of patterns resulting from the same event with temporal relations. The latter integrates multimodal patterns in a Bayesian network to explore pattern co-occurrences and concept interdependence.

Natural language processing experts have not yet applied all the techniques from text to the video's extracted speech. Most approaches to extract information from text and combine this with the information extracted from audio and video are all very simple, such as a simple product between the probabilities of various modalities' classifiers.

Conclusion

This chapter reviewed semantic-multimedia analysis algorithms with special emphasis on visual content. Multimedia datasets are important research tools that provide a means for researchers to evaluate various information extraction strategies. The two parts are not separate, because algorithm performances are intrinsically related to the dataset on which they are evaluated.

Major developments in semantic-multimedia analysis algorithms will probably be related to knowledge-based models and multimodal fusion algorithms. Future applications might boost knowledge-based model research by enforcing a limited application domain (i.e., a constrained knowledge base). Examples of such applications are football game summaries and mobile photo albums.

Multimodal analysis algorithms already have proven to be crucial in semantic multimedia analysis. Large developments are expected in this young research area due to the several problems that wait to be fully explored and to the TRECVID conference series that is pushing forward this research area through a standard evaluation and a rich multimedia dataset.

We believe that semantic-multimedia information analysis for retrieval applications has delivered its first promises and that many novel contributions will be done over the next years. To better understand the field, the conceptual organization by different statistical methods presented here allows readers to easily put into context novel approaches to be published in the future.

References

Adams, W. H. et al. (2003). Semantic indexing of multimedia content using visual, audio and text cues. *EURASIP Journal on Applied Signal Processing, 2*, 170–185.

Akutsu, M., Hamada, A., & Tonomura, Y. (1998). Video handling with music and speech detection. *IEEE Multimedia, 5*(3), 17–25.

Allen, J. F. (1983). Maintaining knowledge about temporal intervals. *Communications of the ACM, 26*(11), 832–843.

Barnard, K., & Forsyth, D. A. (2001). Learning the semantics of words and pictures. In *Proceedings of the International Conference on Computer Vision*, Vancouver, Canada.

Benitez, A. (2005). *Multimedia knowledge: Discovery, classification, browsing, and retrieval* [doctoral thesis]. New York: Columbia University.

Benitez, A. B., & Chang, S. F. (2002). Multimedia knowledge integration, summarization and evaluation. In *Proceedings of the International Workshop on Multimedia Data Mining in conjunction with the International Conference on Knowledge Discovery & Data Mining*, Alberta, Canada.

Benitez, A. B. et al. (2002). Semantics of multimedia in MPEG-7. In *Proceedings of the IEEE International Conference on Image Processing, Rochester*, NY.

Blei, D., & Jordan, M. (2003). Modeling annotated data. In *Proceedings of the ACM SIGIR Conference on Research and Development in Information Retrieval*, Toronto, Canada.

Brunelli, R., Mich, O., & Modena, C. M. (1999). A survey on the automatic indexing of video data. *Journal of Visual Communication and Image Representation, 10*(2), 78–112.

Carneiro, G., & Vasconcelos, N. (2005). Formulating semantic image annotation as a supervised learning problem. In *Proceedings of the IEEE Conference on Computer Vision and Pattern Recognition*, San Diego, CA.

Deerwester, S., Dumais, S. T., Furnas, G. W., Landauer, T. K., & Harshman, R. (1990). Indexing by latent semantic analysis. *Journal of the American Society for Information Science, 41*(6), 391–407.

Duygulu, P., Barnard, K., de Freitas, N., & Forsyth, D. (2002). Object recognition as machine translation: Learning a lexicon for a fixed image vocabulary. In *Proceedings of the European Conference on Computer Vision*, Copenhagen, Denmark.

Ekin, A., Tekalp, A. M., & Mehrotra, R. (2003). Automatic video analysis and summarization. *IEEE Transactions on Image Processing, 12*(7), 796–807.

Feng, S. L., Lavrenko, V., & Manmatha, R. (2004). Multiple Bernoulli relevance models for image and video annotation. In *Proceedings of the IEEE Conference on Computer Vision and Pattern Recognition*, Cambridge, UK.

Forsyth, D., & Ponce, J. (2003). *Computer vision: A modern approach*. Prentice Hall.

Harabagiu, S., et al. (2000). Falcon: Boosting knowledge for answer engines. In *Proceedings of the Text Retrieval Conference*, Gaithersburg, MD.

Hartley, R., & Zisserman, A. (2004). *Multiple view geometry in computer vision* (2nd ed.). Cambridge University Press.

Hastie, T., Tibshirani, R., & Friedman, J. (2001). *The elements of statistical learning: Data mining, inference and prediction*. Springer.

He, X., Zemel, R. S., & Carreira-Perpiñán, M. Á. (2004). Multiscale conditional random fields for image labeling. In *Proceedings of the IEEE International Conference on Computer Vision and Pattern Recognition*, Cambridge, UK.

Hofmann, T., & Puzicha, J. (1998). *Statistical models for co-occurrence data* (No. 1635 A. I. Memo). Massachusetts Institute of Technology.

Jeon, J., Lavrenko, V., & Manmatha, R. (2003). Automatic image annotation and retrieval using cross-media relevance models. In *Proceedings of the ACM SIGIR Conference on Research and Development in Information Retrieval*, Toronto, Canada.

Kumar, S., & Herbert, M. (2003a). Discriminative random fields: A discriminative framework for contextual interaction in classification. In *Proceedings of the IEEE International Conference on Computer Vision*, Nice, France.

Kumar, S., & Herbert, M. (2003b). Man-made structure detection in natural images using causal multiscale random field. In *Proceedings of the IEEE International Conference on Computer Vision and Pattern Recognition*, Madison, WI.

Lafferty, J., McCallum, A., & Pereira, F. (2001). Conditional random fields: Probabilistic models for segmenting and labeling sequence data. In *Proceedings of the International Conference on Machine Learning*, San Francisco.

Lavrenko, V., Manmatha, R., & Jeon, J. (2003). A model for learning the semantics of pictures. In *Proceedings of the Neural Information Processing System Conference*, Vancouver, Canada.

Leonardi, R., Migliotari, P., & Prandini, M. (2004). Semantic indexing of soccer audio-visual sequences: A multimodal approach based on controlled Markov chains. *IEEE Transactions on Circuits Systems and Video Technology, 14*(5), 634–643.

Li, B., & Sezan, I. (2003). Semantic sports video analysis: Approaches and new applications. In *Proceedings of the IEEE International Conference on Image Processing*, Barcelona, Spain.

Li, J., & Wang, J. Z. (2003). Automatic linguistic indexing of pictures by a statistical modeling approach. *IEEE Transactions on Pattern Analysis and Machine Intelligence, 25*(9), 1075–1088.

Lowe, D. (1999). Object recognition from local scale-invariant features. In *Proceedings of the International Conference on Computer Vision*, Kerkyra, Corfu, Greece.

Lu, L., Zhang, H-J., & Jiang, H. (2002). Content analysis for audio classification and segmentation. *IEEE Transactions on Speech and Audio Processing, 10*(7), 293–302.

Luo, Y., & Hwang, J. N. (2003). Video sequence modeling by dynamic Bayesian networks: A systematic approach from coarse-to-fine grains. In *Proceedings of the IEEE International Conference on Image Processing*, Barcelona, Spain.

Marr, D. (1983). *Vision*. San Francisco: W.H. Freeman.

Miller, G. A. (1995). Wordnet: A lexical database for English. *Communications of the ACM, 38*(11), 39–41.

Mori, Y., Takahashi, H., & Oka, R. (1999). Image-to-word transformation based on dividing and vector quantizing images with words. In *Proceedings of the First International Workshop on Multimedia Intelligent Storage and Retrieval Management*, Orlando, FL.

Naphade, M., et al. (1998). A high performance shot boundary detection algorithm using multiple cues. In *Proceedings of the IEEE International Conference on Image Processing*, Chicago.

Naphade, M., & Smith, J. (2003). Learning visual models of semantic concepts. In *Proceedings of the IEEE International Conference on Image Processing*, Barcelona, Spain.

Naphade, M. R., & Huang, T. S. (2000). *Stochastic modeling of soundtrack for efficient segmentation and indexing of video.* In *Proceedings of the Conference on SPIE, Storage and Retrieval for Media Databases*, San Jose, CA.

Naphade, M. R., & Huang, T. S. (2001). A probabilistic framework for semantic video indexing filtering and retrieval. *IEEE Transactions on Multimedia, 3*(1), 141–151.

Natsev, A., Naphade, M., & Smith, J. (2003). Exploring semantic dependencies for scalable concept detection. In *Proceedings of the IEEE International Conference on Image Processing*, Barcelona, Spain.

Pearl, J. (1988). *Probabilistic reasoning in intelligent systems: Networks of plausible inference.* Los Angeles: Morgan Kaufmann Publishers.

Quattoni, A., Collins, M., & Darrell, T. (2004). Conditional random fields for object recognition. In *Proceedings of the Neural Information Processing Systems Conference*, Vancouver, Canada.

Rabiner, L. R. (1989). A tutorial on hidden Markov models and selected applications in speech recognition. *Proceedings of IEEE, 77*(2), 257–286.

Shi, J., & Malik, J. (2000). Normalized cuts and image segmentation. *IEEE Transactions on Pattern Analysis and Machine Intelligence, 22*(8), 888–905.

Smeulders, A. W. M., Worring, M., Santini, S., Gupta, A., & Jain, R. (2000). Content-based image retrieval at the end of the early years. *IEEE Transactions on Pattern Analysis and Machine Intelligence, 22*(12), 1349–1380.

Snoek, C. G. M., & Worring, M. (2005). Multimedia event based video indexing using time intervals. *IEEE Transactions on Multimedia, 7*(4), 638-647.

Snoek, C. G. M., & Worring, M. (2005). Multimodal video indexing: A review of the state-of-the-art. *Multimedia Tools and Applications, 25*(1), 5–35.

Souvannavong, F., Merialdo, B., & Huet, B. (2003). Latent semantic indexing for video content modeling and analysis. In *Proceedings of the TREC Video Retrieval Evaluation Workshop*, Gaithersburg, MD.

Srikanth, M., Varner, J., Bowden, M., & Moldovan, D. (2005). Exploiting ontologies for automatic image annotation. In *Proceedings of the ACM SIGIR Conference on Research and Development in Information Retrieval*, Salvador, Brazil.

Tan, Y-P., Saur, D. D., Kulkarni, S. R., & Ramadge, P. J. (2000). Rapid estimation of camera motion from compressed video with application to video annotation. *IEEE Transactions on Circuits and Systems for Video Technology, 10*(1), 133–146.

Tansley, R. (2000). *The multimedia thesaurus: Adding a semantic layer to multimedia information* [doctoral thesis]. University of Southampton, UK.

TRECVID. (2004). *TREC video retrieval evaluation.* Retrieved November 2005, from http://www-nlpir.nist.gov/projects/trecvid/

Tseng, B. L., Lin, C-Y., Naphade, M., Natsev, A., & Smith, J. (2003). Normalised classifier fusion for semantic visual concept detection. In *Proceedings of the IEEE International Conference on Image Processing*, Barcelona, Spain.

Tzanetakis, G., & Cook, P. (2002). Musical genre classification of audio signals. *IEEE Transactions on Speech and Audio Processing, 10*(5), 293–302.

Vailaya, A., Figueiredo, M., Jain, A., & Zhang, H. (1999). A Bayesian framework for semantic classification of outdoor vacation images. In *Proceedings of the SPIE: Storage and Retrieval for Image and Video Databases VII*, San Jose, CA.

Vailaya, A., Figueiredo, M., Jain, A. K., & Zhang, H. J. (2001). Image classification for content-based indexing. *IEEE Transactions on Image Processing, 10*(1), 117–130.

Vasconcelos, N., & Lippman, A. (1998). A Bayesian framework for semantic content characterization. In *Proceedings of the IEEE Conference on Computer Vision and Pattern Recognition*, Santa Barbara, CA.

Vasconcelos, N., & Lippman, A. (2000). A probabilistic architecture for content-based image retrieval. In *Proceedings of the IEEE Computer Vision and Pattern Recognition, Hilton Head*, SC.

Wang, Y., Liu, Z., & Huang, J-C. (2000). Multimedia content analysis using both audio and visual clues. *IEEE Signal Processing, 17*(6), 12–36.

Westerveld, T., & de Vries, A. P. (2003). Experimental result analysis for a generative probabilistic image retrieval model. In *Proceedings of the ACM SIGIR Conference on Research and Development in Information Retrieval*, Toronto, Canada.

Westerveld, T., de Vries, A. P., Ianeva, T., Boldareva, L., & Hiemstra, D. (2003). Combining information sources for video retrieval. In *Proceedings of the TREC Video Retrieval Evaluation Workshop*, Gaithersburg, MD.

Yavlinsky, A., Schofield, E., & Rüger, S. (2005). Automated image annotation using global features and robust nonparametric density estimation. In *Proceedings of the International Conference on Image and Video Retrieval*, Singapore.

About the Authors

Yu-Jin Zhang (zhang-yj@tsinghua.edu.cn) (PhD, State University of Liège, Belgium) is a professor of image engineering at Tsinghua University, Beijing, China. Previously, he was with Delft University of Technology, The Netherlands. In 2003, he was a visiting professor at National Technological University, Singapore. His research interests are mainly in the area of image engineering, including image processing, image analysis, and image understanding, as well as their applications. He has published nearly 300 research papers and a dozen books, including two monographs: *Image Segmentation* and *Content-Based Visual Information Retrieval* (Science Press); and two edited collections: *Advances in Image and Video Segmentation* and *Semantic-Based Visual Information Retrieval* (IRM Press). He is vice president of the China Society of Image and Graphics and the director of the academic committee of the society. He is deputy editor-in-chief of *Journal of Image and Graphics* and on the editorial boards of several scientific journals. He was the program co-chair of the First International Conference on Image and Graphics (ICIG2000) and the Second International Conference on Image and Graphics (ICIG2002). He is a senior member of IEEE.

* * *

Ioannis Andreadis (iandread@ee.duth.gr) received a diploma degree from the Department of Electrical & Computer Engineering, DUTH, Greece (1983), and an MSc and a PhD from the University of Manchester Institute of Science & Technology, UK (1985 and 1989, respectively). His research interests are mainly in intelligent systems, machine vision, and VLSI-based computing architectures. He joined the Department of Electrical & Computer Engineering, DUTH, in 1993. He is a member of the editorial board of the *Pattern Recognition Journal*, *TEE* and *IEEE*.

Ryan Benton (rbenton@cacs.louisiana.edu) earned his doctorate in computer science from the University of Louisiana (UL) at Lafayette, USA (May 2001). Since 2002, he has been a research scientist at UL Lafayette, where his primary focus is developing solutions for existing industrial problems. From November 2001 to June 2005, he was also the director for research at Star Software Systems Corporation, a UL Lafayette industrial partner. Benton has been involved in developing image retrieval and photogrammetric systems, conducting feasibility studies on automated annotation of satellite imagery and equipment health prognostics, and creating novel image steganalytic techniques.

Eugeny Bovbel (bovbel@bsu.by) was born in Belarus. In 1970, he graduated from the physical faculty of the Belarusian State University. He is the candidate of science (Phys-Math, 1976) and assistant professor of the Radiophysics Department (since 1976). He is also the chairman of teaching and methodical council and deputy dean of the Faculty of Radiophysics and Electronics (1981-1992). At the present time, he teaches a course on statistical radiophysics on the Faculty of Radiophysics. The areas of scientific interest cover statistical signal processing; time-and-frequency analysis; and recently, hidden Markov models and wavelet transforms. He is the author of more than 120 papers. In 2003, he passed the training at the Faculty of Computer Science, École Central de Lyon (France).

Liming Chen (liming.chen@ec-lyon.fr) received a BSc in mathematics and computer science from the Universite de Nantes in 1984. He obtained a master's degree in 1986 and a PhD in computer science from the University of Paris 6. He served for eight years at the Universite de Technologies de Compiegne as associate professor and in 1998 joined École Centrale de Lyon as a professor. He is the author of more than 90 publications in the multimedia indexing field from 1995, and his current research interests include face detection, reconstruction and recognition, image and video analysis and indexing, audio general classification, and music indexing and retrieval. He is a member of the IEEE Computer Society.

Stamatia Dasiopoulou (dasiop@iti.gr) received a diploma in electronic and computer engineering from the Electronic and Computer Engineering Department, Technical University of Crete, Hania, Greece (2003). She is currently working toward a PhD with the Electrical and Computer Engineering Department, Aristotle University of Thessaloniki, Greece. She is also a graduate research assistant with the Informatics and Telematics Institute/Centre for Research and Technology Hellas, Thessaloniki, Greece. Her research interests include knowledge-assisted multimedia analysis, knowledge discovery, multimedia semantic annotation, Semantic Web technologies, and reasoning support for multimedia.

Charalampos Doulaverakis (doulaver@iti.gr) received his diploma in electronic and computer engineering from the Electronic and Computer Engineering Department, Technical University of Crete, Hania, Greece (2003) and is currently an MSc student in advanced computing at Aristotle University of Thessaloniki, Greece. His research interests include multimedia analysis and retrieval, Semantic Web technologies, semantic information retrieval, and MPEG-7 standard. Up to now, he has been research associate in the Informatics and Telematics Institute/Centre for Research and Technology Hellas. Doulaverakis is a member of the Technical Chamber of Greece.

Antonios Gasteratos (agaster@pme.duth.gr) is a lecturer of robotics with the Department of Production and Management, DUTH, Greece. He received a PhD from the Department of Electrical and Computer Engineering, DUTH, Greece (1999). He serves as a reviewer to numerous scientific journals and at international conferences. His research interests are mainly in computer and robot vision and sensory data fusion. He is a member of the IEEE, the IAPR, the EURASIP, the Hellenic Society of Artificial Intelligence (SETN), and the TEE.

Hakim Hacid (hhacid@eric.univ-lyon2.fr) received an MSc in computer science from the University of Lyon 2, France. He is currently a PhD student at the ERIC Laboratory at the University of Lyon 2. His research interests include databases, data mining, and multimedia. For more information, visit http://eric.univ-lyon2.fr/~hhacid/.

Daniel Heesch (daniel.heesch@imperial.ac.uk) is currently a post-doctoral researcher in the Department of Electrical and Electronic Engineering at Imperial College London, working on contextual inference for problems in computer vision. His doctoral research was concerned with the question of how to represent the multiplicity of similarity relationships that may exist among images and how to employ those representations for effective content-based image search. He holds an MSc in computer science from Imperial College London, a BA in biological sciences from Oxford University, and a BSc in mathematical sciences from Open University.

Mohammed Lamine Kherfi (kherfi@uqtr.ca) received MSc and PhD degrees in computer science from the Université de Sherbrooke, Canada (2002 and 2004, respectively). Between 1997 and 2000, he worked as head of the General Computer Services Agency. Previously, he received a BEng in computer science from the Institut National d'Informatique INI, Algeria (1997). Currently, he is a professor with the Department of Mathematics and Computer Science, Université du Québec à Trois-Rivières (Canada). His research interests include image and multimedia retrieval, computer vision, and machine learning. Kherfi has published numerous refereed papers in the field of computer vision and holds an invention patent. He has served as a reviewer for a number of journals and conferences. He was the recipient of six scholarships and awards from the Université de Sherbrooke and several awards from the INI. He received the FQRNT excellence postdoctoral fellowship in 2005.

Yasushi Kiyoki (kiyoki@mdbl.sfc.keio.ac.jp) received his BEng, MEng, and PhD degrees in electrical engineering from Keio University, Japan (1978, 1980, and 1983, respectively). From 1984 to 1996, he was an assistant professor and an associate professor at the Institute of Information Sciences and Electronics, University of Tsukuba. In 1996, he joined the faculty of Keio University, where he is a tenured professor at the Department of Environmental Information. His research addresses multidatabase systems, knowledge-base systems, semantic associative processing, and multimedia database systems. He serves as the editor-in-chief of *Information Modeling and Knowledge Bases* (IOS Press). He also served as the program chair for the 7th International Conference on Database Systems for Advanced Applications.

Ioannis Kompatsiaris (ikom@iti.gr) received a diploma in electrical engineering and a PhD in 3-D model-based image sequence coding from Aristotle University of Thessaloniki (AUTH), Thessaloniki, Greece (1996 and 2001, respectively). He is a senior researcher (Researcher C') with the Informatics and Telematics Institute, Thessaloniki. His research interests include semantic annotation of multimedia content, multimedia information retrieval and knowledge discovery, and MPEG-4 and MPEG-7 standards. He is the co-author of four book chapters, 14 papers in refereed journals, and more than 50 papers in international conferences. He is a member of IEEE and IEE VIE TAP.

Konstantinos Konstantinidis (konkonst@ee.duth.gr) received a BSc and an MSc from the Department of Automatic Control and Systems Engineering, University of Sheffield, UK (1999 and 2000, respectively). He is currently a PhD student with the Department of Electrical and Computer Engineering, Democritus University of Thrace (DUTH), Greece. He is a member of the Technical Chamber of Greece (TEE).

Siddhivinayak Kulkarni (S.Kulkarni@ballarat.edu.au) received a Bachelor of Electronics Engineering from the University of Pune, India (1992) and a Master in Information Technology and PhD from Griffith University, Australia (1997 and 2002). Currently, he is a lecturer in the School of Information Technology and Mathematical Science at the University of Ballarat, Australia. Before moving to the University of Ballarat, Australia, he worked as an assistant professor with the Department of Computer Science and Mathematics, Nipissing University, Canada. He also holds industrial experience in the field of electronics and computer engineering. He has served as a program-organizing committee member for many International Conferences and as a reviewer for prestigious journals. His research interests include content-based image retrieval, neural networks, fuzzy logic, and pattern recognition.

João Magalhães (j.magalhaes@imperial.ac.uk) is currently a PhD candidate with the Department of Computing, Imperial College London, working in the area of semantic-multimedia analysis and indexing. His main research interests are in statistical modeling, information extraction, and related applications. He received both an electrical engineering degree and an MSc in electrical and computer engineering from the Instituto Superior Técnico, Lisbon, Portugal, and in 2002 was awarded the Young Engineer Award by the Portuguese National Engineering Association. He has worked for Siemens R&D in Telecommunication Networks for six years. His research is funded by the Fundação para a Ciência e Tecnologia under the scholarship SFRH/BD/16283/2004/K5N0.

Vasileios Mezaris (bmezaris@iti.gr) received a diploma and a PhD in electrical and computer engineering from the Aristotle University of Thessaloniki, Thessaloniki, Greece (2001 and 2005, respectively). He is a post-doctoral research fellow with the Informatics and Telematics Institute/Centre for Research and Technology Hellas (ITI/CERTH), Thessaloniki, Greece. Prior to this, he was a post-graduate research fellow with ITI/CERTH and a teaching assistant with the Electrical and Computer Engineering Department of the Aristotle University

of Thessaloniki. His research interests include still image segmentation, video segmentation, knowledge-assisted multimedia analysis, and knowledge extraction from multimedia,
content-based, and semantic indexing and retrieval.

Vyacheslav Parshin (vyacheslav.parshin@ec-lyon.fr) graduated from Belorussian State
University in 1996 and obtained a degree in radiophysics. He worked as a research associate
at the aforementioned university until 1999 in the field of speech recognition and neural-
network technologies. Currently, he is a PhD student at École Centrale de Lyon in France.
His research interests include semantic video segmentation and summarization.

Antonio Picariello (antonio.picariello@unina.it) received the Laurea degree in electronics engineering from the University of Napoli, Italy (1991). In 1993, he joined the Istituto
Ricerca Sui Sistemi Informatici Paralleli, the National Research Council, Napoli, Italy. He
received a PhD in computer science and engineering in 1998 from the University of Napoli
"Federico II." In 1999, he joined the Dipartimento di Informatica e Sistemistica, University
of Napoli "Federico II," Italy, where he is currently an associate professor of database. He
has been active in the fields of computer vision, medical image processing and pattern recognition, object-oriented models for image processing, multimedia database, and information retrieval. His current research interests are in knowledge extraction and management,
multimedia integration, and image and video databases. He is a member of the International
Association of Pattern Recognition.

Vijay Raghavan (raghavan@cacs.louisiana.edu) is a distinguished professor of computer
science at the University of Louisiana at Lafayette, USA. Raghavan obtained a BTech degree
in mechanical engineering from the Indian Institute of Technology, Madras, an MBA from
McMaster University, and a PhD in computer science from the University of Alberta. He
specializes in data management, including information retrieval, database data mining, and
data warehousing. Raghavan's research projects have been supported by funding agencies
such as the NSF, DoE, DoD, NASA, and the State of Louisiana Board of Regents. Raghavan has served in various professional capacities and was the recent conference chair of the
International Conference on Data Mining. He is a member of the ACM, the IEEE, and the
Phi Kappa Phi and Upsilon Pi Epsilon Computer Science honor societies. He is currently
an ACM national lecturer.

Stefan Rüger (s.rueger@imperial.ac.uk), currently a reader in multimedia and information
systems with the Department of Computing, Imperial College London, has been working in
the area of multimedia information retrieval since 1999, when he was awarded a prestigious
five-year EPSRC Advanced Research Fellowship. His research is funded by industry, research
councils, and the Commission of European Communities, and he is involved in a number of
international research networks and projects. Rüger received a doctorate in computer science
from the Technical University Berlin for his contributions to the theory of neural networks
and holds a diploma (equivalent to MSc) in physics from the Free University Berlin.

Enver Sangineto (sanginet@dia.uniroma3.it) obtained a "laurea" degree in computer science from the University of Pisa (1995). In 2001, he received a PhD in computer engineering from the University of Rome "La Sapienza," discussing a thesis on *Object Classification through Geometric Abstraction*. From 2001 to the present, he has been working at the Artificial Intelligence Laboratory of the University of Rome "Roma Tre" and with the CRMPA (a research centre of the University of Salerno, Italy). He won a post-doc. position at the University of Salerno in 2005. His research activity concerns AI and computer vision. His main scientific interests concern e-learning, robotics, pattern recognition, and image retrieval. He has published various papers on these topics in important national and international conferences and journals.

Maria Luisa Sapino (mlsapino@di.unito.it) received her master's degree and PhD in computer science at the University of Torino, Italy, where she is currently an associate professor. She initially worked in the area of logic programming and artificial intelligence, specifically interested in the semantics of negation in logic programming and in the abductive extensions of logic programs. Her current research is in the area of heterogeneous and multimedia databases. In particular, she is interested in similarity-based information retrieval, content-based image retrieval, Web accessibility for users who are visually impaired, and multimedia presentations. She has been serving as a reviewer for several international conferences and journals.

Hideyasu Sasaki (h-sasaki-7@alumni.uchicago.edu), a graduate of the University of Tokyo in 1994, received an LLM from the University of Chicago Law School in 1999, an MS and a PhD in cybernetic knowledge engineering (media and governance) (Hons.) from Keio University (2001 and 2003, respectively). He is an associate professor with the Department of Information Science and Engineering, Ritsumeikan University, Japan. He was an assistant professor at Keio University from 2003 to 2005. His research interests include content-based metadata indexing and image retrieval, digital libraries, multimedia databases, and intellectual property law and management. He was admitted to practice as an attorney and counselor at law in the New York State Bar in 2000. He is active as an organizing committee member of the International Conference on Asian Digital Libraries (ICADL) and as an international program committee member of the International Society on Law and Technology (LawTech) and Euro-Japan Conference on Information Modeling and Knowledge Bases (EJC).

Biren Shah (bshah@cacs.louisiana.edu) received a bachelor's degree in computer science from the Veermata Jijabai Technological Institute, Mumbai, India (1999). He received his master's degree and PhD in computer science, both from the University of Louisiana (UL) at Lafayette, USA (2001 and 2005, respectively). From 2001 to 2002, he was a software engineer at Lucent Technologies. He is a research scientist at the UL Lafayette. His research interests include information retrieval, databases, data warehouses, XML, and heterogeneous information management. Shah has served in various professional capacities and has published more than 15 papers in refereed conferences, journals, symposiums, and workshops. He is a member of the ACM.

Michael G. Strintzis (strintzi@eng.auth.gr) received a diploma in electrical engineering from the National Technical University of Athens, Greece (1967), and MA and PhD degrees in electrical engineering from Princeton University, New Jersey (1969 and 1970, respectively). He joined the Electrical Engineering Department, University of Pittsburgh, Pennsylvania, where he served as an assistant professor from 1970 to 1976 and an associate professor from 1976 to 1980. Since 1980, he has been a professor of electrical and computer engineering at the Aristotle University of Thessaloniki, Thessaloniki, Greece. In 1998, he founded the Informatics and Telematics Institute, currently part of the Centre for Research and Technology Hellas, Thessaloniki.

Dzmitry Tsishkou (dzmitry.tsishkou@ec-lyon.fr) received a diploma from Belarussian State University (BSU), Belarus, in 2001. He was an assistant professor in 2002 at BSU. He obtained a PhD in computer science from École Centrale de Lyon, France, in 2005. He is currently a post doc. at École Centrale de Lyon, while doing research at LIRIS. His research interests are in the areas of computer vision, machine learning, and multimedia indexing, especially in face detection and recognition.

Brijesh Verma (B.Verma@cqu.edu.au) is an associate professor in the School of Information Technology at Central Queensland University in Australia. His research interests include computational intelligence and pattern recognition. He has published one book, 10 edited books, three book chapters, 92 journal and conference papers in the areas of computational intelligence, handwriting recognition, face recognition, microcalcification classification, Web document retrieval, and content-based image retrieval. He has supervised five PhD, 17 master's, and six honor students, most of them in the areas of computational intelligence and pattern recognition. He is a co-editor in chief of the *International Journal of Computational Intelligence and Applications*, an associate editor of the *IEEE Transaction on Biomedicine in Information Technology*, and an editorial board member of the *International Journal of Hybrid Intelligent Systems*. He has served on the program/organizing committees of more than 15 conferences, including IEEE IJCNN'06 and IEEE CEC'06. He is a member of IEEE and IEEE Computational Intelligence Society.

Zonghuan Wu (zwu@cacs.louisiana.edu) received both his bachelor's and master's degrees in computer science from Sichuan University (1993 and 1998, respectively). He received his PhD in computer science from the State University of New York at Binghamton. He is a research scientist at the Center for Advanced Computer Studies, University of Louisiana at Lafayette, USA. He specializes in information retrieval, metasearch engines, and Web development. Wu has around two dozen published papers. He also has been enthusiastically promoting academia-to-business technology transfer.

Feng Xu (f-xu02@mails.tsinghua.edu.cn) received a BS in electronic engineering from Lanzhou University, China (2002). She is currently a PhD candidate with the Department of Electronic Engineering, Tsinghua University, Beijing, China. Her research interests include content-based image retrieval, automatic image annotation, and machine learning.

Hun-Woo Yoo (paulyhw@yonsei.ac.kr) is a research professor at the Center for Cognitive Science, Yonsei University, Korea. He received a BS and an MS in electrical engineering from Inha University, Korea, and a PhD in industrial systems and information engineering at Korea University, Korea. From 1994 to 1997, he worked as a research engineer at the Manufacturing Technology Center of LG Electronics. His current research interests include multimedia information retrieval, computer vision, and image processing.

Abdelkader Djamel Zighed (zighed@univ-lyon2.fr) is a full professor at the university Lumière Lyon 2 (France). He has worked in data mining and machine learning for more than 20 years. He organized many conferences related to knowledge engineering and data mining, such as the PKDD conference in 2000 and the ISMIS conference in 2002. He is now developing research in mining complex data. For more information, visit http://www.morgon.univ-lyon2.fr.

Djemel Ziou (djemel.ziou@usherbrooke.ca) received a BEng in computer science from the University of Annaba (Algeria) in 1984, and a PhD in computer science from the Institut National Polytechnique de Lorraine (INPL), France, in 1991. From 1987 to 1993, he served as lecturer in several universities in France. During the same period, he was a researcher in the Centre de Recherche en Informatique de Nancy (CRIN) and the Institut National de Recherche en Informatique et Automatique (INRIA) in France. Presently, he is a full professor with the Department of Computer Science at the University of Sherbrooke in Canada. He has served on numerous conference committees as a member or chair. He heads the laboratory MOIVRE and the consortium CoRIMedia, which he founded. His research interests include image processing, information retrieval, computer vision, and pattern recognition.

Index

M

machine learning 230-232
machine translation model 121
machine translation research 339
Mahalanobis distance 167
manual and automatic image annotation 208
manual annotation of the training set 316
manual classification 193
mathematic model 14
maximum thresholding process 148
meaningful regions 8
measure conversion process 147
MediaNet 344, 345
media ontology 344
medical information systems 65
metadata 136
metadata generation module 144
metric 23
military intelligence 65
million-hour archives 188
minimum thresholding process 147
Minkowski metric 165
MLP as a classifier 267
MLP as a texture feature classifier 262
motivation 213
MPEG-7 190, 212
multi-dimensional scaling 170
multi-level image and video 5
multi-level image retrieval 7
multi-level representation 7
multi-level video abstraction 11
multi-level video organization 10
multi-level video retrieval 9
multi-modal 70
multi-scale random field (MSRF) 343
multimedia 136
multimedia-semantic analysis 336
multimedia analysis 335
multimedia database 136, 231-232
multimodal analysis 348
multimodal data fusion 72
museum 65
music retrieval 275

N

naive method 241
natural language processing 131
nearest-neighbor network 177
negative example (NE) 14
neighborhood graph construction 236
neighborhood graph 235-237
network model 336, 342
neural network 252-256, 263, 267
neuro-fuzzy 261
neuro-fuzzy AND 264
NNk network 177
nonparametric density 337
normalization process 148

O

object relational database 307
observable feature 74
occlusion 196
ontology-based framework 208
ontology-based method 116
ontology-driven analysis 209
ontology-driven framework 208
ontology infrastructure 208, 216
open video 200
optimal parameter 321
optimistic combination 304
optimization with negative feedback 169
optomistic and combinations 305
orientation 195
ostensive relevance feedback 179

P

painting retrieval 275
parameter initialization 174
pathfinder network 178
pattern recognition 14
performance evaluation 151
point access method (PAM) 233
positive example (PE) 14
preprocessing texture images 261
probabilistic approach 171